Studies in Systems, Decision and Control

Volume 147

Series editor

Janusz Kacprzyk, Polish Academy of Sciences, Warsaw, Poland
e-mail: kacprzyk@ibspan.waw.pl

The series "Studies in Systems, Decision and Control" (SSDC) covers both new developments and advances, as well as the state of the art, in the various areas of broadly perceived systems, decision making and control- quickly, up to date and with a high quality. The intent is to cover the theory, applications, and perspectives on the state of the art and future developments relevant to systems, decision making, control, complex processes and related areas, as embedded in the fields of engineering, computer science, physics, economics, social and life sciences, as well as the paradigms and methodologies behind them. The series contains monographs, textbooks, lecture notes and edited volumes in systems, decision making and control spanning the areas of Cyber-Physical Systems, Autonomous Systems, Sensor Networks, Control Systems, Energy Systems, Automotive Systems, Biological Systems, Vehicular Networking and Connected Vehicles, Aerospace Systems, Automation, Manufacturing, Smart Grids, Nonlinear Systems, Power Systems, Robotics, Social Systems, Economic Systems and other. Of particular value to both the contributors and the readership are the short publication timeframe and the world-wide distribution and exposure which enable both a wide and rapid dissemination of research output.

More information about this series at http://www.springer.com/series/13304

George A. Anastassiou

Nonlinearity: Ordinary and Fractional Approximations by Sublinear and Max-Product Operators

 Springer

George A. Anastassiou
Department of Mathematical Sciences
University of Memphis
Memphis, TN
USA

ISSN 2198-4182   ISSN 2198-4190 (electronic)
Studies in Systems, Decision and Control
ISBN 978-3-030-07788-4   ISBN 978-3-319-89509-3 (eBook)
https://doi.org/10.1007/978-3-319-89509-3

Printed on acid-free paper

This Springer imprint is published by the registered company Springer International Publishing AG
part of Springer Nature
The registered company address is: Gewerbestrasse 11, 6330 Cham, Switzerland

Dedicated to My Family.

Preface

Nonlinear mathematics extend the linear mathematics to meet various needs and demands of pure and applied mathematics, among others to cover a great variety of applications in the real world. Approximation by sublinear operators with applications to max-product operators is a new trend in approximation theory. These operators are nonlinear and rational giving very fast and flexible approximations based on limited data.

In this book, we focus more in approximations under the presence of ordinary and various kinds of fractional smoothness, deriving better approximations than without smoothness. We present both the univariate and multivariate cases. The last three chapters contain approximations under the influence of convexity, there the estimates are more elegant and compact with small constants, and the convergence of high speeds. This monograph is the natural evolution of recent author's research work put in a book form for the first time. The presented approaches are original, and chapters are self-contained and can be read independently. This monograph is suitable to be used in related graduate classes and research projects. We exhibit to the maximum our approximation methods to all possible directions.

The motivation to write this monograph came by the following: various issues related to the modelling and analysis of ordinary and fractional-order systems have gained an increased popularity, as witnesses by many books and volumes in Springer's program:

http://www.springer.com/gp/search?query=fractional&submit=Prze%C5%9Blij

and the purpose of our book is to provide a deeper formal analysis on some issues that are relevant to many areas, for instance, decision-making, complex processes, systems modelling and control, and related areas. The above are deeply embedded in the fields of mathematics, engineering, computer science, physics, economics, social and life sciences.

The list of presented topics follows:

approximation by sublinear operators, approximation by max-product operators, conformable fractional approximation by max-product operators,

Caputo fractional approximation by sublinear operators,
Canavati fractional approximation by max-product operators,
iterated fractional approximation by max-product operators,
mixed conformable fractional approximation by sublinear operators,
approximation of fuzzy numbers by max-product operators,
approximation by multivariate sublinear and max-product operators,
approximation by sublinear and max-product operators using convexity,
conformable fractional approximations by max-product operators using convexity,
and approximations by multivariate sublinear and max-product operators under convexity.
An extensive list of references is given per chapter.

The book's results are expected to find applications in many areas of pure and applied mathematics, especially in approximation theory and numerical analysis in both ordinary and fractional sense. As such this monograph is suitable for researchers, graduate students and seminars of the above disciplines, also to be in all science and engineering libraries.

The preparation of the book took place during 2017 at the University of Memphis.

The author likes to thank Prof. Alina Alb Lupas of University of Oradea, Romania, for checking and reading the manuscript.

Memphis, USA   George A. Anastassiou
January 2018

Contents

Chapter 1
Approximation by Positive Sublinear Operators

Here we study the approximation of functions by positive sublinear operators under differentiability. We produce general Jackson type inequalities under initial conditions. We apply these to a series of well-known Max-product operators. So our approach is quantitative by producing inequalities with their right hand sides involving the modulus of continuity of a high order derivative of the function under approximation. It follows [3].

1.1 Introduction

The main motivation here is the monograph by B. Bede, L. Coroianu and S. Gal [5], 2016.

Let $N \in \mathbb{N}$, the well-known Bernstein polynomials [7] are positive linear operators, defined by the formula

$$B_N(f)(x) = \sum_{k=0}^{N} \binom{N}{k} x^k (1-x)^{N-k} f\left(\frac{k}{N}\right), x \in [0,1], f \in C([0,1]).$$

(1.1)

T. Popoviciu in [7], 1935, proved for $f \in C([0,1])$ that

$$|B_N(f)(x) - f(x)| \leq \frac{5}{4}\omega_1\left(f, \frac{1}{\sqrt{N}}\right), \forall x \in [0,1],$$

(1.2)

where

$$\omega_1(f, \delta) = \sup_{\substack{x,y\in[0,1]:\\|x-y|\leq\delta}} |f(x) - f(y)|, \delta > 0,$$

(1.3)

© Springer International Publishing AG, part of Springer Nature 2018
G. A. Anastassiou, *Nonlinearity: Ordinary and Fractional Approximations by Sublinear and Max-Product Operators*, Studies in Systems, Decision and Control 147, https://doi.org/10.1007/978-3-319-89509-3_1

is the first modulus of continuity.

G.G. Lorentz in [6], 1986, p. 21, proved for $f \in C^1([0, 1])$ that

$$|B_N(f)(x) - f(x)| \leq \frac{3}{4\sqrt{N}}\omega_1\left(f', \frac{1}{\sqrt{N}}\right), \forall x \in [0, 1], \qquad (1.4)$$

In [5], p. 10, the authors introduced the basic Max-product Bernstein operators,

$$B_N^{(M)}(f)(x) = \frac{\bigvee_{k=0}^{N} p_{N,k}(x) f\left(\frac{k}{N}\right)}{\bigvee_{k=0}^{N} p_{N,k}(x)}, N \in \mathbb{N}, \qquad (1.5)$$

where $\bigvee$ stands for maximum, and $p_{N,k}(x) = \binom{N}{k} x^k (1-x)^{N-k}$ and $f : [0, 1] \rightarrow \mathbb{R}_+ = [0, \infty)$.

These are nonlinear and piecewise rational operators.

The authors in [5] studied similar such nonlinear operators such as: the Max-product Favard–Szász–Mirakjan operators and their truncated version, the Max-product Baskakov operators and their truncated version, also many other similar specific operators. The study in [5] is based on presented there general theory of sublinear operators. These Max-product operators tend to converge faster to the on hand function.

So we mention from [5], p. 30, that for $f : [0, 1] \rightarrow \mathbb{R}_+$ continuous, we have the estimate

$$\left|B_N^{(M)}(f)(x) - f(x)\right| \leq 12\omega_1\left(f, \frac{1}{\sqrt{N+1}}\right), \text{ for all } N \in \mathbb{N}, x \in [0, 1], \quad (1.6)$$

Also from [5], p. 36, we mention that for $f : [0, 1] \rightarrow \mathbb{R}_+$ being concave function we get that

$$\left|B_N^{(M)}(f)(x) - f(x)\right| \leq 2\omega_1\left(f, \frac{1}{N}\right), \text{ for all } x \in [0, 1], \qquad (1.7)$$

a much faster convergence.

In this chapter we expand the study in [5] by considering smoothness of functions, which is not done in [5]. So our inequalities are with respect to $\omega_1\left(f^{(n)}, \delta\right), \delta > 0, n \in \mathbb{N}$.

We present at first some general related theory of sublinear operators and then we apply it to specific as above Max-product operators.

1.2 Main Results

Let $I \subset \mathbb{R}$ be a bounded or unbounded interval, $n \in \mathbb{N}$, and

$$CB_+^n (I) = \left\{ f : I \to \mathbb{R}_+ : f^{(i)} \text{ is continuous and bounded on } I, \text{ all } i = 0, 1, \ldots, n \right\}.$$

(1.8)

Let $f \in CB_+^n (I)$, and any $x, y \in I$. By Taylor's formula we have

$$f(y) = \sum_{i=0}^{n} f^{(i)} (x) \frac{(y-x)^i}{i!} + \frac{1}{(n-1)!} \int_x^y (y-t)^{n-1} \left(f^{(n)} (t) - f^{(n)} (x) \right) dt.$$

(1.9)

We define for

$$f \in CB_+ (I) = \{ f : I \to \mathbb{R}_+ : f \text{ is continuous and bounded on } I \},$$

the first modulus of continuity

$$\omega_1 (f, \delta) = \sup_{\substack{x, y \in I: \\ |x-y| \le \delta}} |f(x) - f(y)|,$$

where $0 < \delta \le diameter (I)$.

We call the remainder in (1.9) as

$$R_n (x, y) = \frac{1}{(n-1)!} \int_x^y (y-t)^{n-1} \left(f^{(n)} (t) - f^{(n)} (x) \right) dt, \forall\, x, y \in I. \quad (1.10)$$

By [1], p. 217, and [2], p. 194, Chap. 7, (7.27) there, we get

$$|R_n (x, y)| \le \frac{\omega_1 \left(f^{(n)}, \delta \right)}{n!} |x - y|^n \left(1 + \frac{|x-y|}{(n+1)\delta} \right), \forall\, x, y \in I, \delta > 0. \quad (1.11)$$

We may rewrite (1.11) as

$$|R_n (x, y)| \le \frac{\omega_1 \left(f^{(n)}, \delta \right)}{n!} \left[|x-y|^n + \frac{|x-y|^{n+1}}{(n+1)\delta} \right], \forall\, x, y \in I, \delta > 0. \quad (1.12)$$

That is

$$\left| f(y) - \sum_{i=0}^{n} f^{(i)} (x) \frac{(y-x)^i}{i!} \right| \le \frac{\omega_1 \left(f^{(n)}, \delta \right)}{n!} \left[|x-y|^n + \frac{|x-y|^{n+1}}{(n+1)\delta} \right], \quad (1.13)$$

$\forall\, x, y \in I, \delta > 0.$

Furthermore it holds

$$|f(y) - f(x)| \leq \sum_{i=1}^{n} |f^{(i)}(x)| \frac{|y - x|^i}{i!} + \frac{\omega_1\left(f^{(n)}, \delta\right)}{n!} \left[|x - y|^n + \frac{|x - y|^{n+1}}{(n+1)\delta} \right],$$
$$(1.14)$$

$\forall\, x, y \in I, \delta > 0.$

In case of $f^{(i)}(x) = 0$, for $i = 1, \ldots, n$; for a specific $x \in I$, we get

$$|f(y) - f(x)| \leq \frac{\omega_1\left(f^{(n)}, \delta\right)}{n!} \left[|x - y|^n + \frac{|x - y|^{n+1}}{(n+1)\delta} \right], \forall\, y \in I, \delta > 0. \quad (1.15)$$

In case of $n = 1$, we derive

$$|f(y) - f(x)| \leq |f'(x)| |y - x| + \omega_1\left(f', \delta\right) \left[|x - y| + \frac{(x - y)^2}{2\delta} \right], \quad (1.16)$$

$\forall\, x, y \in I, \delta > 0.$

In case of $n = 1$ and $f'(x) = 0$, for a specific $x \in I$, we get that

$$|f(y) - f(x)| \leq \omega_1\left(f', \delta\right) \left[|x - y| + \frac{(x - y)^2}{2\delta} \right], \forall\, y \in I, \delta > 0. \quad (1.17)$$

Call $C_+(I) = \{f : I \to \mathbb{R}_+ : f \text{ is continuous on } I\}$.

Let $L_N : C_+(I) \to CB_+(I), n, N \in \mathbb{N}$ be a sequence of operators satisfying the following properties (see also [5], p. 17):

(i) (positive homogeneous)

$$L_N(\alpha f) = \alpha L_N(f), \forall\, \alpha \geq 0, f \in C_+(I), \quad (1.18)$$

(ii) (Monotonicity)

$$\text{if } f, g \in C_+(I) \text{ satisfy } f \leq g, \text{ then } L_N(f) \leq L_N(g), \forall\, N \in \mathbb{N}, \quad (1.19)$$

and

(iii) (Subadditivity)

$$L_N(f + g) \leq L_N(f) + L_N(g), \forall\, f, g \in C_+(I). \quad (1.20)$$

We call L_N positive sublinear operators.

In particular we will study the restrictions $L_N|_{CB_+^n(I)} : CB_+^n(I) \to CB_+(I)$.

As in [5], p. 17, we get that for $f, g \in CB_+(I)$,

$$|L_N(f)(x) - L_N(g)(x)| \leq L_N(|f - g|)(x), \forall\, x \in I. \quad (1.21)$$

Furthermore, also from [5], p. 17, we have

$$|L_N(f)(x) - f(x)| \le L_N(|f(\cdot) - f(x)|)(x) + |f(x)||L_N(1)(x) - 1|, \forall x \in I. \tag{1.22}$$

Using (1.14) into (1.22) we obtain

$$|L_N(f)(x) - f(x)| \le f(x)|L_N(1)(x) - 1| + \sum_{i=1}^{n} \frac{\left|f^{(i)}(x)\right|}{i!} L_N(|\cdot - x|^i)(x) +$$

$$\frac{\omega_1\left(f^{(n)}, \delta\right)}{n!}\left[L_N(|\cdot - x|^n)(x) + \frac{L_N(|\cdot - x|^{n+1})(x)}{(n+1)\delta}\right], \forall x \in I, \delta > 0. \tag{1.23}$$

If $L_N(1) = 1$ and $f^{(i)}(x) = 0$, $i = 1, \ldots, n$; x is fixed in I, we derive that

$$|L_N(f)(x) - f(x)| \le \frac{\omega_1\left(f^{(n)}, \delta\right)}{n!}\left[L_N(|\cdot - x|^n)(x) + \frac{L_N(|\cdot - x|^{n+1})(x)}{(n+1)\delta}\right], \tag{1.24}$$

$\delta > 0$.

We assume and choose

$$\delta = \left(L_N(|\cdot - x|^{n+1})(x)\right)^{\frac{1}{n+1}} > 0. \tag{1.25}$$

Therefore we get

$$|L_N(f)(x) - f(x)| \le \frac{\omega_1\left(f^{(n)}, \left(L_N(|\cdot - x|^{n+1})(x)\right)^{\frac{1}{n+1}}\right)}{n!} \cdot$$

$$\left[L_N(|\cdot - x|^n)(x) + \frac{\left(L_N(|\cdot - x|^{n+1})(x)\right)^{\frac{n}{n+1}}}{(n+1)}\right]. \tag{1.26}$$

Using (1.16) into (1.22) we also obtain

$$|L_N(f)(x) - f(x)| \le f(x)|L_N(1)(x) - 1| + \left|f'(x)\right|L_N(|\cdot - x|)(x) +$$

$$\omega_1\left(f', \delta\right)\left[L_N(|\cdot - x|)(x) + \frac{L_N((\cdot - x)^2)(x)}{2\delta}\right], \forall x \in I, \delta > 0. \tag{1.27}$$

Assuming $L_N(1) = 1$ and $f'(x) = 0$, for a specific $x \in I$, we get from (1.27), that ($n = 1$ case)

$$|L_N(f)(x) - f(x)| \le \omega_1(f', \delta) \left[L_N(|\cdot - x|)(x) + \frac{L_N((\cdot - x)^2)(x)}{2\delta} \right], \delta > 0.$$

$$(1.28)$$

Assume and choose

$$\delta = \sqrt{L_N((\cdot - x)^2)(x)} > 0 \qquad (1.29)$$

then it holds

$$|L_N(f)(x) - f(x)| \le \omega_1\left(f', \sqrt{L_N((\cdot - x)^2)(x)}\right) \cdot$$

$$\left[L_N(|\cdot - x|)(x) + \frac{\sqrt{L_N((\cdot - x)^2)(x)}}{2} \right], \forall N \in \mathbb{N}. \qquad (1.30)$$

We present Hölder's inequality for positive sublinear operators

Theorem 1.1 *Let $L : C_+(I) \to CB_+(I)$, be a positive sublinear operator and $f, g \in C_+(I)$, furthermore let $p, q > 1 : \frac{1}{p} + \frac{1}{q} = 1$. Assume that $L((f(\cdot))^p)(s_*)$, $L((g(\cdot))^q)(s_*) > 0$ for some $s_* \in I$. Then*

$$L(f(\cdot)g(\cdot))(s_*) \le \left(L((f(\cdot))^p)(s_*)\right)^{\frac{1}{p}} \left(L((g(\cdot))^q)(s_*)\right)^{\frac{1}{q}}. \qquad (1.31)$$

Proof Let $a, b \ge 0$, $p, q > 1 : \frac{1}{p} + \frac{1}{q} = 1$. The Young's inequality says

$$ab \le \frac{a^p}{p} + \frac{b^q}{q}. \qquad (1.32)$$

Then

$$\frac{f(s)}{\left(L((f(\cdot))^p)(s_*)\right)^{\frac{1}{p}}} \cdot \frac{g(s)}{\left(L((g(\cdot))^q)(s_*)\right)^{\frac{1}{q}}} \le$$

$$\frac{(f(s))^p}{p\left(L((f(\cdot))^p)(s_*)\right)} + \frac{(g(s))^q}{q\left(L((g(\cdot))^q)(s_*)\right)}, \forall s \in I. \qquad (1.33)$$

Hence it holds

$$\frac{L(f(\cdot)g(\cdot))(s_*)}{\left(L((f(\cdot))^p)(s_*)\right)^{\frac{1}{p}} \left(L((g(\cdot))^q)(s_*)\right)^{\frac{1}{q}}} \le \qquad (1.34)$$

$$\frac{(L((f(\cdot))^p))(s_*)}{p\left(L((f(\cdot))^p)(s_*)\right)} + \frac{(L((g(\cdot))^q))(s_*)}{q\left(L((g(\cdot))^q)(s_*)\right)} = \frac{1}{p} + \frac{1}{q} = 1, \text{ for } s_* \in I,$$

proving the claim. ∎

By (1.25) and (1.31) and $L_N (1) = 1$, we obtain

$$L_N \left(|\cdot - x|^n \right) (x) \leq \left(L_N \left(|\cdot - x|^{n+1} \right) (x) \right)^{\frac{n}{n+1}}, \tag{1.35}$$

in case of $n = 1$ we derive

$$L_N \left(|\cdot - x| \right) (x) \leq \sqrt{\left(L_N \left((\cdot - x)^2 \right) (x) \right)}. \tag{1.36}$$

We have proved the following result.

Theorem 1.2 *Let* $\left(L_N \right)_{N \in \mathbb{N}}$ *be a sequence of positive sublinear operators from* $C_+ (I)$ *into* $C B_+ (I)$, *and* $f \in C B_+^n (I)$, *where* $n \in \mathbb{N}$ *and* $I \subset \mathbb{R}$ *a bounded or unbounded interval. Assume* $L_N (1) = 1$, $\forall N \in \mathbb{N}$, *and* $f^{(i)} (x) = 0$, $i = 1, \ldots, n$, *for a fixed* $x \in I$. *Furthermore assume that* $L_N \left(|\cdot - x|^{n+1} \right) (x) > 0$, $\forall N \in \mathbb{N}$.
Then

$$|L_N (f) (x) - f (x)| \leq \frac{\omega_1 \left(f^{(n)}, \left(L_N \left(|\cdot - x|^{n+1} \right) (x) \right)^{\frac{1}{n+1}} \right)}{n!} \cdot$$

$$\left[L_N \left(|\cdot - x|^n \right) (x) + \frac{\left(L_N \left(|\cdot - x|^{n+1} \right) (x) \right)^{\frac{n}{n+1}}}{(n+1)} \right], \forall N \in \mathbb{N}. \tag{1.37}$$

We give ($n = 1$ case)

Corollary 1.3 *Let* $(L_N)_{N \in \mathbb{N}}$ *be a sequence of positive sublinear operators from* $C_+ (I)$ *into* $C B_+ (I)$, *and* $f \in C B_+^1 (I)$, *and* $I \subset \mathbb{R}$ *a bounded or unbounded interval. Assume* $L_N (1) = 1$, $\forall N \in \mathbb{N}$, *and* $f' (x) = 0$, *for a fixed* $x \in I$. *Furthermore assume that* $L_N \left((\cdot - x)^2 \right) (x) > 0$, $\forall N \in \mathbb{N}$.
Then

$$|L_N (f) (x) - f (x)| \leq \omega_1 \left(f', \sqrt{\left(L_N \left((\cdot - x)^2 \right) (x) \right)} \right) \cdot$$

$$\left[L_N \left(|\cdot - x| \right) (x) + \frac{\sqrt{\left(L_N \left((\cdot - x)^2 \right) (x) \right)}}{2} \right], \forall N \in \mathbb{N}. \tag{1.38}$$

Remark 1.4 (i) To Theorem 1.2: Assuming $f^{(n)}$ is uniformly continuous on I, and $L_N \left(|\cdot - x|^{n+1} \right) (x) \to 0$, as $N \to \infty$, using (1.35), we get that $(L_N (f)) (x) \to f (x)$, as $N \to \infty$.

(ii) To Corollary 1.3: Assuming f' is uniformly continuous on I, and $L_N \left((\cdot - x)^2 \right) (x) \to 0$, as $N \to \infty$, using (1.36), we get that $(L_N (f)) (x) \to f (x)$, as $N \to \infty$.

(iii) The right hand sides of (1.37), (1.38) are finite.

We also give the basic result ($n = 0$ case).

Theorem 1.5 *Let* $(L_N)_{N\in\mathbb{N}}$ *be a sequence of positive sublinear operators from* $C_+(I)$ *into* $CB_+(I)$, *and* $f \in CB_+(I)$, *where* $I \subset \mathbb{R}$ *a bounded or unbounded interval. Assume that* $L_N(|\cdot - x|)(x) > 0$, *for some fixed* $x \in I$, $\forall N \in \mathbb{N}$. *Then*

(1)

$$|L_N(f)(x) - f(x)| \leq f(x)|L_N(1)(x) - 1| +$$

$$[L_N(1)(x) + 1]\omega_1(f, L_N(|\cdot - x|)(x)), \forall N \in \mathbb{N}, \tag{1.39}$$

(2) when $L_N(1) = 1$, *we get*

$$|L_N(f)(x) - f(x)| \leq 2\omega_1(f, L_N(|\cdot - x|)(x)), \forall N \in \mathbb{N}. \tag{1.40}$$

Proof From [5], p. 17, we get

$$|L_N(f)(x) - f(x)| \leq f(x)|L_N(1)(x) - 1| +$$

$$\left[L_N(1)(x) + \frac{1}{\delta}L_N(|\cdot - x|)(x)\right]\omega_1(f, \delta), \tag{1.41}$$

where $\delta > 0$.

In (1.41) we choose $\delta = L_N(|\cdot - x|)(x) > 0$. ∎

Remark 1.6 (To Theorem 1.5) Here $x \in I$ is fixed.

(i) Assume $L_N(1)(x) \to 1$, as $N \to \infty$, and $L_N(|\cdot - x|)(x) \to 0$, as $N \to \infty$, given that f is uniformly continuous we get that $L_n(f)(x) \to f(x)$, as $N \to \infty$ (use of (1.39)). Notice here that $L_N(1)(x)$ is bounded.

(ii) Assume that $L_N(1) = 1$, and $L_N(|\cdot - x|)(x) \to 0$, as $N \to \infty$, and f is uniformly continuous on I, then $L_n(f)(x) \to f(x)$, as $N \to \infty$ (use of (1.40)).

(iii) The right hand sides of (1.39) and (1.40) are finite.

(iv) Variants of Theorem 1.5 have been applied extensively in [4, 5].

1.3 Applications

Here we give applications to Theorem 1.2 and Corollary 1.3.

Remark 1.7 We start with the Max-product Bernstein operators

$$B_N^{(M)}(f)(x) = \frac{\bigvee_{k=0}^{N} p_{N,k}(x) f\left(\frac{k}{N}\right)}{\bigvee_{k=0}^{N} p_{N,k}(x)}, \forall N \in \mathbb{N}, \tag{1.42}$$

$p_{N,k}(x) = \binom{N}{k} x^k (1-x)^{N-1}$, $x \in [0, 1]$, $\bigvee$ stands for maximum, and $f \in C_+([0, 1]) = \{f : [0, 1] \to \mathbb{R}_+ \text{ is continuous}\}$.

Clearly $B_N^{(M)}$ is a positive sublinear operator from $C_+([0, 1])$ into itself with $B_N^{(M)}(1) = 1$. Furthermore we notice that

$$B_N^{(M)}(|\cdot - x|^m)(x) = \frac{\bigvee_{k=0}^{N} p_{N,k}(x) |\frac{k}{N} - x|^m}{\bigvee_{k=0}^{N} p_{N,k}(x)} > 0, \tag{1.43}$$

$\forall x \in (0, 1)$ for any $m \in \mathbb{N}$, $\forall N \in \mathbb{N}$.

By [5], p. 31, we have

$$B_N^{(M)}(|\cdot - x|)(x) \le \frac{6}{\sqrt{N+1}}, \forall x \in [0, 1], N \in \mathbb{N}. \tag{1.44}$$

Notice that $|\cdot - x|^{m-1} \le 1$, therefore

$$|\cdot - x|^m = |\cdot - x| |\cdot - x|^{m-1} \le |\cdot - x| \, m \in \mathbb{N},$$

hence by (1.19),

$$B_N^{(M)}(|\cdot - x|^m)(x) \le B_N^{(M)}(|\cdot - x|)(x),$$

that is

$$B_N^{(M)}(|\cdot - x|^m)(x) \le \frac{6}{\sqrt{N+1}}, \forall x \in [0, 1], m, N \in \mathbb{N}. \tag{1.45}$$

Denote by

$$C_+^n([0, 1]) = \{f : [0, 1] \to \mathbb{R}_+, n\text{-times continuously differentiable}\}, n \in \mathbb{N}.$$

We give

Theorem 1.8 *Let* $f \in C_+^n([0, 1])$, *a fixed* $x \in (0, 1)$ *such that* $f^{(i)}(x) = 0$, $i = 1, \ldots, n$. *Then*

$$\left| B_N^{(M)}(f)(x) - f(x) \right| \le \frac{\omega_1\left(f^{(n)}, \left(\frac{6}{\sqrt{N+1}}\right)^{\frac{1}{n+1}}\right)}{n!} \cdot$$

$$\left[\frac{6}{\sqrt{N+1}} + \frac{1}{(n+1)}\left(\frac{6}{\sqrt{N+1}}\right)^{\frac{n}{n+1}}\right], \forall N \in \mathbb{N}. \tag{1.46}$$

We get $B_N^{(M)}(f)(x) \to f(x)$, *as* $N \to \infty$.

Proof By Theorem 1.2. ∎

The case $n = 1$ follows:

Corollary 1.9 *Let $f \in C_+^1 ([0, 1])$, a fixed $x \in (0, 1)$ such that $f'(x) = 0$. Then*

$$\left| B_N^{(M)} (f)(x) - f(x) \right| \leq \omega_1 \left(f', \frac{\sqrt{6}}{\sqrt[4]{N+1}} \right) \cdot$$

$$\left[\frac{6}{\sqrt{N+1}} + \frac{\sqrt{6}}{2 \left(\sqrt[4]{N+1} \right)} \right], \forall N \in \mathbb{N}. \tag{1.47}$$

We make

Remark 1.10 Let $f \in C^2 ([a, b], \mathbb{R}_+)$, then

$$|f(x) - f(y)| \leq \|f'\|_\infty |x - y| \, \forall \, x, y \in [a, b], \tag{1.48}$$

and

$$\left| f'(x) - f'(y) \right| \leq \|f''\|_\infty |x - y| \, \forall \, x, y \in [a, b]. \tag{1.49}$$

That is f, f' are Lipschitz type functions.

Next we provide examples so that

$$\|f''\|_\infty \leq \|f'\|_\infty. \tag{1.50}$$

(i) Let $f(x) = \sin x$, $f'(x) = \cos x$, $f''(x) = -\sin x$, here $\|f''\|_\infty = \|f'\|_\infty = 1$, for $x \in [0, \pi]$. Notice also that for $x = \frac{\pi}{2}$, we have $f'\left(\frac{\pi}{2}\right) = \cos\frac{\pi}{2} = 0$.

(ii) Let $x \in [0, \pi]$, $f(x) = (x-1)^3 + 1$, $f'(x) = 3(x-1)^2$, $f''(x) = 6(x-1)$, $f'(1) = 0$. Notice that $\|f'\|_\infty = 3(\pi-1)^2$, $\|f''\|_\infty = 6(\pi-1)$, by $|x - 1| \leq \pi - 1$. Because $6(\pi-1) \leq 3(\pi-1)^2$, we get that $\|f''\|_\infty \leq \|f'\|_\infty$.

So over Lipschitz classes of functions with Lipschitz derivatives we would like to compare (1.6) to (1.47).

Thus we some calculations we get that

$$\frac{\sqrt{6}}{\sqrt{N+1}} \left[\frac{6}{\sqrt{N+1}} + \frac{\sqrt{6}}{2 \left(\sqrt[4]{N+1} \right)} \right] \leq \frac{12}{\sqrt{N+1}}, \tag{1.51}$$

true $\forall N \in \mathbb{N}$, $N \geq 7$.

Similarly, we get that

$$\frac{1}{n!} \left(\frac{6}{\sqrt{N+1}} \right)^{\frac{1}{n+1}} \left[\frac{6}{\sqrt{N+1}} + \frac{1}{(n+1)} \left(\frac{6}{\sqrt{N+1}} \right)^{\frac{n}{n+1}} \right] \leq \frac{12}{\sqrt{N+1}}, \tag{1.52}$$

for large enough $N \in \mathbb{N}$.

Therefore (1.46) and (1.47), over differentiability, can give better estimates and speeds than (1.6).

We continue with

Remark 1.11 Here we focus on the truncated Favard–Szász–Mirakjan operators

$$T_N^{(M)}(f)(x) = \frac{\bigvee_{k=0}^N s_{N,k}(x) f\left(\frac{k}{N}\right)}{\bigvee_{k=0}^N s_{N,k}(x)}, x \in [0,1], \ N \in \mathbb{N}, f \in C_+([0,1]),$$

(1.53)

$s_{N,k}(x) = \frac{(Nx)^k}{k!}$, see also [5], p. 11.

By Theorem 3.2.5, [5], p. 178, we get that

$$\left| T_N^{(M)}(f)(x) - f(x) \right| \le 6\omega_1\left(f, \frac{1}{\sqrt{N}}\right) \forall N \in \mathbb{N}, \text{ any } x \in [0,1].$$

(1.54)

Also from [5], p. 178-179, we get that

$$T_N^{(M)}(|\cdot - x|)(x) = \frac{\bigvee_{k=0}^N \frac{(Nx)^k}{k!} \left|\frac{k}{N} - x\right|}{\bigvee_{k=0}^N \frac{(Nx)^k}{k!}} \le \frac{3}{\sqrt{N}}, \forall x \in [0,1], \ N \in \mathbb{N}.$$

(1.55)

Clearly it holds, for $m \in \mathbb{N}$ that

$$T_N^{(M)}(|\cdot - x|^m)(x) \le T_N^{(M)}(|\cdot - x|)(x),$$

and

$$T_N^{(M)}(|\cdot - x|^m)(x) \le \frac{3}{\sqrt{N}}, \forall x \in [0,1], \ N \in \mathbb{N}, m \in \mathbb{N}.$$

(1.56)

The operators $T_N^{(M)}$ are positive sublinear from $C_+([0,1])$ into itself with $T_N^{(M)}(1) = 1$. Also it holds

$$T_N^{(M)}(|\cdot - x|^m)(x) = \frac{\bigvee_{k=0}^N \frac{(Nx)^k}{k!} \left|\frac{k}{N} - x\right|^m}{\bigvee_{k=0}^N \frac{(Nx)^k}{k!}} > 0,$$

(1.57)

$\forall x \in (0,1]$, for any $m \in \mathbb{N}$, $\forall N \in \mathbb{N}$.

We give

Theorem 1.12 *Let* $f \in C_+^n([0,1])$, x *fixed in* $(0,1]$ *such that* $f^{(i)}(x) = 0$, $i = 1, \ldots, n$. *Then*

$$\left| T_N^{(M)}(f)(x) - f(x) \right| \le \frac{\omega_1\left(f^{(n)}, \left(\frac{3}{\sqrt{N}}\right)^{\frac{1}{n+1}}\right)}{n!}.$$

$$\left[\frac{3}{\sqrt{N}} + \frac{1}{(n+1)}\left(\frac{3}{\sqrt{N}}\right)^{\frac{n}{n+1}}\right], \forall N \in \mathbb{N}. \tag{1.58}$$

Proof By Theorem 1.2. ∎

The case $n = 1$ follows:

Corollary 1.13 *Let* $f \in C_+^1([0, 1])$, $x \in (0, 1] : f'(x) = 0$. *Then*

$$\left|T_N^{(M)}(f)(x) - f(x)\right| \le \omega_1\left(f', \frac{\sqrt{3}}{\sqrt[4]{N}}\right)\left[\frac{3}{\sqrt{N}} + \frac{\sqrt{3}}{2\sqrt[4]{N}}\right], \forall N \in \mathbb{N}. \tag{1.59}$$

From (1.58) and/or (1.59) we get that $T_N^{(M)}(f)(x) \to f(x)$, as $N \to \infty$.
We make

Remark 1.14 We compare (1.58), (1.59) to (1.54). We have

$$\frac{\sqrt{3}}{\sqrt{N}}\left[\frac{3}{\sqrt{N}} + \frac{\sqrt{3}}{2\sqrt[4]{N}}\right] \le \frac{6}{\sqrt{N}} \tag{1.60}$$

$$\Leftrightarrow$$

$$\frac{1}{\sqrt{N}} \le \frac{3\sqrt{3}}{6},$$

true for large enough $N \in \mathbb{N}$.
Also we find that

$$\frac{1}{n!}\left(\frac{3}{\sqrt{N}}\right)^{\frac{1}{n+1}}\left[\frac{3}{\sqrt{N}} + \frac{1}{(n+1)}\left(\frac{3}{\sqrt{N}}\right)^{\frac{n}{n+1}}\right] \le \frac{6}{\sqrt{N}} \tag{1.61}$$

$$\Leftrightarrow$$

$$\frac{1}{^{2(n+1)}\sqrt{N}} \le \frac{2n! - \frac{1}{(n+1)}}{\sqrt[n+1]{3}}, \tag{1.62}$$

true for large enough $N \in \mathbb{N}$.
Therefore (1.58), (1.59), over differentiability, give better estimates and speeds than (1.54).

We make

Remark 1.15 Next we study the truncated Max-product Baskakov operators (see [5], p. 11)

$$U_N^{(M)}(f)(x) = \frac{\bigvee_{k=0}^N b_{N,k}(x) f\left(\frac{k}{N}\right)}{\bigvee_{k=0}^N b_{N,k}(x)}, x \in [0, 1], \ f \in C_+([0, 1]), \ N \in \mathbb{N},$$

(1.63)

where

$$b_{N,k}(x) = \binom{N+k-1}{k} \frac{x^k}{(1+x)^{N+k}}.$$

(1.64)

From [5], pp. 217–218, we get $(x \in [0, 1])$

$$\left(U_N^{(M)}(|\cdot - x|)\right)(x) = \frac{\bigvee_{k=0}^N b_{N,k}(x) \left|\frac{k}{N} - x\right|}{\bigvee_{k=0}^N b_{N,k}(x)} \leq \frac{2\sqrt{3}\left(\sqrt{2}+2\right)}{\sqrt{N+1}} N \geq 2, N \in \mathbb{N}.$$

(1.65)

Let $f \in C_+([0, 1])$, then (by [5], p. 217):

$$\left|U_N^{(M)}(f)(x) - f(x)\right| \leq 24\omega_1\left(f, \frac{1}{\sqrt{N+1}}\right) N \in \mathbb{N}, N \geq 2, x \in [0, 1].$$

(1.66)

See here that

$$\left|\frac{k}{N} - x\right| \leq 1 \forall x \in [0, 1].$$

Let $m \in \mathbb{N}$, clearly then it holds

$$\left(U_N^{(M)}(|\cdot - x|^m)\right)(x) \leq \frac{2\sqrt{3}\left(\sqrt{2}+2\right)}{\sqrt{N+1}}, N \geq 2, N \in \mathbb{N}.$$

(1.67)

Also it holds $U_N^{(M)}(1)(x) = 1$, and $U_N^{(M)}$ are positive sublinear operators from $C_+([0, 1])$ into itself. Also it holds

$$U_N^{(M)}\left(|\cdot - x|^m\right)(x) > 0,$$

$\forall x \in (0, 1]$, for any $m \in \mathbb{N}$, $\forall N \in \mathbb{N}$.

We give

Theorem 1.16 *Let* $f \in C_+^n([0, 1])$, $x \in (0, 1]$ *fixed, such that* $f^{(i)}(x) = 0$, $i = 1, \ldots, n$, $n \in \mathbb{N}$. *Then*

$$\left|U_N^{(M)}(f)(x) - f(x)\right| \leq \frac{\omega_1\left(f^{(n)}, \left(\frac{2\sqrt{3}(\sqrt{2}+2)}{\sqrt{N+1}}\right)^{\frac{1}{n+1}}\right)}{n!}.$$

$$\left[\frac{2\sqrt{3}\left(\sqrt{2}+2\right)}{\sqrt{N+1}} + \frac{1}{(n+1)} \left(\frac{2\sqrt{3}\left(\sqrt{2}+2\right)}{\sqrt{N+1}} \right)^{\frac{n}{n+1}} \right], \forall N \in \mathbb{N} - \{1\}. \quad (1.68)$$

Proof By Theorem 1.2. ∎

The case $n = 1$ follows:

Corollary 1.17 *Let* $f \in C_+^1 \left([0, 1]\right)$, $x \in (0, 1]$ *fixed:* $f'(x) = 0$. *Then*

$$\left| U_N^{(M)} (f)(x) - f(x) \right| \leq \omega_1 \left(f', \left(\frac{2\sqrt{3}\left(\sqrt{2}+2\right)}{\sqrt{N+1}} \right)^{\frac{1}{2}} \right) \cdot$$

$$\left[\frac{2\sqrt{3}\left(\sqrt{2}+2\right)}{\sqrt{N+1}} + \frac{1}{2} \left(\frac{2\sqrt{3}\left(\sqrt{2}+2\right)}{\sqrt{N+1}} \right)^{\frac{1}{2}} \right], \forall N \in \mathbb{N} - \{1\}. \quad (1.69)$$

From (1.68) and/or (1.69) we get that $U_N^{(M)} (f)(x) \to f(x)$, as $N \to \infty$.

Remark 1.18 Next we compare (1.68), (1.69) to (1.66). We notice that

$$\left(\frac{2\sqrt{3}\left(\sqrt{2}+2\right)}{\sqrt{N+1}} \right)^{\frac{1}{2}} \left[\frac{2\sqrt{3}\left(\sqrt{2}+2\right)}{\sqrt{N+1}} + \frac{1}{2} \left(\frac{2\sqrt{3}\left(\sqrt{2}+2\right)}{\sqrt{N+1}} \right)^{\frac{1}{2}} \right] \leq \frac{24}{\sqrt{N+1}} \quad (1.70)$$

$$\Leftrightarrow$$

$$\frac{1}{\sqrt[4]{N+1}} \leq \frac{24 - \sqrt{3}\left(\sqrt{2}+2\right)}{\sqrt{2\sqrt{3}\left(\sqrt{2}+2\right)\left(2\sqrt{3}\left(\sqrt{2}+2\right)\right)}},$$

true for large enough $N \in \mathbb{N} - \{1\}$.

We also observe that

$$\frac{1}{n!} \left(\frac{2\sqrt{3}\left(\sqrt{2}+2\right)}{\sqrt{N+1}} \right)^{\frac{1}{n+1}} \left[\frac{2\sqrt{3}\left(\sqrt{2}+2\right)}{\sqrt{N+1}} + \frac{\left(\frac{2\sqrt{3}\left(\sqrt{2}+2\right)}{\sqrt{N+1}} \right)^{\frac{n}{n+1}}}{(n+1)} \right] \leq \frac{24}{\sqrt{N+1}} \quad (1.71)$$

$$\Leftrightarrow$$

$$\frac{1}{\sqrt[2(n+1)]{N+1}} \leq \frac{n!}{\sqrt[n+1]{2\sqrt{3}\left(\sqrt{2}+2\right)}}\left[\frac{12 - \frac{\sqrt{3}(\sqrt{2}+2)}{(n+1)!}}{\sqrt{3}\left(\sqrt{2}+2\right)}\right], \qquad (1.72)$$

true for large enough $N \in \mathbb{N} - \{1\}$.

Therefore (1.68), (1.69), over differentiability, give better estimates and speeds than (1.66).

We continue with

Remark 1.19 Here we study Max-product Meyer-Köning and Zeller operators (see [5], p. 11) defined by

$$Z_N^{(M)}(f)(x) = \frac{\bigvee_{k=0}^{\infty} s_{N,k}(x) f\left(\frac{k}{N+k}\right)}{\bigvee_{k=0}^{\infty} s_{N,k}(x)}, \forall N \in \mathbb{N}, f \in C_+([0,1]), \qquad (1.73)$$

$$s_{N,k}(x) = \binom{N+k}{k} x^k, x \in [0,1].$$

By [5], p. 248, we obtain

$$\left|Z_N^{(M)}(f)(x) - f(x)\right| \leq 18\omega_1\left(f, \frac{(1-x)\sqrt{x}}{\sqrt{N}}\right), N \geq 4, x \in [0,1]. \quad (1.74)$$

By [5], p. 253, we get that

$$Z_N^{(M)}(|\cdot - x|)(x) \leq \frac{8\left(1+\sqrt{5}\right)}{3} \frac{\sqrt{x}(1-x)}{\sqrt{N}} \forall x \in [0,1]. \qquad (1.75)$$

Let $m \in \mathbb{N}$, then

$$Z_N^{(M)}\left(|\cdot - x|^m\right)(x) = \frac{\bigvee_{k=0}^{N} s_{N,k}(x)\left|\frac{k}{N+k} - x\right|^m}{\bigvee_{k=0}^{N} s_{N,k}(x)} \leq Z_N^{(M)}(|\cdot - x|)(x), \quad (1.76)$$

so that

$$Z_N^{(M)}\left(|\cdot - x|^m\right)(x) \leq \frac{8\left(1+\sqrt{5}\right)}{3} \frac{\sqrt{x}(1-x)}{\sqrt{N}} =: \rho(x), \qquad (1.77)$$

$\forall x \in [0,1]$, $N \geq 4$, $\forall m \in \mathbb{N}$.

Also it holds $Z_N^{(M)}(1) = 1$, and $Z_N^{(M)}$ are positive sublinear operators from $C_+([0,1])$ into itself. Also it holds

$$Z_N^{(M)}\left(|\cdot - x|^m\right)(x) > 0, \qquad (1.78)$$

$\forall x \in (0,1)$, for any $m \in \mathbb{N}$, $\forall N \in \mathbb{N}$.

We give

Theorem 1.20 *Let $f \in C_+^n ([0, 1])$, $n \in \mathbb{N}$, $x \in (0, 1)$, $f^{(i)} (x) = 0$, $i = 1, \ldots, n$.*
Then

$$\left| Z_N^{(M)} (f) (x) - f (x) \right| \leq \frac{\omega_1 \left(f^{(n)}, (\rho (x))^{\frac{1}{n+1}} \right)}{n!} \left[\rho (x) + \frac{(\rho (x))^{\frac{n}{n+1}}}{(n + 1)} \right], \quad (1.79)$$

$\forall N \geq 4$, $N \in \mathbb{N}$.

Proof By Theorem 1.2. ∎

The case $n = 1$ follows:

Corollary 1.21 *Let $f \in C_+^1 ([0, 1])$, $x \in (0, 1)$: $f' (x) = 0$. Then*

$$\left| Z_N^{(M)} (f) (x) - f (x) \right| \leq \omega_1 \left(f', \sqrt{\rho (x)} \right) \left[\rho (x) + \frac{\sqrt{\rho (x)}}{2} \right], \quad (1.80)$$

$\forall N \geq 4$, $N \in \mathbb{N}$.
From (1.79), (1.80) we get that $Z_N^{(M)} (f) (x) \to f (x)$, as $N \to \infty$.

We finish with

Remark 1.22 Next we compare (1.79), (1.80) to (1.74).
We notice that

$$\sqrt{\rho (x)} \left[\rho (x) + \frac{\sqrt{\rho (x)}}{2} \right] \leq \frac{18 (1 - x) \sqrt{x}}{\sqrt{N}} \quad (1.81)$$

$$\Leftrightarrow$$

$$\frac{1}{\sqrt{N}} \leq \frac{3}{8 \left(1 + \sqrt{5} \right) \sqrt{x} (1 - x)} \left(\frac{27 - 2 \left(1 + \sqrt{5} \right)}{4 \left(1 + \sqrt{5} \right)} \right)^2,$$

true for large enough $N \geq 4$, $N \in \mathbb{N}$, $x \in (0, 1)$.
We also observe that

$$\frac{(\rho (x))^{\frac{1}{n+1}}}{n!} \left[\rho (x) + \frac{(\rho (x))^{\frac{n}{n+1}}}{n + 1} \right] \leq \frac{18 (1 - x) \sqrt{x}}{\sqrt{N}} \quad (1.82)$$

$$\Leftrightarrow$$

$$\frac{1}{\sqrt{N}} \leq \frac{3}{8\left(1+\sqrt{5}\right)\sqrt{x}\,(1-x)} \left(\frac{27n!}{4\left(1+\sqrt{5}\right)} - \frac{1}{n+1}\right)^{n+1}, \qquad (1.83)$$

true for large enough $N \geq 4$, $N \in \mathbb{N}$, $x \in (0, 1)$.

Therefore (1.79), (1.80), under differentiability, perform better than (1.74).

References

1. G. Anastassiou, *Moments in Probability and Approximation Theory*, Pitman Research Notes in Mathematics Series (Longman Group, New York, 1993)
2. G. Anastassiou, *Intelligent Computations: Abstract Fractional Calculus, Inequalities, Approximations* (Springer, Heidelberg, 2018)
3. G. Anastassiou, *Approximation by Sublinear Operators* (2017, submitted)
4. G. Anastassiou, L. Coroianu, S. Gal, Approximation by a nonlinear Cardaliagnet–Euvrard neural network operator of max-product kind. J. Comput. Anal. Appl. **12**(2), 396–406 (2010)
5. B. Bede, L. Coroianu, S. Gal, *Approximation by Max-Product Type Operators* (Springer, Heidelberg, 2016)
6. G.G. Lorentz, *Bernstein Polynomials*, 2nd edn. (Chelsea Publishing Company, New York, 1986)
7. T. Popoviciu, Sur l'approximation de fonctions convexes d'order superieur. Mathematica (Cluj) **10**, 49–54 (1935)

Chapter 2
High Order Approximation by Max-Product Operators

Here we study the approximation of functions by a great variety of Max-Product operators under differentiability. These are positive sublinear operators. Our study is based on our general results about positive sublinear operators. We produce Jackson type inequalities under initial conditions. So our approach is quantitative by producing inequalities with their right hand sides involving the modulus of continuity of a high order derivative of the function under approximation. We improve known related results which do not use smoothness of functions. It follows [2].

2.1 Introduction

The main motivation here comes from the monograph by B. Bede, L. Coroianu and S. Gal [4], 2016.

We mention the interpolation Hermite-Fejer polynomials on Kuibyshev knots of the first kind (see [4], p. 4): Let $f : [-1, 1] \to \mathbb{R}$ and based on the knots $x_{N,k} = \cos\left(\frac{(2(N-k)+1)}{2(N+1)}\pi\right) \in (-1, 1)$, $k \in \{0, ..., N\}$, $-1 < x_{N,0} < x_{N,1} < ... < x_{N,N} < 1$, which are the roots of the first kind Chebyshev polynomial $T_{N+1}(x) = \cos((N + 1) \arccos x)$, we define (see Fejér [5])

$$H_{2N+1}(f)(x) = \sum_{k=0}^{N} h_{N,k}(x) f\left(x_{N,k}\right), \qquad (2.1)$$

where

$$h_{N,k}(x) = \left(1 - x \cdot x_{N,k}\right)\left(\frac{T_{N+1}(x)}{(N+1)\left(x - x_{N,k}\right)}\right)^2, \qquad (2.2)$$

the fundamental interpolation polynomials.

© Springer International Publishing AG, part of Springer Nature 2018
G. A. Anastassiou, *Nonlinearity: Ordinary and Fractional Approximations by Sublinear and Max-Product Operators*, Studies in Systems, Decision and Control 147, https://doi.org/10.1007/978-3-319-89509-3_2

Denoting $A_{N+1}(f) = \|H_{2N+1} - f\|_\infty$, Fejer [5] proved that $\lim_{N \to \infty} A_{N+1}(f) = 0$, for all $f \in C([-1,1])$.

Popoviciu ([7]) also proved that $A_{N+1}(f) = O\left(\omega_1\left(f, \frac{1}{\sqrt{N+1}}\right)\right)$, and Moldovan ([6]) improved it to

$$A_{N+1}(f) = O\left(\omega_1\left(f, \frac{\ln(N+1)}{N+1}\right)\right).$$

Here $\omega_1(f,\delta) = \sup_{\substack{x,y \in [-1,1]: \\ |x-y| \le \delta}} |f(x) - f(y)|$, $\delta > 0$, is the first modulus of continuity.

The Max-product interpolation Hermite-Fejér operators on Chebyshev knots of the first kind (see p. 12 of [4]) are defined by

$$H_{2N+1}^{(M)}(f)(x) = \frac{\bigvee_{k=0}^{N} h_{N,k}(x) f(x_{N,k})}{\bigvee_{k=0}^{N} h_{N,k}(x)}, \quad \forall\, N \in \mathbb{N}, \tag{2.3}$$

where $f : [-1,1] \to \mathbb{R}_+$ is continuous.

By [4], p. 286 we get that

$$\left| H_{2N+1}^{(M)}(f)(x) - f(x) \right| \le 14\omega_1\left(f, \frac{1}{N+1}\right), \quad \forall\, N \in \mathbb{N}, \text{ any } x \in [-1,1]. \tag{2.4}$$

Call

$$E_N(x) := H_{2N+1}^{(M)}(|\cdot - x|)(x) = \frac{\bigvee_{k=0}^{N} h_{N,k}(x) |x_{N,k} - x|}{\bigvee_{k=0}^{N} h_{N,k}(x)}, \quad x \in [-1,1]. \tag{2.5}$$

Then by [4], p. 287 we obtain that

$$E_N(x) \le \frac{2\pi}{N+1}, \quad \forall\, x \in [-1,1], \ N \in \mathbb{N}. \tag{2.6}$$

For $m \in \mathbb{N}$, we get

$$H_{2N+1}^{(M)}(|\cdot - x|^m)(x) = \frac{\bigvee_{k=0}^{N} h_{N,k}(x) |x_{N,k} - x|^m}{\bigvee_{k=0}^{N} h_{N,k}(x)} =$$

$$\frac{\bigvee_{k=0}^{N} h_{N,k}(x) |x_{N,k} - x| |x_{N,k} - x|^{m-1}}{\bigvee_{k=0}^{N} h_{N,k}(x)} \le 2^{m-1} \frac{\bigvee_{k=0}^{N} h_{N,k}(x) |x_{N,k} - x|}{\bigvee_{k=0}^{N} h_{N,k}(x)} \tag{2.7}$$

$$\le \frac{2^m \pi}{N+1}, \quad \forall\, x \in [-1,1], \ N \in \mathbb{N}.$$

Hence it holds

$$H_{2N+1}^{(M)} \left(|\cdot - x|^m \right) (x) \leq \frac{2^m \pi}{N+1}, \forall\, x \in [-1, 1],\ m \in \mathbb{N}, \forall\, N \in \mathbb{N}. \tag{2.8}$$

Clearly it holds

$$H_{2N+1}^{(M)} \left(|\cdot - x|^m \right) (x) > 0, \tag{2.9}$$

$\forall\, x \in [-1, 1] : x \neq x_{N,k}, \forall\, N \in \mathbb{N}$, any $k \in \{0, 1, ..., N\}$; any $m \in \mathbb{N}$.

Furthermore we have

$$H_{2N+1}^{(M)} (1) (x) = 1, \forall\, x \in [-1, 1],$$

and $H_{2N+1}^{(M)}$ maps continuous functions to continuous functions over $[-1, 1]$ and for any $x \in \mathbb{R}$ we have $\bigvee_{k=0}^{N} h_{N,k}(x) > 0$.

We also have $h_{N,k}(x_{N,k}) = 1$, and $h_{N,k}(x_{N,j}) = 0$, if $k \neq j$, furthermore it holds $H_{2N+1}^{(M)} (f) (x_{N,j}) = f(x_{N,j})$, for all $j \in \{0, ..., N\}$, see [4], p. 282.

In this chapter we will improve (2.4) by assuming differentiability of f. Similar improvements, using the differentiability of f, will be presented for Max-product Lagrange interpolation operators, Max-product truncated sampling operators and Max-product Neural network operators.

2.2 Main Results

Let $I \subset \mathbb{R}$ be a bounded or unbounded interval, $n \in \mathbb{N}$, and

$$CB_+^n (I) = \left\{ f : I \rightarrow \mathbb{R}_+ : f^{(i)} \text{ is continuous and bounded on } I, \text{ for both } i = 0, n \right\}. \tag{2.10}$$

We define for

$$f \in CB_+ (I) = \{ f : I \rightarrow \mathbb{R}_+ : f \text{ is continuous and bounded on } I \}, \tag{2.11}$$

the first modulus of continuity

$$\omega_1 (f, \delta) = \sup_{\substack{x, y \in I: \\ |x-y| \leq \delta}} |f(x) - f(y)|, \tag{2.12}$$

where $0 < \delta \leq diameter\,(I)$.

Call $C_+ (I) = \{ f : I \rightarrow \mathbb{R}_+ : f \text{ is continuous on } I \}$.

Let $L_N : C_+ (I) \rightarrow CB_+ (I), n, N \in \mathbb{N}$ be a sequence of operators satisfying the following properties (see also [4], p. 17):

(i) (positive homogeneous)

$$L_N(\alpha f) = \alpha L_N(f), \forall \alpha \geq 0, f \in C_+(I), \tag{2.13}$$

(ii) (Monotonicity)

$$\text{if } f, g \in C_+(I) \text{ satisfy } f \leq g, \text{ then } L_N(f) \leq L_N(g), \forall N \in \mathbb{N}, \tag{2.14}$$

and

(iii) (Subadditivity)

$$L_N(f + g) \leq L_N(f) + L_N(g), \forall f, g \in C_+(I). \tag{2.15}$$

We call L_N positive sublinear operators.

In particular we will study the restrictions $L_N|_{CB_+^n(I)} : CB_+^n(I) \to CB_+(I)$.

The operators $H_{2N+1}^{(M)}$ are positive sublinear operators. From [1] we will be using the following result:

Theorem 2.1 ([1]) *Let* $(L_N)_{N \in \mathbb{N}}$ *be a sequence of positive sublinear operators from* $C_+(I)$ *into* $CB_+(I)$, *and* $f \in CB_+^n(I)$, *where* $n \in \mathbb{N}$ *and* $I \subset \mathbb{R}$ *a bounded or unbounded interval. Assume* $L_N(1) = 1$, $\forall N \in \mathbb{N}$, *and* $f^{(i)}(x) = 0$, $i = 1, ..., n$, *for a fixed* $x \in I$, *and* $\delta > 0$. *Then*

$$|L_N(f)(x) - f(x)| \leq \frac{\omega_1\left(f^{(n)}, \delta\right)}{n!} \left[L_N\left(|\cdot - x|^n\right)(x) + \frac{L_N\left(|\cdot - x|^{n+1}\right)(x)}{(n+1)\delta} \right], \tag{2.16}$$

$\forall N \in \mathbb{N}$.

We give

Theorem 2.2 *Let* $f \in C^n([-1, 1], \mathbb{R}_+)$, *with* $f^{(i)}(x) = 0$, $i = 1, ..., n \in \mathbb{N}$, *for some fixed* $x \in [-1, 1]$, $N \in \mathbb{N}$. *Then*

$$\left| H_{2N+1}^{(M)}(f)(x) - f(x) \right| \leq$$

$$\frac{1}{n!}\omega_1\left(f^{(n)}, 2\sqrt[n+1]{\frac{\pi}{N+1}}\right) \left[\frac{2^n \pi}{N+1} + \frac{2^n}{(n+1)}\left(\sqrt[n+1]{\frac{\pi}{N+1}}\right)^n \right]. \tag{2.17}$$

When $x = x_{N,k}$, *the left hand side of (2.17) is zero.*

Proof Here we are using (2.8) and (2.16), namely we have

$$H_{2N+1}^{(M)} \left(|\cdot - x|^m \right) (x) \leq \frac{2^m \pi}{N+1}, \forall\, x \in [-1, 1], \, m \in \mathbb{N}, \forall\, N \in \mathbb{N}. \tag{2.18}$$

and

$$\left| H_{2N+1}^{(M)} (f) (x) - f(x) \right| \leq \frac{\omega_1 \left(f^{(n)}, \delta \right)}{n!} \left[\frac{2^n \pi}{N+1} + \frac{1}{(n+1)\delta} \left(\frac{2^{n+1}\pi}{N+1} \right) \right], \tag{2.19}$$

$\delta > 0$.

Then choose

$$\delta := 2 \sqrt[n+1]{\frac{\pi}{N+1}}, \tag{2.20}$$

we get

$$\delta^{n+1} = 2^{n+1} \frac{\pi}{N+1}, \tag{2.21}$$

and

$$\left| H_{2N+1}^{(M)} (f) (x) - f(x) \right| \leq \frac{\omega_1 \left(f^{(n)}, 2 \sqrt[n+1]{\frac{\pi}{N+1}} \right)}{n!} \left[\frac{2^n \pi}{N+1} + \frac{1}{(n+1)} \delta^n \right] =$$

$$\frac{1}{n!} \omega_1 \left(f^{(n)}, 2 \sqrt[n+1]{\frac{\pi}{N+1}} \right) \left[\frac{2^n \pi}{N+1} + \frac{1}{(n+1)} 2^n \left(\sqrt[n+1]{\frac{\pi}{N+1}} \right)^n \right], \tag{2.22}$$

proving the claim. ■

It follows the $n = 1$ case.

Corollary 2.3 *Let* $f \in C^1 ([-1, 1], \mathbb{R}_+)$ *, with* $f'(x) = 0$*, for some fixed* $x \in [-1, 1]$*,* $N \in \mathbb{N}$*. Then*

$$\left| H_{2N+1}^{(M)} (f) (x) - f(x) \right| \leq$$

$$\omega_1 \left(f', 2 \sqrt{\frac{\pi}{N+1}} \right) \left[\frac{2\pi}{N+1} + \sqrt{\frac{\pi}{N+1}} \right]. \tag{2.23}$$

From (2.17) and/or (2.23), as $N \to \infty$*, we get that* $H_{2N+1}^{(M)} (f) (x) \to f(x)$*.*

Proof By (2.17). ■

We make

Remark 2.4 We compare (2.23) to (2.4). We prove that (2.23) gives a sharper estimate and speed than (2.4). We observe that

$$\frac{2\sqrt{\pi}}{\sqrt{N+1}} \left(\frac{2\pi}{N+1} + \frac{\sqrt{\pi}}{\sqrt{N+1}} \right) \le \frac{14}{N+1} \qquad (2.24)$$

$$\Leftrightarrow$$

$$\frac{1}{\sqrt{N+1}} \le \frac{7-\pi}{2\pi\sqrt{\pi}}, \qquad (2.25)$$

true for large enough $N \in \mathbb{N}$.

We also make

Remark 2.5 Here we compare (2.17) to (2.4). We prove that (2.17) gives a better estimate and speed than (2.4). We see that

$$\frac{2 \sqrt[n+1]{\pi}}{n! \sqrt[n+1]{N+1}} \left[\frac{2^n \pi}{N+1} + \frac{2^n}{(n+1)} \left(\frac{\sqrt[n+1]{\pi}}{\sqrt[n+1]{N+1}} \right)^n \right] \le \frac{14}{N+1}, \qquad (2.26)$$

$$\Leftrightarrow$$

$$\frac{1}{\sqrt[n+1]{N+1}} \le \left(7 - \frac{2^n \pi}{(n+1)!} \right) \frac{n!}{2^n \pi \sqrt[n+1]{\pi}}, \qquad (2.27)$$

true for large enough $N \in \mathbb{N}$.
About notice that $7 - \frac{2^n \pi}{(n+1)!} > 0$.

We continue with

Remark 2.6 Here we deal with Lagrange interpolation polynomials on Chebyshev knots of second kind plus the endpoints ± 1 (see [4], p. 5). These polynomials are linear operators attached to $f : [-1, 1] \to \mathbb{R}$ and to the knots $x_{N,k} = \cos \left(\left(\frac{N-k}{N-1} \right) \pi \right) \in [-1, 1]$, $k = 1, ..., N$, $N \in \mathbb{N}$, which are the roots of $\omega_N(x) = \sin(N-1) t \sin t$, $x = \cos t$. Notice that $x_{N,1} = -1$ and $x_{N,N} = 1$. Their formula is given by ([4], p. 5)

$$L_N(f)(x) = \sum_{k=1}^{N} l_{N,k}(x) f(x_{N,k}), \qquad (2.28)$$

where

$$l_{N,k}(x) = \frac{(-1)^{k-1} \omega_N(x)}{(1 + \delta_{k,1} + \delta_{k,N})(N-1)(x - x_{N,k})}, \qquad (2.29)$$

$N \ge 2$, $k = 1, ..., N$, and $\omega_N(x) = \prod_{k=1}^{N}(x - x_{N,k})$ and $\delta_{i,j}$ denotes the Kronecher's symbol, that is $\delta_{i,j} = 1$, if $i = j$, and $\delta_{i,j} = 0$, if $i \ne j$. Then (see [4], p. 5)

$$\|L_N(f) - f\|_{\infty,[-1,1]} \le c\omega_1 \left(f, \frac{1}{N} \right) \ln N. \qquad (2.30)$$

The Max-product Lagrange interpolation operators on Chebyshev knots of second kind, plus the endpoints ± 1, are defined by ([4], p. 12)

$$L_N^{(M)}(f)(x) = \frac{\bigvee_{k=1}^{N} l_{N,k}(x) f(x_{N,k})}{\bigvee_{k=1}^{N} l_{N,k}(x)}, \quad x \in [-1, 1], \tag{2.31}$$

where $f : [-1, 1] \to \mathbb{R}_+$ is continuous.

First we see that $L_N^{(M)}(f)(x)$ is well defined and continuous for any $x \in [-1, 1]$. Following [4], p. 289, because $\sum_{k=1}^{N} l_{N,k}(x) = 1$, $\forall x \in \mathbb{R}$, for any x there exists $k \in \{1, ..., N\} : l_{N,k}(x) > 0$, hence $\bigvee_{k=1}^{N} l_{N,k}(x) > 0$. We have that $l_{N,k}(x_{N,k}) = 1$, and $l_{N,k}(x_{N,j}) = 0$, if $k \neq j$. Furthermore it holds $L_N^{(M)}(f)(x_{N,j}) = f(x_{N,j})$, all $j \in \{1, ..., N\}$, and $L_N^{(M)}(1) = 1$.

Call $I_N^+(x) = \{k \in \{1, ..., N\}; l_{N,k}(x) > 0\}$, then $I_N^+(x) \neq \emptyset$.

So for $f \in CB_+([-1, 1])$ we get

$$L_N^{(M)}(f)(x) = \frac{\bigvee_{k \in I_N^+(x)} l_{N,k}(x) f(x_{N,k})}{\bigvee_{k \in I_N^+(x)} l_{N,k}(x)} \geq 0. \tag{2.32}$$

By [4], p. 295, we have:

Let $f \in C([-1, 1], \mathbb{R}_+)$, $N \in \mathbb{N}$, $N \geq 3$, N is odd, then

$$\left| L_N^{(M)}(f)(x) - f(x) \right| \leq 4\omega_1\left(f, \frac{1}{N-1}\right), \quad \forall x \in [-1, 1]. \tag{2.33}$$

Notice here that $|x_{N,k} - x| \leq 2$, $\forall x \in [-1, 1]$.

By [4], p. 297, we get that

$$L_N^{(M)}(|\cdot - x|)(x) = \frac{\bigvee_{k=1}^{N} l_{N,k}(x) |x_{N,k} - x|}{\bigvee_{k=1}^{N} l_{N,k}(x)} =$$

$$\frac{\bigvee_{k \in I_N^+(x)} l_{N,k}(x) |x_{N,k} - x|}{\bigvee_{k \in I_N^+(x)} l_{N,k}(x)} \leq \frac{\pi^2}{6(N-1)}, \tag{2.34}$$

$N \geq 3$, $\forall x \in (-1, 1)$, N is odd.

We get that ($m \in \mathbb{N}$)

$$L_N^{(M)}(|\cdot - x|^m)(x) = \frac{\bigvee_{k \in I_N^+(x)} l_{N,k}(x) |x_{N,k} - x|^m}{\bigvee_{k \in I_N^+(x)} l_{N,k}(x)} \leq \frac{2^{m-1}\pi^2}{6(N-1)}, \tag{2.35}$$

$N \geq 3$ odd, $\forall\, x \in (-1, 1)$.

We present

Theorem 2.7 *Let* $f \in C^n\left([-1,1], \mathbb{R}_+\right)$, $n \in \mathbb{N}$, $x \in [-1,1]$, $f^{(i)}(x) = 0$, $i = 1, ..., n$. *Here* $N \in \mathbb{N}$, $N \geq 3$ *is odd. Then*

$$\left| L_N^{(M)}(f)(x) - f(x) \right| \leq$$

$$\frac{\omega_1\left(f^{(n)}, \left(\frac{2^n \pi^2}{6(N-1)}\right)^{\frac{1}{n+1}}\right)}{n!} \left[\frac{2^{n-1} \pi^2}{6(N-1)} + \frac{1}{(n+1)} \left(\frac{2^n \pi^2}{6(N-1)}\right)^{\frac{n}{n+1}} \right]. \tag{2.36}$$

Proof When $x = \pm 1$, the left hand side of (2.36) is zero, hence (2.36) is trivially true. Let now $x \in (-1, 1)$, by Theorem 2.1 and (2.16), (2.35), we obtain

$$\left| L_N^{(M)}(f)(x) - f(x) \right| \leq$$

$$\frac{\omega_1\left(f^{(n)}, \delta\right)}{n!} \left[\frac{2^{n-1} \pi^2}{6(N-1)} + \frac{1}{(n+1)\delta} \frac{2^n \pi^2}{6(N-1)} \right] \tag{2.37}$$

(setting $\delta := \left(\frac{2^n \pi^2}{6(N-1)}\right)^{\frac{1}{n+1}}$, i.e. $\delta^{n+1} = \frac{2^n \pi^2}{6(N-1)}$)

$$= \frac{\omega_1\left(f^{(n)}, \left(\frac{2^n \pi^2}{6(N-1)}\right)^{\frac{1}{n+1}}\right)}{n!} \left[\frac{2^{n-1} \pi^2}{6(N-1)} + \frac{1}{(n+1)} \left(\frac{2^n \pi^2}{6(N-1)}\right)^{\frac{n}{n+1}} \right],$$

proving the claim. ∎

The case $n = 1$ follows:

Corollary 2.8 *Let* $f \in C^1\left([-1,1], \mathbb{R}_+\right)$, $x \in [-1,1]$, $f'(x) = 0$. *Here* $N \in \mathbb{N}$, $N \geq 3$ *is odd. Then*

$$\left| L_N^{(M)}(f)(x) - f(x) \right| \leq$$

$$\omega_1\left(f', \frac{\pi}{\sqrt{3(N-1)}}\right) \left[\frac{\pi^2}{6(N-1)} + \frac{\pi}{2\sqrt{3(N-1)}} \right]. \tag{2.38}$$

By (2.36) and/or (2.38), we get that $L_N^{(M)}(f)(x) \to f(x)$, *as* $N \to \infty$.

We make

Remark 2.9 Here we compare (2.38) to (2.33), and we prove that (2.38) gives better estimates and speeds than (2.33). We observe that

$$\frac{\pi}{\sqrt{3(N-1)}}\left[\frac{\pi^2}{6(N-1)}+\frac{\pi}{2\sqrt{3(N-1)}}\right]\leq\frac{4}{N-1} \tag{2.39}$$

$$\Leftrightarrow$$

$$\frac{1}{\sqrt{3(N-1)}}\leq\frac{24-\pi^2}{\pi^3}, \tag{2.40}$$

true for large enough $N\geq 3$ odd.

Remark 2.10 Here we compare (2.36) to (2.33), and we prove that (2.36) gives better estimates and speeds that (2.33). We see that

$$\frac{1}{n!}\left(\frac{2^n\pi^2}{6(N-1)}\right)^{\frac{1}{n+1}}\left[\frac{2^{n-1}\pi^2}{6(N-1)}+\frac{1}{(n+1)}\left(\frac{2^n\pi^2}{6(N-1)}\right)^{\frac{n}{n+1}}\right]\leq\frac{4}{N-1} \tag{2.41}$$

$$\Leftrightarrow$$

$$\frac{1}{N-1}\leq\frac{3}{2^{n-1}\pi^2}\left(\frac{24(n+1)!-2^n\pi^2}{(n+1)2^{n-1}\pi^2}\right)^{n+1}, \tag{2.42}$$

true for large enough N (odd) ≥ 3.

We continue with

Remark 2.11 From [4], p. 297, we have: Let $f\in C([-1,1],\mathbb{R}_+)$, $N\geq 4$, $N\in\mathbb{N}$, N even. Then

$$\left|L_N^{(M)}(f)(x)-f(x)\right|\leq 28\omega_1\left(f,\frac{1}{N-1}\right),\quad\forall x\in[-1,1]. \tag{2.43}$$

From [4], p. 298, we get

$$L_N^{(M)}(|\cdot-x|)(x)\leq\frac{4\pi^2}{3(N-1)}=\frac{2^2\pi^2}{3(N-1)},\quad\forall x\in(-1,1). \tag{2.44}$$

Hence ($m\in\mathbb{N}$)

$$L_N^{(M)}(|\cdot-x|^m)(x)\leq\frac{2^{m+1}\pi^2}{3(N-1)},\quad\forall x\in(-1,1). \tag{2.45}$$

We present

Theorem 2.12 *Let* $f \in C^n([-1, 1], \mathbb{R}_+)$, $n \in \mathbb{N}$, $x \in [-1, 1]$, $f^{(i)}(x) = 0$, $i = 1, ..., n$. *Here* $N \in \mathbb{N}$, $N \geq 4$, N *is even. Then*

$$\left| L_N^{(M)}(f)(x) - f(x) \right| \leq$$

$$\frac{\omega_1 \left(f^{(n)}, \left(\frac{2^{n+2}\pi^2}{3(N-1)} \right)^{\frac{1}{n+1}} \right)}{n!} \left[\frac{2^{n+1}\pi^2}{3(N-1)} + \frac{1}{(n+1)} \left(\frac{2^{n+2}\pi^2}{3(N-1)} \right)^{\frac{n}{n+1}} \right]. \qquad (2.46)$$

Proof When $x = \pm 1$, the left hand side of (2.46) is zero, thus (2.46) is trivially true. Let now $x \in (-1, 1)$, by Theorem 2.1 and (2.16), (2.45), we obtain

$$\left| L_N^{(M)}(f)(x) - f(x) \right| \leq$$

$$\frac{\omega_1 \left(f^{(n)}, \delta \right)}{n!} \left[\frac{2^{n+1}\pi^2}{3(N-1)} + \frac{1}{(n+1)\delta} \frac{2^{n+2}\pi^2}{3(N-1)} \right] \qquad (2.47)$$

(setting $\delta := \left(\frac{2^{n+2}\pi^2}{3(N-1)} \right)^{\frac{1}{n+1}}$, i.e. $\delta^{n+1} = \frac{2^{n+2}\pi^2}{3(N-1)}$)

$$= \frac{\omega_1 \left(f^{(n)}, \left(\frac{2^{n+2}\pi^2}{3(N-1)} \right)^{\frac{1}{n+1}} \right)}{n!} \left[\frac{2^{n+1}\pi^2}{3(N-1)} + \frac{1}{(n+1)} \left(\frac{2^{n+2}\pi^2}{3(N-1)} \right)^{\frac{n}{n+1}} \right],$$

proving the claim.   ■

The case $n = 1$ follows:

Corollary 2.13 *Let* $f \in C^1([-1, 1], \mathbb{R}_+)$, $x \in [-1, 1]$, $f'(x) = 0$. *Here* $N \in \mathbb{N}$, $N \geq 4$, N *is even. Then*

$$\left| L_N^{(M)}(f)(x) - f(x) \right| \leq$$

$$\omega_1 \left(f', \frac{2\pi\sqrt{2}}{\sqrt{3(N-1)}} \right) \left[\frac{4\pi^2}{3(N-1)} + \frac{\pi\sqrt{2}}{\sqrt{3(N-1)}} \right]. \qquad (2.48)$$

By (2.46) and/or (2.48), we get that $L_N^{(M)}(f)(x) \to f(x)$, *as* $N \to \infty$.

We make

Remark 2.14 Here we compare (2.48) to (2.43). We prove that (2.48) gives better estimates and speeds that (2.43). Indeed we have

$$\frac{2\pi\sqrt{2}}{\sqrt{3}\,(N-1)}\left[\frac{4\pi^2}{3\,(N-1)}+\frac{\pi\sqrt{2}}{\sqrt{3}\,(N-1)}\right]\le\frac{28}{N-1} \qquad (2.49)$$

$$\Leftrightarrow$$

$$\frac{1}{N-1}\le\frac{3}{8\pi^2}\left(\frac{42-2\pi^2}{2\pi^2}\right)^2, \qquad (2.50)$$

true for large enough $N\ge 4$, even.

We make

Remark 2.15 Here we compare (2.46) to (2.43). We prove that (2.46) gives better estimates and speeds that (2.43). We observe that

$$\frac{1}{n!}\left(\frac{2^{n+2}\pi^2}{3\,(N-1)}\right)^{\frac{1}{n+1}}\left[\frac{2^{n+1}\pi^2}{3\,(N-1)}+\frac{1}{(n+1)}\left(\frac{2^{n+2}\pi^2}{3\,(N-1)}\right)^{\frac{n}{n+1}}\right]\le\frac{28}{N-1} \qquad (2.51)$$

$$\Leftrightarrow$$

$$\frac{1}{N-1}\le\left(\frac{3}{2^{n+2}\pi^2}\right)\left[\frac{42\,(n+1)!-2^{n+1}\pi^2}{(n+1)\,2^n\pi^2}\right]^{n+1}, \qquad (2.52)$$

true for large enough $N\ge 4$, N even.

We continue with

Remark 2.16 The sampling truncated linear operators (see [4], p. 7) are defined by

$$W_N\,(f)\,(x)=\sum_{k=0}^{N}\frac{\sin\,(Nx-k\pi)}{Nx-k\pi}f\left(\frac{k\pi}{N}\right),\ \forall\,x\in[0,\pi], \qquad (2.53)$$

and

$$T_N\,(f)\,(x)=\sum_{k=0}^{N}\frac{\sin^2\,(Nx-k\pi)}{(Nx-k\pi)^2}f\left(\frac{k\pi}{N}\right),\ \forall\,x\in[0,\pi];\ f\in C\,([0,\pi]\,,\mathbb{R})\,, \qquad (2.54)$$

and they are used as approximators.

Here we deal with the Max-product truncated sampling operators (see [4], p. 13) defined by

$$W_N^{(M)}\,(f)\,(x)=\frac{\bigvee_{k=0}^{N}\frac{\sin(Nx-k\pi)}{Nx-k\pi}f\left(\frac{k\pi}{N}\right)}{\bigvee_{k=0}^{N}\frac{\sin(Nx-k\pi)}{Nx-k\pi}},\ x\in[0,\pi], \qquad (2.55)$$

$f:[0,\pi]\to\mathbb{R}_+$, continuous,

and

$$T_N^{(M)}(f)(x) = \frac{\bigvee_{k=0}^N \frac{\sin^2(Nx-k\pi)}{(Nx-k\pi)^2} f\left(\frac{k\pi}{N}\right)}{\bigvee_{k=0}^N \frac{\sin^2(Nx-k\pi)}{(Nx-k\pi)^2}}, \quad x \in [0, \pi], \tag{2.56}$$

$f : [0, \pi] \to \mathbb{R}_+$, continuous.

Following [4], p. 343, and making the convention $\frac{\sin(0)}{0} = 1$ and denoting $s_{N,k}(x) = \frac{\sin(Nx-k\pi)}{Nx-k\pi}$, we get that $s_{N,k}\left(\frac{k\pi}{N}\right) = 1$, and $s_{N,k}\left(\frac{j\pi}{N}\right) = 0$, if $k \neq j$, furthermore $W_N^{(M)}(f)\left(\frac{j\pi}{N}\right) = f\left(\frac{j\pi}{N}\right)$, for all $j \in \{0, ..., N\}$.

Clearly $W_N^{(M)}(f)$ is a well-defined function for all $x \in [0, \pi]$, and it is continuous on $[0, \pi]$, also $W_N^{(M)}(1) = 1$.

By [4], p. 344, $W_N^{(M)}$ are positive sublinear operators.

Call $I_N^+(x) = \{k \in \{0, 1, ..., N\}; s_{N,k}(x) > 0\}$, and set $x_{N,k} := \frac{k\pi}{N}$, $k \in \{0, 1, ..., N\}$.

We see that

$$W_N^{(M)}(f)(x) = \frac{\bigvee_{k \in I_N^+(x)} s_{N,k}(x) f(x_{N,k})}{\bigvee_{k \in I_N^+(x)} s_{N,k}(x)}. \tag{2.57}$$

We call

$$F_N(x) := W_N^{(M)}(|\cdot - x|)(x) = \frac{\bigvee_{k=0}^N s_{N,k}(x)|x_{N,k} - x|}{\bigvee_{k=0}^N s_{N,k}(x)}$$

$$= \frac{\bigvee_{k \in I_N^+(x)} s_{N,k}(x)|x_{N,k} - x|}{\bigvee_{k \in I_N^+(x)} s_{N,k}(x)}. \tag{2.58}$$

By Theorem 8.2.8 ([4], p. 345) we get: Let $f \in C([0, \pi], \mathbb{R}_+)$. Then

$$\left| W_N^{(M)}(f)(x) - f(x) \right| \leq 4\omega_1\left(f, \frac{1}{N}\right)_{[0,\pi]}, \quad \forall N \in \mathbb{N}, x \in [0, \pi]. \tag{2.59}$$

We have that ([4], p. 346)

$$F_N(x) \leq \frac{\pi}{2N}. \tag{2.60}$$

Notice also $|x_{N,k} - x| \leq \pi, \forall x \in [0, \pi]$.

Therefore ($m \in \mathbb{N}$) it holds

$$W_N^{(M)}(|\cdot - x|^m)(x) \leq \frac{\pi^{m-1}\pi}{2N} = \frac{\pi^m}{2N}. \tag{2.61}$$

We present

Theorem 2.17 *Let $f \in C^n([0, \pi], \mathbb{R}_+)$, $x \in [0, \pi]$ fixed, $f^{(i)}(x) = 0, i = 1, ..., n$. Then*

$$\left| W_N^{(M)}\left(f\right)(x) - f\left(x\right) \right| \leq \frac{\omega_1\left(f^{(n)}, \left(\frac{\pi^{n+1}}{2N}\right)^{\frac{1}{n+1}}\right)}{n!} \left[\frac{\pi^n}{2N} + \frac{\left(\frac{\pi^{n+1}}{2N}\right)^{\frac{n}{n+1}}}{n+1}\right] = \quad (2.62)$$

$$\frac{1}{n!}\omega_1\left(f^{(n)}, \frac{\pi}{\sqrt[n+1]{2N}}\right)\left[\frac{\pi^n}{2N} + \frac{1}{n+1}\left(\frac{\pi}{\sqrt[n+1]{2N}}\right)^n\right]. \quad (2.63)$$

Proof Using Theorem 2.1, (2.16) and (2.61), we get

$$\left| W_N^{(M)}\left(f\right)(x) - f\left(x\right) \right| \leq \frac{\omega_1\left(f^{(n)}, \delta\right)}{n!}\left[\frac{\pi^n}{2N} + \frac{1}{(n+1)\delta}\left(\frac{\pi^{n+1}}{2N}\right)\right] \quad (2.64)$$

(choosing $\delta := \left(\frac{\pi^{n+1}}{2N}\right)^{\frac{1}{n+1}}$, i.e. $\delta^{n+1} = \frac{\pi^{n+1}}{2N}$)

$$= \frac{1}{n!}\omega_1\left(f^{(n)}, \left(\frac{\pi^{n+1}}{2N}\right)^{\frac{1}{n+1}}\right)\left[\frac{\pi^n}{2N} + \frac{1}{(n+1)}\left(\frac{\pi^{n+1}}{2N}\right)^{\frac{n}{n+1}}\right],$$

proving the claim. ∎

The case $n = 1$ follows:

Corollary 2.18 *Let* $f \in C^1\left([0, \pi], \mathbb{R}_+\right)$, $x \in [0, \pi]$ *fixed,* $f'\left(x\right) = 0$. *Then*

$$\left| W_N^{(M)}\left(f\right)(x) - f\left(x\right) \right| \leq \omega_1\left(f', \frac{\pi}{\sqrt{2N}}\right)\left[\frac{\pi}{2N} + \frac{\pi}{2\sqrt{2N}}\right]. \quad (2.65)$$

By (2.62)–(2.63) and/or (2.65), we get that $W_N^{(M)}\left(f\right)(x) \to f\left(x\right)$, *as* $N \to +\infty$.

We make

Remark 2.19 Here we compare (2.65) to (2.59) and we prove that (2.65) gives better estimates and speeds that (2.59). Indeed we have

$$\frac{\pi}{\sqrt{2N}}\left(\frac{\pi}{2N} + \frac{\pi}{2\sqrt{2N}}\right) \leq \frac{4}{N} \quad (2.66)$$

$$\Leftrightarrow$$

$$\frac{1}{\sqrt{2N}} \leq \frac{16 - \pi^2}{2\pi^2}, \quad (2.67)$$

true for large enough $N \in \mathbb{N}$.

We also make

Remark 2.20 Here we compare (2.62)–(2.63) to (2.59), and we prove that (2.62)–(2.63) gives better estimates and speeds that (2.59). We observe that

$$\frac{1}{n!}\left(\frac{\pi^{n+1}}{2N}\right)^{\frac{1}{n+1}}\left[\frac{\pi^n}{2N}+\frac{\left(\frac{\pi^{n+1}}{2N}\right)^{\frac{n}{n+1}}}{(n+1)}\right]\leq\frac{4}{N} \tag{2.68}$$

$$\Leftrightarrow$$

$$\frac{1}{\sqrt[n+1]{2N}}\leq\frac{8\,(n+1)!-\pi^{n+1}}{(n+1)\,\pi^{n+1}}, \tag{2.69}$$

true for large enough $N \in \mathbb{N}$.

Notice here that $8\,(n+1)!-\pi^{n+1}>0, \forall\, n\in\mathbb{N}$.

We continue with

Remark 2.21 Here we study $T_N\,(f)\,(x)$, see (2.54).

By Theorem 8.2.13, [4], p. 352, we get: Let $f\in C\,([0,\pi],\mathbb{R}_+)$, then

$$\left|T_N^{(M)}\,(f)\,(x)-f\,(x)\right|\leq 4\omega_1\left(f,\frac{1}{N}\right)_{[0,\pi]}, \quad \forall\, N\in\mathbb{N}, x\in[0,\pi]. \tag{2.70}$$

By [4], p. 352, we get

$$T_N^{(M)}\,(|\cdot-x|)\,(x)\leq\frac{\pi}{2N}, \tag{2.71}$$

hence ($m\in\mathbb{N}$) we find

$$T_N^{(M)}\,(|\cdot-x|^m)\,(x)\leq\frac{\pi^m}{2N}. \tag{2.72}$$

Here again $x_{N,k}=\frac{k\pi}{N}, k\in\{0,1,...,N\}$.

The operators $T_N^{(M)}$ are positive sublinear operators, mapping $C\,([0,\pi],\mathbb{R}_+)$ into itself, and $T_N^{(M)}\,(1)=1$. So we can apply again Theorem 2.1. We obtain the same results as before with $W_N^{(M)}$, we state them:

Theorem 2.22 *Let $f\in C^1\,([0,\pi],\mathbb{R}_+), x\in[0,\pi]$ fixed, $f'\,(x)=0$. Then*

$$\left|T_N^{(M)}\,(f)\,(x)-f\,(x)\right|\leq\omega_1\left(f',\frac{\pi}{\sqrt{2N}}\right)\left[\frac{\pi}{2N}+\frac{\pi}{2\sqrt{2N}}\right], \quad \forall\, N\in\mathbb{N}. \tag{2.73}$$

Theorem 2.23 *Let $f\in C^n\,([0,\pi],\mathbb{R}_+), x\in[0,\pi]$ fixed, $f^{(i)}\,(x)=0, i=1,...,n$. Then*

$$\left|T_N^{(M)}\,(f)\,(x)-f\,(x)\right|\leq$$

$$\frac{\omega_1\left(f^{(n)}, \frac{\pi}{n+\sqrt[1]{2N}}\right)}{n!}\left[\frac{\pi^n}{2N} + \frac{\left(\frac{\pi}{n+\sqrt[1]{2N}}\right)^n}{n+1}\right], \ \forall N \in \mathbb{N}. \tag{2.74}$$

Clearly (2.73) and (2.74) can perform better than (2.70), the same study as for $W_N^{(M)}$. Furthermore we derive $T_N^{(M)}(f)(x) \rightarrow f(x)$, as $N \rightarrow +\infty$.

We continue with

Remark 2.24 Let $b : \mathbb{R} \rightarrow \mathbb{R}_+$ be a centered (it takes a global maximum at 0) bell-shaped function, with compact support $[-T, T]$, $T > 0$ (that is $b(x) > 0$ for all $x \in (-T, T)$) and $I = \int_{-T}^{T} b(x)\,dx > 0$.

The Cardaliaguet-Euvrard neural network operators are defined by (see [3])

$$C_{N,\alpha}(f)(x) = \sum_{k=-N^2}^{N^2} \frac{f\left(\frac{k}{N}\right)}{IN^{1-\alpha}} b\left(N^{1-\alpha}\left(x - \frac{k}{N}\right)\right), \tag{2.75}$$

$0 < \alpha < 1$, $N \in \mathbb{N}$ and $f : \mathbb{R} \rightarrow \mathbb{R}$ is continuous and bounded or uniformly continuous on $\mathbb{R}$.

$CB(\mathbb{R})$ denotes the continuous and bounded function on $\mathbb{R}$, and

$$CB_+(\mathbb{R}) = \{f : \mathbb{R} \rightarrow [0, \infty); \ f \in CB(\mathbb{R})\}.$$

The corresponding max-product Cardaliaguet-Euvrard neural network operators will be given by

$$C_{N,\alpha}^{(M)}(f)(x) = \frac{\bigvee_{k=-N^2}^{N^2} b\left(N^{1-\alpha}\left(x - \frac{k}{N}\right)\right) f\left(\frac{k}{N}\right)}{\bigvee_{k=-N^2}^{N^2} b\left(N^{1-\alpha}\left(x - \frac{k}{N}\right)\right)}, \tag{2.76}$$

$f \in \mathbb{R}$, $f \in CB_+(\mathbb{R})$, see also [3].

Next we follow [3].

For any $x \in \mathbb{R}$, denoting

$$J_{T,N}(x) = \left\{k \in \mathbb{Z}; \ -N^2 \le k \le N^2, N^{1-\alpha}\left(x - \frac{k}{N}\right) \in (-T, T)\right\},$$

we can write

$$C_{N,\alpha}^{(M)}(f)(x) = \frac{\bigvee_{k \in J_{T,N}(x)} b\left(N^{1-\alpha}\left(x - \frac{k}{N}\right)\right) f\left(\frac{k}{N}\right)}{\bigvee_{k \in J_{T,N}(x)} b\left(N^{1-\alpha}\left(x - \frac{k}{N}\right)\right)}, \tag{2.77}$$

$x \in \mathbb{R}$, $N > \max\left\{T + |x|, T^{-\frac{1}{\alpha}}\right\}$, where $J_{T,N}(x) \ne \emptyset$. Indeed, we have $\bigvee_{k \in J_{T,N}(x)} b\left(N^{1-\alpha}\left(x - \frac{k}{N}\right)\right) > 0$, $\forall x \in \mathbb{R}$ and $N > \max\left\{T + |x|, T^{-\frac{1}{\alpha}}\right\}$.

We have that $C_{N,\alpha}^{(M)}(1)(x) = 1, \forall\, x \in \mathbb{R}$ and $N > \max\left\{T + |x|, T^{-\frac{1}{\alpha}}\right\}$.

See in [3] there: Lemma 2.1, Corollary 2.2 and Remarks.
We need

Theorem 2.25 ([3]) *Let $b(x)$ be a centered bell-shaped function, continuous and with compact support $[-T, T]$, $T > 0$, $0 < \alpha < 1$ and $C_{N,\alpha}^{(M)}$ be defined as in (2.76).*

(i) If $|f(x)| \le c$ for all $x \in \mathbb{R}$ then $\left|C_{N,\alpha}^{(M)}(f)(x)\right| \le c$, for all $x \in \mathbb{R}$ and $N > \max\left\{T + |x|, T^{-\frac{1}{\alpha}}\right\}$ and $C_{N,\alpha}^{(M)}(f)(x)$ is continuous at any point $x \in \mathbb{R}$, for all $N > \max\left\{T + |x|, T^{-\frac{1}{\alpha}}\right\}$;

(ii) If $f, g \in CB_+(\mathbb{R})$ satisfy $f(x) \le g(x)$ for all $x \in \mathbb{R}$, then $C_{N,\alpha}^{(M)}(f)(x) \le C_{N,\alpha}^{(M)}(g)(x)$ for all $x \in \mathbb{R}$ and $N > \max\left\{T + |x|, T^{-\frac{1}{\alpha}}\right\}$;

(iii) $C_{N,\alpha}^{(M)}(f + g)(x) \le C_{N,\alpha}^{(M)}(f)(x) + C_{N,\alpha}^{(M)}(g)(x)$ for all $f, g \in CB_+(\mathbb{R})$, $x \in \mathbb{R}$ and $N > \max\left\{T + |x|, T^{-\frac{1}{\alpha}}\right\}$;

(iv) For all $f, g \in CB_+(\mathbb{R})$, $x \in \mathbb{R}$ and $N > \max\left\{T + |x|, T^{-\frac{1}{\alpha}}\right\}$, we have

$$\left|C_{N,\alpha}^{(M)}(f)(x) - C_{N,\alpha}^{(M)}(g)(x)\right| \le C_{N,\alpha}^{(M)}(|f - g|)(x);$$

(v) $C_{N,\alpha}^{(M)}$ is positive homogeneous, that is $C_{N,\alpha}^{(M)}(\lambda f)(x) = \lambda C_{N,\alpha}^{(M)}(f)(x)$ for all $\lambda \ge 0$, $x \in \mathbb{R}$, $N > \max\left\{T + |x|, T^{-\frac{1}{\alpha}}\right\}$ and $f \in CB_+(\mathbb{R})$.

We make

Remark 2.26 We have

$$E_{N,\alpha}(x) := C_{N,\alpha}^{(M)}(|\cdot - x|)(x) = \frac{\bigvee_{k \in J_{T,N}(x)} b\left(N^{1-\alpha}\left(x - \frac{k}{N}\right)\right)\left|x - \frac{k}{N}\right|}{\bigvee_{k \in J_{T,N}(x)} b\left(N^{1-\alpha}\left(x - \frac{k}{N}\right)\right)}, \quad (2.78)$$

$\forall\, x \in \mathbb{R}$, and $N > \max\left\{T + |x|, T^{-\frac{1}{\alpha}}\right\}$.

By (2.77), $C_{N,\alpha}^{(M)}$ satisfies

$$C_{N,\alpha}^{(M)}(f \vee g)(x) = C_{N,\alpha}^{(M)}(f)(x) \vee C_{N,\alpha}^{(M)}(g)(x),$$

$\forall\, f, g \in CB_+(\mathbb{R})$, $x \in \mathbb{R}$, $N > \max\left\{T + |x|, T^{-\frac{1}{\alpha}}\right\}$.

Notice that

$$\bigvee_{k \in J_{T,N}(x)} b\left(N^{1-\alpha}\left(x - \frac{k}{N}\right)\right) = \bigvee_{k=-N^2}^{N^2} b\left(N^{1-\alpha}\left(x - \frac{k}{N}\right)\right). \quad (2.79)$$

By [3], Lemma 3.1 there, we have: Let $b(x)$ be a centered bell-shaped function, continuous and with compact support $[-T, T]$, $T > 0$ and $0 < \alpha < 1$. Then for any $j \in \mathbb{Z}$ with $-N^2 \le j \le N^2$, all $x \in \left[\frac{j}{N}, \frac{j+1}{N}\right]$ and $N > \max\left\{T + |x|, T^{-\frac{1}{\alpha}}\right\}$, we have

$$\bigvee_{k=-N^2}^{N^2} b\left(N^{1-\alpha}\left(x - \frac{k}{N}\right)\right) =$$

$$\max\left\{b\left(N^{1-\alpha}\left(x - \frac{j}{N}\right)\right), b\left(N^{1-\alpha}\left(x - \frac{j+1}{N}\right)\right)\right\} > 0. \qquad (2.80)$$

Lemma 3.1 ([3]), is valid only for all $x \in [-N, N]$.

We mention from [3] the following:

Theorem 2.27 ([3]) *Let $b(x)$ be a centered bell-shaped function, continuous and with compact support $[-T, T]$, $T > 0$ and $0 < \alpha < 1$. In addition, suppose that the following requirements are fulfilled:*

(i) There exist $0 < m_1 \le M_1 < \infty$ such that $m_1 (T - x) \le b(x) \le M_1 (T - x)$, $\forall\, x \in [0, T]$;

(ii) There exist $0 < m_2 \le M_2 < \infty$ such that $m_2 (x + T) \le b(x) \le M_2 (x + T)$, $\forall\, x \in [-T, 0]$.

Then for all $f \in CB_+(\mathbb{R})$, $x \in \mathbb{R}$ and for all $N \in \mathbb{N}$ satisying $N > \max\left\{T + |x|, \left(\frac{2}{T}\right)^{\frac{1}{\alpha}}\right\}$, we have the estimate

$$\left|C_{N,\alpha}^{(M)}(f)(x) - f(x)\right| \le c\omega_1\left(f, N^{\alpha-1}\right)_{\mathbb{R}}, \qquad (2.81)$$

where

$$c := 2\left(\max\left\{\frac{TM_2}{2m_2}, \frac{TM_1}{2m_1}\right\} + 1\right),$$

and

$$\omega_1(f, \delta)_{\mathbb{R}} := \sup_{\substack{x, y \in \mathbb{R}: \\ |x-y| \le \delta}} |f(x) - f(y)|.$$

We make

Remark 2.28 In [3], was proved that

$$E_{N,\alpha}(x) \le \max\left\{\frac{TM_2}{2m_2}, \frac{TM_1}{2m_1}\right\} N^{\alpha-1}, \ \forall\, N > \max\left\{T + |x|, \left(\frac{2}{T}\right)^{\frac{1}{\alpha}}\right\}. \qquad (2.82)$$

That is

$$C_{N,\alpha}^{(M)} \left(|\cdot - x| \right) (x) \leq \max \left\{ \frac{TM_2}{2m_2}, \frac{TM_1}{2m_1} \right\} N^{\alpha-1}, \ \forall \ N > \max \left\{ T + |x|, \left(\frac{2}{T} \right)^{\frac{1}{\alpha}} \right\}.$$
(2.83)

From (2.78) we have that $\left| x - \frac{k}{N} \right| \leq \frac{T}{N^{1-\alpha}}$.

Hence $(m \in \mathbb{N})$ $(\forall \ x \in \mathbb{R}$ and $N > \max \left\{ T + |x|, \left(\frac{2}{T} \right)^{\frac{1}{\alpha}} \right\})$

$$C_{N,\alpha}^{(M)} \left(|\cdot - x|^m \right) (x) = \frac{\bigvee_{k \in J_{T,N}(x)} b \left(N^{1-\alpha} \left(x - \frac{k}{N} \right) \right) \left| x - \frac{k}{N} \right|^m}{\bigvee_{k \in J_{T,N}(x)} b \left(N^{1-\alpha} \left(x - \frac{k}{N} \right) \right)} \leq \qquad (2.84)$$

$$\left(\frac{T}{N^{1-\alpha}} \right)^{m-1} \max \left\{ \frac{TM_2}{2m_2}, \frac{TM_1}{2m_1} \right\} N^{\alpha-1}, \ \forall \ N > \max \left\{ T + |x|, \left(\frac{2}{T} \right)^{\frac{1}{\alpha}} \right\}.$$

Then $(m \in \mathbb{N})$ it holds

$$C_{N,\alpha}^{(M)} \left(|\cdot - x|^m \right) (x) \leq$$

$$T^{m-1} \max \left\{ \frac{TM_2}{2m_2}, \frac{TM_1}{2m_1} \right\} \frac{1}{N^{m(1-\alpha)}}, \ \forall \ N > \max \left\{ T + |x|, \left(\frac{2}{T} \right)^{\frac{1}{\alpha}} \right\}. \quad (2.85)$$

Call

$$\lambda := \max \left\{ \frac{TM_2}{2m_2}, \frac{TM_1}{2m_1} \right\} > 0. \qquad (2.86)$$

Consequently $(m \in \mathbb{N})$ we derive

$$C_{N,\alpha}^{(M)} \left(|\cdot - x|^m \right) (x) \leq \frac{\lambda T^{m-1}}{N^{m(1-\alpha)}}, \ \forall \ N > \max \left\{ T + |x|, \left(\frac{2}{T} \right)^{\frac{1}{\alpha}} \right\}. \qquad (2.87)$$

We need

Theorem 2.29 *Let $b(x)$ be a centered bell-shaped function, continuous and with compact support $[-T, T]$, $T > 0$, $0 < \alpha < 1$ and $C_{N,\alpha}^{(M)}$ be defined as in (2.76).*

Let $f \in CB_+^n (\mathbb{R})$, $n \in \mathbb{N}$. Let $x \in \mathbb{R} : f^{(i)} (x) = 0$, $i = 1, ..., n$, and $\delta > 0$. Then

$$\left| C_{N,\alpha}^{(M)} (f) (x) - f (x) \right| \leq$$

$$\frac{\omega_1 \left(f^{(n)}, \delta \right)_{\mathbb{R}}}{n!} \left[C_{N,\alpha}^{(M)} \left(|\cdot - x|^n \right) (x) + \frac{C_{N,\alpha}^{(M)} \left(|\cdot - x|^{n+1} \right) (x)}{(n+1) \delta} \right], \qquad (2.88)$$

$$\forall \ N \in \mathbb{N} : N > \max \left\{ T + |x|, T^{-\frac{1}{\alpha}} \right\}.$$

Proof By [1], we get that

$$|f(x) - f(y)| \le \frac{\omega_1\left(f^{(n)}, \delta\right)_{\mathbb{R}}}{n!} \left[|x - y|^n + \frac{|x - y|^{n+1}}{(n+1)\delta}\right], \ \forall \, y \in \mathbb{R}, \delta > 0. \tag{2.89}$$

Using Theorem 2.25 and $C_{N,\alpha}^{(M)}(1) = 1$, we get

$$\left|C_{N,\alpha}^{(M)}(f)(x) - f(x)\right| \le C_{N,\alpha}^{(M)}\left(|f(\cdot) - f(x)|\right)(x) \le$$

$$\frac{\omega_1\left(f^{(n)}, \delta\right)_{\mathbb{R}}}{n!}\left[C_{N,\alpha}^{(M)}\left(|\cdot - x|^n\right)(x) + \frac{C_{N,\alpha}^{(M)}\left(|\cdot - x|^{n+1}\right)(x)}{(n+1)\delta}\right], \tag{2.90}$$

$\forall \, N \in \mathbb{N} : N > \max\left\{T + |x|, T^{-\frac{1}{\alpha}}\right\}.$ ∎

We give

Theorem 2.30 *Same assumptions as in Theorem 2.27. Let* $f \in CB_+^n(\mathbb{R})$, $n \in \mathbb{N}$, $x \in \mathbb{R} : f^{(i)}(x) = 0$, $i = 1, \dots, n$. *Then*

$$\left|C_{N,\alpha}^{(M)}(f)(x) - f(x)\right| \le$$

$$\frac{\omega_1\left(f^{(n)}, \left(\frac{\lambda T^n}{N^{(n+1)(1-\alpha)}}\right)^{\frac{1}{n+1}}\right)_{\mathbb{R}}}{n!}\left[\frac{\lambda T^{n-1}}{N^{n(1-\alpha)}} + \frac{1}{(n+1)}\left(\frac{\lambda T^n}{N^{(n+1)(1-\alpha)}}\right)^{\frac{n}{n+1}}\right], \tag{2.91}$$

$\forall \, N > \max\left\{T + |x|, \left(\frac{2}{T}\right)^{\frac{1}{\alpha}}\right\}.$

Proof We use (2.88) and we choose

$$\delta := \left(\frac{\lambda T^n}{N^{(n+1)(1-\alpha)}}\right)^{\frac{1}{n+1}}, \tag{2.92}$$

i.e. $\delta^{n+1} = \frac{\lambda T^n}{N^{(n+1)(1-\alpha)}}$. Hence

$$\left|C_{N,\alpha}^{(M)}(f)(x) - f(x)\right| \overset{((2.87),\,(2.88))}{\le}$$

$$\frac{1}{n!}\omega_1\left(f^{(n)}, \left(\frac{\lambda T^n}{N^{(n+1)(1-\alpha)}}\right)^{\frac{1}{n+1}}\right)_{\mathbb{R}}\left[\frac{\lambda T^{n-1}}{N^{n(1-\alpha)}} + \frac{1}{(n+1)\delta}\frac{\lambda T^n}{N^{(n+1)(1-\alpha)}}\right] =$$

$$\frac{1}{n!}\omega_1\left(f^{(n)}, \left(\frac{\lambda T^n}{N^{(n+1)(1-\alpha)}}\right)^{\frac{1}{n+1}}\right)_{\mathbb{R}}\left[\frac{\lambda T^{n-1}}{N^{n(1-\alpha)}} + \frac{1}{(n+1)}\left(\frac{\lambda T^n}{N^{(n+1)(1-\alpha)}}\right)^{\frac{n}{n+1}}\right], \tag{2.93}$$

$\forall\, N > \max\left\{T + |x|, \left(\frac{2}{T}\right)^{\frac{1}{\alpha}}\right\}$, proving the claim. ∎

It follows the case $n = 1$.

Corollary 2.31 *Same assumptions as in Theorem 2.27. Let $f \in CB_+^1(\mathbb{R})$, $x \in \mathbb{R}$:*
$f'(x) = 0$. Then

$$\left|C_{N,\alpha}^{(M)}(f)(x) - f(x)\right| \le$$

$$\omega_1\left(f', \sqrt{\frac{\lambda T}{N^{2(1-\alpha)}}}\right)_{\mathbb{R}}\left[\frac{\lambda}{N^{1-\alpha}} + \frac{1}{2}\sqrt{\frac{\lambda T}{N^{2(1-\alpha)}}}\right], \tag{2.94}$$

$\forall\, N > \max\left\{T + |x|, \left(\frac{2}{T}\right)^{\frac{1}{\alpha}}\right\}$.

By (2.91) and/or (2.94) we get that $C_{N,\alpha}^{(M)}(f)(x) \to f(x)$, as $N \to +\infty$.

We make

Remark 2.32 We prove that (2.94) performs better than (2.81).
Indeed we have that

$$\sqrt{\frac{\lambda T}{N^{2(1-\alpha)}}}\left[\frac{\lambda}{N^{1-\alpha}} + \frac{1}{2}\sqrt{\frac{\lambda T}{N^{2(1-\alpha)}}}\right] \le \frac{2(\lambda + 1)}{N^{1-\alpha}}, \tag{2.95}$$

$$\Leftrightarrow$$

$$\frac{1}{N^{(1-\alpha)}} \le \frac{2(\lambda + 1)}{\lambda\left(\sqrt{\lambda T} + \frac{T}{2}\right)}, \tag{2.96}$$

true $\forall\, N > \max\left\{T + |x|, \left(\frac{2}{T}\right)^{\frac{1}{\alpha}}\right\}$, large enough.

We also make

Remark 2.33 We prove that (2.91) performs better than (2.81). We observe that

$$\frac{1}{n!}\left(\frac{\lambda T^n}{N^{(n+1)(1-\alpha)}}\right)^{\frac{1}{n+1}}\left[\frac{\lambda T^{n-1}}{N^{n(1-\alpha)}} + \frac{1}{(n+1)}\left(\frac{\lambda T^n}{N^{(n+1)(1-\alpha)}}\right)^{\frac{n}{n+1}}\right] \le \frac{2(\lambda + 1)}{N^{1-\alpha}} \tag{2.97}$$

$$\Leftrightarrow$$

$$\frac{1}{N^{n(1-\alpha)}} \le \frac{2(\lambda + 1)}{\left[\frac{\lambda^{n+1}\sqrt{\lambda T}^{\frac{n^2+n-1}{n+1}}}{n!} + \frac{\lambda T^n}{(n+1)!}\right]}, \tag{2.98}$$

true $\forall\, N > \max\left\{T + |x|, \left(\frac{2}{T}\right)^{\frac{1}{\alpha}}\right\}$, large enough.

Here using Theorem 2.1 we extend the domain of the application results of [1].

Remark 2.34 We start with the Max-product Bernstein operators ([4], p. 10)

$$B_N^{(M)}(f)(x) = \frac{\bigvee_{k=0}^{N} p_{N,k}(x) f\left(\frac{k}{N}\right)}{\bigvee_{k=0}^{N} p_{N,k}(x)}, \; \forall N \in \mathbb{N}, \tag{2.99}$$

$p_{N,k}(x) = \binom{N}{k} x^k (1-x)^{N-1}$, $x \in [0, 1]$, $\bigvee$ stands for maximum, and $f \in C_+([0, 1]) = \{f : [0, 1] \to \mathbb{R}_+ \text{ is continuous}\}$.
    From [1] we get

$$B_N^{(M)}\left(|\cdot - x|^m\right)(x) \le \frac{6}{\sqrt{N+1}}, \forall x \in [0, 1], m, N \in \mathbb{N}. \tag{2.100}$$

Denote by

$$C_+^n([0, 1]) = \{f : [0, 1] \to \mathbb{R}_+, n\text{-times continuously differentiable}\}, \; n \in \mathbb{N}.$$

We give

Theorem 2.35 *Let* $f \in C_+^n([0, 1])$, *a fixed* $x \in [0, 1]$ *such that* $f^{(i)}(x) = 0$, $i = 1, ..., n$. *Then*

$$\left| B_N^{(M)}(f)(x) - f(x) \right| \le \frac{\omega_1\left(f^{(n)}, \left(\frac{6}{\sqrt{N+1}}\right)^{\frac{1}{n+1}}\right)}{n!} \cdot$$

$$\left[\frac{6}{\sqrt{N+1}} + \frac{1}{(n+1)} \left(\frac{6}{\sqrt{N+1}}\right)^{\frac{n}{n+1}} \right], \; \forall N \in \mathbb{N}. \tag{2.101}$$

We get $B_N^{(M)}(f)(x) \to f(x)$, *as* $N \to \infty$.

Proof Use of (2.16) for $\delta = \left(\frac{6}{\sqrt{N+1}}\right)^{\frac{1}{n+1}}$. ∎

We continue with

Remark 2.36 Here we focus on the truncated Favard-Szász-Mirakjan operators

$$K_N^{(M)}(f)(x) = \frac{\bigvee_{k=0}^{N} s_{N,k}(x) f\left(\frac{k}{N}\right)}{\bigvee_{k=0}^{N} s_{N,k}(x)}, \; x \in [0, 1], \; N \in \mathbb{N}, f \in C_+([0, 1]), \tag{2.102}$$

$s_{N,k}(x) = \frac{(Nx)^k}{k!}$, see [4], p. 11.

From [1] we get

$$K_N^{(M)}\left(|\cdot - x|^m\right)(x) \le \frac{3}{\sqrt{N}}, \ \forall\, x \in [0, 1], \ N \in \mathbb{N}, m \in \mathbb{N}. \qquad (2.103)$$

We give

Theorem 2.37 *Let* $f \in C_+^n\left([0, 1]\right)$, x *fixed in* $[0, 1]$ *such that* $f^{(i)}(x) = 0$, $i = 1, ..., n$. *Then*

$$\left|K_N^{(M)}(f)(x) - f(x)\right| \le \frac{\omega_1\left(f^{(n)}, \left(\frac{3}{\sqrt{N}}\right)^{\frac{1}{n+1}}\right)}{n!} \cdot$$

$$\left[\frac{3}{\sqrt{N}} + \frac{1}{(n+1)}\left(\frac{3}{\sqrt{N}}\right)^{\frac{n}{n+1}}\right], \ \forall\, N \in \mathbb{N}. \qquad (2.104)$$

Proof Use of (2.16) for $\delta = \left(\frac{3}{\sqrt{N}}\right)^{\frac{1}{n+1}}$. ∎

We make

Remark 2.38 Next we study the truncated Max-product Baskakov operators (see [4], p. 11)

$$U_N^{(M)}(f)(x) = \frac{\bigvee_{k=0}^{N} b_{N,k}(x)\, f\left(\frac{k}{N}\right)}{\bigvee_{k=0}^{N} b_{N,k}(x)}, \ x \in [0, 1], \ f \in C_+\left([0, 1]\right), \ N \in \mathbb{N},$$

$$\qquad (2.105)$$

where

$$b_{N,k}(x) = \binom{N+k-1}{k} \frac{x^k}{(1+x)^{N+k}}.$$

We give

Theorem 2.39 *Let* $f \in C_+^n\left([0, 1]\right)$, $x \in [0, 1]$ *fixed, such that* $f^{(i)}(x) = 0$, $i = 1, ..., n$, $n \in \mathbb{N}$. *Then*

$$\left|U_N^{(M)}(f)(x) - f(x)\right| \le \frac{\omega_1\left(f^{(n)}, \left(\frac{2\sqrt{3}(\sqrt{2}+2)}{\sqrt{N+1}}\right)^{\frac{1}{n+1}}\right)}{n!} \cdot \qquad (2.106)$$

$$\left[\frac{2\sqrt{3}\left(\sqrt{2}+2\right)}{\sqrt{N+1}} + \frac{1}{(n+1)}\left(\frac{2\sqrt{3}\left(\sqrt{2}+2\right)}{\sqrt{N+1}}\right)^{\frac{n}{n+1}}\right], \ \forall\, N \in \mathbb{N} - \{1\}.$$

Proof Use of (2.16) for $\delta = \left(\frac{2\sqrt{3}(\sqrt{2}+2)}{\sqrt{N+1}}\right)^{\frac{1}{n+1}}$, we use that (see [1])

$$\left(U_N^{(M)}\left(|\cdot - x|^m\right)\right)(x) \leq \frac{2\sqrt{3}\left(\sqrt{2}+2\right)}{\sqrt{N+1}}, N \geq 2, N \in \mathbb{N}. \qquad (2.107)$$

■

We make

Remark 2.40 Here we study Max-product Meyer-Köning and Zeller operators (see [4], p. 11) defined by

$$Z_N^{(M)}(f)(x) = \frac{\bigvee_{k=0}^{\infty} s_{N,k}(x) f\left(\frac{k}{N+k}\right)}{\bigvee_{k=0}^{\infty} s_{N,k}(x)}, \ \forall \, N \in \mathbb{N}, \, f \in C_+([0,1]), \qquad (2.108)$$

$$s_{N,k}(x) = \binom{N+k}{k} x^k, \, x \in [0,1].$$

From [1] we get

$$Z_N^{(M)}\left(|\cdot - x|^m\right)(x) \leq \frac{8\left(1+\sqrt{5}\right)}{3} \frac{\sqrt{x}\,(1-x)}{\sqrt{N}} =: \rho(x), \qquad (2.109)$$

$\forall \, x \in [0,1], \, N \geq 4, \, \forall \, m \in \mathbb{N}.$

We finish with

Theorem 2.41 *Let* $f \in C_+^n([0,1])$, $n \in \mathbb{N}$, $x \in [0,1]$, $f^{(i)}(x) = 0$, $i = 1, ..., n$. *Then*

$$\left| Z_N^{(M)}(f)(x) - f(x) \right| \leq \frac{\omega_1\left(f^{(n)}, (\rho(x))^{\frac{1}{n+1}}\right)}{n!} \left[\rho(x) + \frac{(\rho(x))^{\frac{n}{n+1}}}{(n+1)}\right], \qquad (2.110)$$

$\forall \, N \geq 4, \, N \in \mathbb{N}.$

Proof Use of (2.16) with $\delta = (\rho(x))^{\frac{1}{n+1}}$. ■

References

1. G. Anastassiou, *Approximation by Sublinear Operators* (2017, submitted).
2. G. Anastassiou, *Approximation by Max-Product Operators* (2017, submitted)
3. G. Anastassiou, L. Coroianu, S. Gal, Approximation by a nonlinear Cardaliaguet-Euvrard neural network operator of max-product kind. J. Comput. Anal. Appl. **12**(2), 396–406 (2010)
4. B. Bede, L. Coroianu, S. Gal, *Approximation by Max-Product Type Operators* (Springer, Heidelberg, New York, 2016)
5. L. Fejér, *Über Interpolation*, Göttingen Nachrichten, (1916), pp. 66–91

6. E. Moldovan, *Observations sur certains procédé s d'interpolation généralisés*, (Romanian, Russian and French summaries). Acad. Republicii Pop. Romane Bul. Stiint. Sect. Stiint. Mat. Fiz. **6**, 477–482 (1954)
7. T. Popoviciu, *On the proof of Weierstrass' theorem with the interpolation polynomials*, Acad. Republicii Pop. Romane, Lucrarile sesiunii generale stiintifice din 2-12 Iunie 1950, vol. 1–4, pp. 1664–1667 (1950) (in Romanian)

Chapter 3
Conformable Fractional Approximations Using Max-Product Operators

Here we study the approximation of functions by a big variety of Max-Product operators under conformable fractional differentiability. These are positive sublinear operators. Our study is based on our general results about positive sublinear operators. We produce Jackson type inequalities under conformable fractional initial conditions. So our approach is quantitative by producing inequalities with their right hand sides involving the modulus of continuity of a high order conformable fractional derivative of the function under approximation. It follows [3].

3.1 Introduction

The main motivation here is the monograph by B. Bede, L. Coroianu and S. Gal [5], 2016.

Let $N \in \mathbb{N}$, the well-known Bernstein polynomials [8] are positive linear operators, defined by the formula

$$B_N(f)(x) = \sum_{k=0}^{N} \binom{N}{k} x^k (1-x)^{N-k} f\left(\frac{k}{N}\right), \quad x \in [0, 1], f \in C([0, 1]).$$
(3.1)

T. Popoviciu in [9], 1935, proved for $f \in C([0, 1])$ that

$$|B_N(f)(x) - f(x)| \le \frac{5}{4}\omega_1\left(f, \frac{1}{\sqrt{N}}\right), \quad \forall x \in [0, 1],$$
(3.2)

where

$$\omega_1(f, \delta) = \sup_{\substack{x,y\in[0,1]: \\ |x-y|\le\delta}} |f(x) - f(y)|, \quad \delta > 0,$$
(3.3)

is the first modulus of continuity.

© Springer International Publishing AG, part of Springer Nature 2018
G. A. Anastassiou, *Nonlinearity: Ordinary and Fractional Approximations
by Sublinear and Max-Product Operators*, Studies in Systems, Decision
and Control 147, https://doi.org/10.1007/978-3-319-89509-3_3

G.G. Lorentz in [8], 1986, p. 21, proved for $f \in C^1([0,1])$ that

$$|B_N(f)(x) - f(x)| \leq \frac{3}{4\sqrt{N}} \omega_1\left(f', \frac{1}{\sqrt{N}}\right), \forall x \in [0,1], \quad (3.4)$$

In [5], p. 10, the authors introduced the basic Max-product Bernstein operators,

$$B_N^{(M)}(f)(x) = \frac{\bigvee_{k=0}^{N} p_{N,k}(x) f\left(\frac{k}{N}\right)}{\bigvee_{k=0}^{N} p_{N,k}(x)}, \quad N \in \mathbb{N}, \quad (3.5)$$

where $\bigvee$ stands for maximum, and $p_{N,k}(x) = \binom{N}{k} x^k (1-x)^{N-k}$ and $f : [0,1] \to \mathbb{R}_+ = [0, \infty)$.

These are nonlinear and piecewise rational operators.

The authors in [5] studied similar such nonlinear operators such as: the Max-product Favard-Szász-Mirakjan operators and their truncated version, the Max-product Baskakov operators and their truncated version, also many other similar specific operators. The study in [5] is based on presented there general theory of sublinear operators. These Max-product operators tend to converge faster to the on hand function.

So we mention from [5], p. 30, that for $f : [0,1] \to \mathbb{R}_+$ continuous, we have the estimate

$$\left|B_N^{(M)}(f)(x) - f(x)\right| \leq 12\omega_1\left(f, \frac{1}{\sqrt{N+1}}\right), \text{ for all } N \in \mathbb{N}, \ x \in [0,1], \quad (3.6)$$

Also from [5], p. 36, we mention that for $f : [0,1] \to \mathbb{R}_+$ being concave function we get that

$$\left|B_N^{(M)}(f)(x) - f(x)\right| \leq 2\omega_1\left(f, \frac{1}{N}\right), \text{ for all } x \in [0,1], \quad (3.7)$$

a much faster convergence.

In this chapter we expand the study in [5] by considering conformable fractional smoothness of functions. So our inequalities are with respect to $\omega_1\left(D_\alpha^n f, \delta\right)$, $\delta > 0$, $n \in \mathbb{N}$, where $D_\alpha^n f$ is the nth order conformable α-fractional derivative, $\alpha \in (0,1]$, see [1, 7].

We present at first some background and general related theory of sublinear operators and then we apply it to specific as above Max-product operators.

3.2 Background

We make

Definition 3.1 Let $f : [0, \infty) \to \mathbb{R}$ and $\alpha \in (0, 1]$. We say that f is an α-fractional continuous function, iff $\forall \varepsilon > 0 \exists \delta > 0 :$ for any $x, y \in [0, \infty)$ such that $|x^\alpha - y^\alpha| \leq \delta$ we get that $|f(x) - f(y)| \leq \varepsilon$.

We give

Theorem 3.2 *Over* $[a, b] \subseteq [0, \infty)$, $\alpha \in [0, 1]$, *a* α-fractional continuous function *is a uniformly continuous function and vice versa, a uniformly continuous function is an* α-fractional continuous function.

(Theorem 3.2 is not valid over $[0, \infty)$.)

Note 3.3 *Let* $x, y \in [a, b] \subseteq [0, \infty)$, *and* $g(x) = x^\alpha$, $0 < \alpha \leq 1$, *then* $g'(x) = \alpha x^{\alpha-1} = \frac{\alpha}{x^{1-\alpha}}$, *for* $x \in (0, \infty)$. *Since* $a \leq x \leq b$, *then* $\frac{1}{x} \geq \frac{1}{b} > 0$ *and* $\frac{\alpha}{x^{1-\alpha}} \geq \frac{\alpha}{b^{1-\alpha}} > 0$.

Assume $y > x$. *By the mean value theorem we get*

$$y^\alpha - x^\alpha = \frac{\alpha}{\xi^{1-\alpha}}(y - x), \quad where \ \xi \in (x, y). \tag{3.8}$$

A similar to (3.8) equality when $x > y$ *is true.*
Then we obtain

$$\frac{\alpha}{b^{1-\alpha}}|y - x| \leq |y^\alpha - x^\alpha| = \frac{\alpha}{\xi^{1-\alpha}}|y - x|. \tag{3.9}$$

Thus, it holds

$$\frac{\alpha}{b^{1-\alpha}}|y - x| \leq |y^\alpha - x^\alpha|. \tag{3.10}$$

Proof of Theorem 3.2
($\Rightarrow$) Assume that f is α-fractional continuous function on $[a, b] \subseteq [0, \infty)$. It means $\forall \ \varepsilon > 0 \ \exists \ \delta > 0 :$ whenever $x, y \in [a, b] : |x^\alpha - y^\alpha| \leq \delta$, then $|f(x) - f(y)| \leq \varepsilon$. Let for $\{x_n\}_{n \in \mathbb{N}} \in [a, b] : \{x_n \to \lambda \in [a, b] \Leftrightarrow x_n^\alpha \to \lambda^\alpha\}$, it implies $f(x_n) \to f(\lambda)$, therefore f is continuous in λ. Therefore f is uniformly continuous over $[a, b]$.
For the converse we use the following criterion:

Lemma 3.4 *A necessary and sufficient condition that the function* f *is not* α-fractional continuous ($\alpha \in (0, 1]$) over $[a, b] \subseteq [0, \infty)$ *is that there exist* $\varepsilon_0 > 0$, *and two sequences* $X = (x_n)$, $Y = (y_n)$ *in* $[a, b]$ *such that if* $n \in \mathbb{N}$, *then* $|x_n^\alpha - y_n^\alpha| \leq \frac{1}{n}$ *and* $|f(x_n) - f(y_n)| > \varepsilon_0$. ∎

Proof Obvious.

(Proof of Theorem 3.2 continuous) ($\Leftarrow$) Uniform continuity implies α-fractional continuity on $[a, b] \subseteq [0, +\infty)$. Indeed: let f uniformly continuous on $[a, b]$, hence f continuous on $[a, b]$. Assume that f is not α-fractional continuous on $[a, b]$. Then by Lemma 3.4 there exist $\varepsilon_0 > 0$, and two sequences $X = (x_n)$, $Y = (y_n)$ in $[a, b]$ such that if $n \in \mathbb{N}$, then $\left|x_n^\alpha - y_n^\alpha\right| \leq \frac{1}{n}$ and

$$|f(x_n) - f(y_n)| > \varepsilon_0. \tag{3.11}$$

Since $[a, b]$ is compact, the sequences $\{x_n\}, \{y_n\}$ are bounded. By the Bolzano-Weierstrass theorem, there is a subsequence $\{x_{n(k)}\}$ of $\{x_n\}$ which converges to an element z. Since $[a, b]$ is closed, the limit $z \in [a, b]$, and f is continuous at z.

We have also that

$$\frac{\alpha}{b^{1-\alpha}} |x_n - y_n| \leq \left|x_n^\alpha - y_n^\alpha\right| \leq \frac{1}{n}, \tag{3.12}$$

hence

$$|x_n - y_n| \leq \frac{b^{1-\alpha}}{\alpha n}. \tag{3.13}$$

It is clear that the corresponding subsequence $\left(y_{n(k)}\right)$ of Y also converges to z. Hence $f\left(x_{n(k)}\right) \to f(z)$, and $f\left(y_{n(k)}\right) \to f(z)$. Therefore, when k is sufficiently large we have $\left|f\left(x_{n(k)}\right) - f\left(y_{n(k)}\right)\right| < \varepsilon_0$, contradicting (3.11). ■

We need

Definition 3.5 Let $[a, b] \subseteq [0, \infty)$, $\alpha \in [0, 1]$. We define the α-fractional modulus of continuity:

$$\omega_1^\alpha(f, \delta) := \sup_{\substack{x, y \in [a, b]: \\ |x^\alpha - y^\alpha| \leq \delta}} |f(x) - f(y)|, \quad \delta > 0. \tag{3.14}$$

The same definition holds over $[0, \infty)$.

Properties:

(1) $\omega_1^\alpha(f, 0) = 0$.

(2) $\omega_1^\alpha(f, \delta) \to 0$ as $\delta \downarrow 0$, iff f is in the set of all α-fractional continuous functions, denoted as $f \in C_\alpha([a, b], \mathbb{R}) (= C([a, b], \mathbb{R}))$.

Proof ($\Rightarrow$) Let $\omega_1^\alpha(f, \delta) \to 0$ as $\delta \downarrow 0$. Then $\forall \varepsilon > 0$, $\exists \delta > 0$ with $\omega_1^\alpha(f, \delta) \leq \varepsilon$. I.e. $\forall x, y \in [a, b] : |x^\alpha - y^\alpha| \leq \delta$ we get $|f(x) - f(y)| \leq \varepsilon$. That is $f \in C_\alpha([a, b], \mathbb{R})$.

($\Leftarrow$) Let $f \in C_\alpha([a, b], \mathbb{R})$. Then $\forall \varepsilon > 0$, $\exists \delta > 0$: whenever $|x^\alpha - y^\alpha| \leq \delta$, $x, y \in [a, b]$, it implies $|f(x) - f(y)| \leq \varepsilon$. I.e. $\forall \varepsilon > 0$, $\exists \delta > 0 : \omega_1^\alpha(f, \delta) \leq \varepsilon$. That is $\omega_1^\alpha(f, \delta) \to 0$, as $\delta \downarrow 0$. ■

(3) ω_1^α is ≥ 0 and non-decreasing on $\mathbb{R}_+$.

(4) ω_1^α is subadditive:

$$\omega_1^\alpha (f, t_1 + t_2) \leq \omega_1^\alpha (f, t_1) + \omega_1^\alpha (f, t_2). \tag{3.15}$$

Proof If $|x^\alpha - y^\alpha| \leq t_1 + t_2$ $(x, y \in [a, b])$, there is a point $z \in [a, b]$ for which $|x^\alpha - z^\alpha| \leq t_1$, $|y^\alpha - z^\alpha| \leq t_2$, and $|f(x) - f(y)| \leq |f(x) - f(z)| + |f(z) - f(y)| \leq \omega_1^\alpha (f, t_1) + \omega_1^\alpha (f, t_2)$, implying $\omega_1^\alpha (f, t_1 + t_2) \leq \omega_1^\alpha (f, t_1) + \omega_1^\alpha (f, t_2)$. ∎

(5) ω_1^α is continuous on $\mathbb{R}_+$.

Proof We get

$$\left| \omega_1^\alpha (f, t_1 + t_2) - \omega_1^\alpha (f, t_1) \right| \leq \omega_1^\alpha (f, t_2). \tag{3.16}$$

By properties (2), (3), (4), we get that $\omega_1^\alpha (f, t)$ is continuous at each $t \geq 0$. ∎

(6) Clearly it holds

$$\omega_1^\alpha (f, t_1 + \dots + t_n) \leq \omega_1^\alpha (f, t_1) + \dots + \omega_1^\alpha (f, t_n), \tag{3.17}$$

for $t = t_1 = \dots = t_n$, we obtain

$$\omega_1^\alpha (f, nt) \leq n\omega_1^\alpha (f, t). \tag{3.18}$$

(7) Let $\lambda \geq 0$, $\lambda \notin \mathbb{N}$, we get

$$\omega_1^\alpha (f, \lambda t) \leq (\lambda + 1) \omega_1^\alpha (f, t). \tag{3.19}$$

Proof Let $n \in \mathbb{Z}_+ : n \leq \lambda < n + 1$, we see that

$$\omega_1^\alpha (f, \lambda t) \leq \omega_1^\alpha (f, (n+1) t) \leq (n+1) \omega_1^\alpha (f, t) \leq (\lambda + 1) \omega_1^\alpha (f, t). \qquad \blacksquare$$

Properties (1), (3), (4), (6), (7) are valid also for ω_1^α defined over $[0, \infty)$.

We notice that $\omega_1^\alpha (f, \delta)$ is finite when f is uniformly continuous on $[a, b]$. If $f : [0, \infty) \to \mathbb{R}$ is bounded then $\omega_1^\alpha (f, \delta)$ is again finite.

We need

Definition 3.6 ([1, 7]) Let $f : [0, \infty) \to \mathbb{R}$. The conformable α-fractional derivative for $\alpha \in (0, 1]$ is given by

$$D_\alpha f(t) := \lim_{\varepsilon \to 0} \frac{f(t + \varepsilon t^{1-\alpha}) - f(t)}{\varepsilon}, \tag{3.20}$$

$$D_\alpha f(0) = \lim_{t \to 0+} D_\alpha f(t). \tag{3.21}$$

If f is differentiable, then

$$D_\alpha f(t) = t^{1-\alpha} f'(t),$$ (3.22)

where f' is the usual derivative.

We define $D_\alpha^n f = D_\alpha^{n-1}(D_\alpha f)$.

If $f : [0, \infty) \to \mathbb{R}$ is α-differentiable at $t_0 > 0$, $\alpha \in (0, 1]$, then f is continuous at t_0, see [7].

We will use

Theorem 3.7 (see [4]) *(Taylor formula) Let $\alpha \in (0, 1]$ and $n \in \mathbb{N}$. Suppose f is $(n + 1)$ times conformable α-fractional differentiable on $[0, \infty)$, and $s, t \in [0, \infty)$, and $D_\alpha^{n+1} f$ is assumed to be continuous on $[0, \infty)$. Then we have*

$$f(t) = \sum_{k=0}^n \frac{1}{k!} \left(\frac{t^\alpha - s^\alpha}{\alpha} \right)^k D_\alpha^k f(s) + \frac{1}{n!} \int_s^t \left(\frac{t^\alpha - \tau^\alpha}{\alpha} \right)^n D_\alpha^{n+1} f(\tau) \tau^{a-1} d\tau.$$
(3.23)

The case $n = 0$ follows.

Corollary 3.8 *Let $\alpha \in (0, 1]$. Suppose f is α-fractional differentiable on $[0, \infty)$, and $s, t \in [0, \infty)$. Assume that $D_\alpha f$ is continuous on $[0, \infty)$. Then*

$$f(t) = f(s) + \int_s^t D_\alpha f(\tau) \tau^{a-1} d\tau.$$ (3.24)

Note: Theorem 3.7 and Corollary 3.8 are also true for $f : [a, b] \to \mathbb{R}$, $[a, b] \subseteq [0, \infty)$, $s, t \in [a, b]$.

Proof of Corollary 3.8

Denote $I_\alpha^s(f)(t) := \int_s^t x^{\alpha-1} f(x) \, dx$. By [7] we get that

$$D_\alpha I_\alpha^s(f)(t) = f(t), \text{ for } t \geq s,$$ (3.25)

where f is any continuous function in the domain of I_α, $\alpha \in (0, 1)$.

Assume that $D_\alpha f$ is continuous, then

$$D_\alpha I_\alpha^s(D_\alpha f)(t) = (D_\alpha f)(t), \forall t \geq s.$$ (3.26)

Then, by [6], there exists a constant c such that

$$I_\alpha^s(D_\alpha f)(t) = f(t) + c.$$ (3.27)

Hence

$$0 = I_\alpha^s(D_\alpha f)(s) = f(s) + c,$$ (3.28)

then $c = -f(s)$.

Therefore

$$I_\alpha^s (D_\alpha f)(t) = f(t) - f(s) = \int_s^t (D_\alpha f)(\tau) \tau^{\alpha-1} d\tau. \tag{3.29}$$

The same proof applies for any $s \geq t$. ■

3.3 Main Results

We give

Theorem 3.9 *Let $\alpha \in (0, 1]$ and $n \in \mathbb{Z}_+$. Suppose f is $(n + 1)$ times conformable α-fractional differentiable on $[0, \infty)$, and $s, t \in [0, \infty)$, and $D_\alpha^{n+1} f$ is assumed to be continuous on $[0, \infty)$ and bounded. Then*

$$\left| f(t) - \sum_{k=0}^{n+1} \frac{1}{k!} \left(\frac{t^\alpha - s^\alpha}{\alpha} \right)^k D_\alpha^k f(s) \right| \leq \frac{\omega_1^\alpha \left(D_\alpha^{n+1} f, \delta \right)}{\alpha^{n+1} (n+1)!} \left| t^\alpha - s^\alpha \right|^{n+1} \left[1 + \frac{\left| t^\alpha - s^\alpha \right|}{(n+2)\delta} \right], \tag{3.30}$$

$\forall s, t \in [0, \infty), \delta > 0$.

Note: Theorem 3.9 is valid also for $f : [a, b] \to \mathbb{R}, [a, b] \subseteq \mathbb{R}_+$, any $s, t \in [a, b]$.

Proof We have that

$$\frac{1}{n!} \int_s^t \left(\frac{t^\alpha - \tau^\alpha}{\alpha} \right)^n D_\alpha^{n+1} f(s) \tau^{\alpha-1} d\tau = \frac{D_\alpha^{n+1} f(s)}{n!} \int_s^t \left(\frac{t^\alpha - \tau^\alpha}{\alpha} \right)^n \tau^{\alpha-1} d\tau$$

(by $\frac{d\tau^\alpha}{d\tau} = \alpha \tau^{\alpha-1} \Rightarrow d\tau^\alpha = \alpha \tau^{\alpha-1} d\tau \Rightarrow \frac{1}{\alpha} d\tau^\alpha = \tau^{\alpha-1} d\tau$)

$$= \frac{D_\alpha^{n+1} f(s)}{\alpha^{n+1} n!} \int_s^t (t^\alpha - \tau^\alpha)^n d\tau^\alpha \tag{3.31}$$

(by $t \leq \tau \leq s \Rightarrow t^\alpha \leq \tau^\alpha (=: z) \leq s^\alpha$)

$$= \frac{D_\alpha^{n+1} f(s)}{\alpha^{n+1} n!} \int_{s^\alpha}^{t^\alpha} (t^\alpha - z)^n dz = \frac{D_\alpha^{n+1} f(s)}{\alpha^{n+1} n!} \frac{(t^\alpha - s^\alpha)^{n+1}}{n+1}$$

$$= \frac{D_\alpha^{n+1} f(s)}{(n+1)!} \left(\frac{t^\alpha - s^\alpha}{\alpha} \right)^{n+1}. \tag{3.32}$$

Therefore it holds

$$\frac{1}{n!} \int_s^t \left(\frac{t^\alpha - \tau^\alpha}{\alpha}\right)^n D_\alpha^{n+1} f(s) \tau^{\alpha-1} d\tau = \frac{D_\alpha^{n+1} f(s)}{(n+1)!} \left(\frac{t^\alpha - s^\alpha}{\alpha}\right)^{n+1}. \tag{3.33}$$

By (3.23) and (3.24) we get:

$$f(t) = \sum_{k=0}^{n+1} \frac{1}{k!} \left(\frac{t^\alpha - s^\alpha}{\alpha}\right)^k D_\alpha^k f(s) + \tag{3.34}$$

$$\frac{1}{n!} \int_s^t \left(\frac{t^\alpha - \tau^\alpha}{\alpha}\right)^n \left(D_\alpha^{n+1} f(\tau) - D_\alpha^{n+1} f(s)\right) \tau^{\alpha-1} d\tau.$$

Call the remainder as

$$R_n(s, t) := \frac{1}{n!} \int_s^t \left(\frac{t^\alpha - \tau^\alpha}{\alpha}\right)^n \left(D_\alpha^{n+1} f(\tau) - D_\alpha^{n+1} f(s)\right) \tau^{\alpha-1} d\tau. \tag{3.35}$$

We estimate $R_n(s, t)$.
   Cases:
   (1) Let $t \geq s$. Then

$$|R_n(s, t)| \leq \frac{1}{n!} \int_s^t \left(\frac{t^\alpha - \tau^\alpha}{\alpha}\right)^n \left|D_\alpha^{n+1} f(\tau) - D_\alpha^{n+1} f(s)\right| \tau^{\alpha-1} d\tau$$

$$\leq \frac{1}{\alpha n!} \int_s^t \left(\frac{t^\alpha - \tau^\alpha}{\alpha}\right)^n \omega_1^\alpha \left(D_\alpha^{n+1} f, \tau^\alpha - s^\alpha\right) d\tau^\alpha \tag{3.36}$$

$$= \frac{1}{\alpha^{n+1} n!} \int_s^t (t^\alpha - \tau^\alpha)^n \omega_1^\alpha \left(D_\alpha^{n+1} f, \frac{\delta(\tau^\alpha - s^\alpha)}{\delta}\right) d\tau^\alpha$$

$(\delta > 0)$

$$\leq \frac{\omega_1^\alpha \left(D_\alpha^{n+1} f, \delta\right)}{\alpha^{n+1} n!} \int_s^t (t^\alpha - \tau^\alpha)^n \left(1 + \frac{\tau^\alpha - s^\alpha}{\delta}\right) d\tau^\alpha$$

$$= \frac{\omega_1^\alpha \left(D_\alpha^{n+1} f, \delta\right)}{\alpha^{n+1} n!} \int_{s^\alpha}^{t^\alpha} (t^\alpha - z)^n \left(1 + \frac{z - s^\alpha}{\delta}\right) dz$$

$$= \frac{\omega_1^\alpha \left(D_\alpha^{n+1} f, \delta\right)}{\alpha^{n+1} n!} \left[\int_{s^\alpha}^{t^\alpha} (t^\alpha - z)^n dz + \frac{1}{\delta} \int_{s^\alpha}^{t^\alpha} (t^\alpha - z)^{(n+1)-1} (z - s^\alpha)^{2-1} dz\right]$$

$$= \frac{\omega_1^\alpha \left(D_\alpha^{n+1} f, \delta\right)}{\alpha^{n+1} n!} \left[\frac{(t^\alpha - s^\alpha)^{n+1}}{n+1} + \frac{1}{\delta} \frac{\Gamma(n+1) \Gamma(2)}{\Gamma(n+3)} (t^\alpha - s^\alpha)^{n+2}\right] \tag{3.37}$$

$$= \frac{\omega_1^\alpha \left(D_\alpha^{n+1} f, \delta \right)}{\alpha^{n+1} n!} \left[\frac{(t^\alpha - s^\alpha)^{n+1}}{n+1} + \frac{1}{\delta} \frac{n!}{(n+2)!} (t^\alpha - s^\alpha)^{n+2} \right]$$

$$= \frac{\omega_1^\alpha \left(D_\alpha^{n+1} f, \delta \right)}{\alpha^{n+1} n!} \left[\frac{(t^\alpha - s^\alpha)^{n+1}}{n+1} + \frac{1}{\delta} \frac{(t^\alpha - s^\alpha)^{n+2}}{(n+1)(n+2)} \right] \qquad (3.38)$$

$$= \frac{\omega_1^\alpha \left(D_\alpha^{n+1} f, \delta \right)}{\alpha^{n+1} (n+1)!} (t^\alpha - s^\alpha)^{n+1} \left[1 + \frac{(t^\alpha - s^\alpha)}{(n+2)\delta} \right].$$

We have proved that (case of $t \geq s$)

$$|R_n(s,t)| \leq \frac{\omega_1^\alpha \left(D_\alpha^{n+1} f, \delta \right)}{\alpha^{n+1} (n+1)!} (t^\alpha - s^\alpha)^{n+1} \left[1 + \frac{(t^\alpha - s^\alpha)}{(n+2)\delta} \right], \qquad (3.39)$$

where $\delta > 0$.

(2) case of $t \leq s$: We have

$$|R_n(s,t)| \leq \frac{1}{n!} \left| \int_t^s \left(\frac{t^\alpha - \tau^\alpha}{\alpha} \right)^n \left(D_\alpha^{n+1} f(\tau) - D_\alpha^{n+1} f(s) \right) \tau^{\alpha-1} d\tau \right| =$$

$$\frac{1}{n!} \left| \int_t^s \left(\frac{\tau^\alpha - t^\alpha}{\alpha} \right)^n \left(D_\alpha^{n+1} f(\tau) - D_\alpha^{n+1} f(s) \right) \tau^{\alpha-1} d\tau \right|$$

$$\leq \frac{1}{\alpha n!} \int_t^s \left(\frac{\tau^\alpha - t^\alpha}{\alpha} \right)^n \left| D_\alpha^{n+1} f(\tau) - D_\alpha^{n+1} f(s) \right| d\tau^\alpha \qquad (3.40)$$

$$= \frac{1}{\alpha^{n+1} n!} \int_t^s (\tau^\alpha - t^\alpha)^n \, \omega_1^\alpha \left(D_\alpha^{n+1} f, s^\alpha - \tau^\alpha \right) d\tau^\alpha$$

$(\delta > 0)$

$$\leq \frac{\omega_1^\alpha \left(D_\alpha^{n+1} f, \delta \right)}{\alpha^{n+1} n!} \int_t^s (\tau^\alpha - t^\alpha)^n \left(1 + \frac{s^\alpha - \tau^\alpha}{\delta} \right) d\tau^\alpha$$

$$= \frac{\omega_1^\alpha \left(D_\alpha^{n+1} f, \delta \right)}{\alpha^{n+1} n!} \int_{t^\alpha}^{s^\alpha} (z - t^\alpha)^n \left(1 + \frac{s^\alpha - z}{\delta} \right) dz$$

$$= \frac{\omega_1^\alpha \left(D_\alpha^{n+1} f, \delta \right)}{\alpha^{n+1} n!} \left[\int_{t^\alpha}^{s^\alpha} (z - t^\alpha)^n \, dz + \frac{1}{\delta} \int_{t^\alpha}^{s^\alpha} (s^\alpha - z)^{2-1} (z - t^\alpha)^{(n+1)-1} \, dz \right]$$

$$\qquad (3.41)$$

$$= \frac{\omega_1^\alpha \left(D_\alpha^{n+1} f, \delta \right)}{\alpha^{n+1} n!} \left[\frac{(s^\alpha - t^\alpha)^{n+1}}{n+1} + \frac{1}{\delta} \frac{\Gamma(2) \Gamma(n+1)}{\Gamma(n+3)} (s^\alpha - t^\alpha)^{n+2} \right]$$

$$= \frac{\omega_1^\alpha \left(D_\alpha^{n+1} f, \delta \right)}{\alpha^{n+1} n!} \left[\frac{(s^\alpha - t^\alpha)^{n+1}}{n+1} + \frac{1}{\delta} \frac{n!}{(n+2)!} (s^\alpha - t^\alpha)^{n+2} \right]$$

$$= \frac{\omega_1^\alpha \left(D_\alpha^{n+1} f, \delta \right)}{\alpha^{n+1} n!} \left[\frac{(s^\alpha - t^\alpha)^{n+1}}{n+1} + \frac{1}{\delta} \frac{(s^\alpha - t^\alpha)^{n+2}}{(n+1)(n+2)} \right] \qquad (3.42)$$

$$= \frac{\omega_1^\alpha \left(D_\alpha^{n+1} f, \delta \right)}{\alpha^{n+1} (n+1)!} (s^\alpha - t^\alpha)^{n+1} \left[1 + \frac{(s^\alpha - t^\alpha)}{(n+2)\delta} \right].$$

We have proved that $(t \le s)$

$$|R_n (s, t)| \le \frac{\omega_1^\alpha \left(D_\alpha^{n+1} f, \delta \right)}{\alpha^{n+1} (n+1)!} (s^\alpha - t^\alpha)^{n+1} \left[1 + \frac{(s^\alpha - t^\alpha)}{(n+2)\delta} \right], \qquad (3.43)$$

$\delta > 0$.

Conclusion: We have proved that $(\delta > 0)$

$$|R_n (s, t)| \le \frac{\omega_1^\alpha \left(D_\alpha^{n+1} f, \delta \right)}{\alpha^{n+1} (n+1)!} |t^\alpha - s^\alpha|^{n+1} \left[1 + \frac{|t^\alpha - s^\alpha|}{(n+2)\delta} \right], \ \forall s, t \in [0, \infty).$$
$$(3.44)$$

The proof of the theorem now is complete.　∎

We proved that

Theorem 3.10 *Let* $\alpha \in (0, 1], n \in \mathbb{N}$. *Suppose* f *is* n *times conformable* α-*fractional differentiable on* $[a, b] \subseteq [0, \infty)$, *and let any* $s, t \in [a, b]$. *Assume that* $D_\alpha^n f$ *is continuous on* $[a, b]$. *Then*

$$\left| f(t) - \sum_{k=0}^n \frac{1}{k!} \left(\frac{t^\alpha - s^\alpha}{\alpha} \right)^k D_\alpha^k f(s) \right| \le \frac{\omega_1^\alpha \left(D_\alpha^n f, \delta \right)}{\alpha^n n!} |t^\alpha - s^\alpha|^n \left[1 + \frac{|t^\alpha - s^\alpha|}{(n+1)\delta} \right],$$
$$(3.45)$$

where $\delta > 0$.

Proof By Theorem 3.9.　∎

Corollary 3.11 ($n = 1$ *case of Theorem 3.10*) *Let* $\alpha \in (0, 1]$. *Suppose* f *is* α-*conformable fractional differentiable on* $[a, b] \subseteq [0, \infty)$, *and let any* $s, t \in [a, b]$. *Assume that* $D_\alpha f$ *is continuous on* $[a, b]$. *Then*

$$\left| f(t) - f(s) - \left(\frac{t^\alpha - s^\alpha}{\alpha} \right) D_\alpha f(s) \right| \le \frac{\omega_1^\alpha (D_\alpha f, \delta)}{\alpha} |t^\alpha - s^\alpha| \left[1 + \frac{|t^\alpha - s^\alpha|}{2\delta} \right],$$
$$(3.46)$$

where $\delta > 0$.

Corollary 3.12 (*to Theorem 3.10*) *Same assumptions as in Theorem 3.10. For specific* $s \in [a, b]$ *assume that* $D_\alpha^k f(s) = 0$, $k = 1, ..., n$. *Then*

$$|f(t) - f(s)| \leq \frac{\omega_1^\alpha (D_\alpha^n f, \delta)}{\alpha^n n!} |t^\alpha - s^\alpha|^n \left[1 + \frac{|t^\alpha - s^\alpha|}{(n+1)\delta} \right], \delta > 0. \quad (3.47)$$

The case $n = 1$ follows:

Corollary 3.13 (to Corollary 3.12) *For specific $s \in [a, b]$ assume that $D_\alpha f(s) = 0$. Then*

$$|f(t) - f(s)| \leq \frac{\omega_1^\alpha (D_\alpha f, \delta)}{\alpha} |t^\alpha - s^\alpha| \left[1 + \frac{|t^\alpha - s^\alpha|}{2\delta} \right], \delta > 0. \quad (3.48)$$

We make

Remark 3.14 For $0 < \alpha \leq 1, t, s \geq 0$, we have

$$2^{\alpha-1} (x^\alpha + y^\alpha) \leq (x + y)^\alpha \leq x^\alpha + y^\alpha. \quad (3.49)$$

Assume that $t > s$, then

$$t = t - s + s \Rightarrow t^\alpha = (t - s + s)^\alpha \leq (t - s)^\alpha + s^\alpha,$$

hence $t^\alpha - s^\alpha \leq (t - s)^\alpha$.

Similarly, when $s > t \Rightarrow s^\alpha - t^\alpha \leq (s - t)^\alpha$.

Therefore it holds

$$|t^\alpha - s^\alpha| \leq |t - s|^\alpha, \forall t, s \in [0, \infty). \quad (3.50)$$

Corollary 3.15 (to Theorem 3.10) *Same assumptions as in Theorem 3.10. For specific $s \in [a, b]$ assume that $D_\alpha^k f(s) = 0, k = 1, ..., n$. Then*

$$|f(t) - f(s)| \leq \frac{\omega_1^\alpha (D_\alpha^n f, \delta)}{\alpha^n n!} |t - s|^{n\alpha} \left[1 + \frac{|t - s|^\alpha}{(n+1)\delta} \right], \delta > 0, \quad (3.51)$$

$\forall t \in [a, b] \subseteq [0, \infty)$.

Corollary 3.16 (to Corollary 3.11) *Same assumptions as in Corollary 3.11. For specific $s \in [a, b]$ assume that $D_\alpha f(s) = 0$. Then*

$$|f(t) - f(s)| \leq \frac{\omega_1^\alpha (D_\alpha f, \delta)}{\alpha} |t - s|^\alpha \left[1 + \frac{|t - s|^\alpha}{2\delta} \right], \delta > 0, \quad (3.52)$$

$\forall t \in [a, b] \subseteq [0, \infty)$.

We need

Definition 3.17 Here $C_+ ([a, b]) := \{f : [a, b] \subseteq [0, \infty) \to \mathbb{R}_+$, continuous functions$\}$. Let $L_N : C_+ ([a, b]) \to C_+ ([a, b])$, operators, $\forall N \in \mathbb{N}$, such that

(i)

$$L_N\left(\alpha f\right) = \alpha L_N\left(f\right), \; \forall \alpha \geq 0, \forall f \in C_+\left([a, b]\right), \tag{3.53}$$

(ii) if $f, g \in C_+\left([a, b]\right) : f \leq g$, then

$$L_N\left(f\right) \leq L_N\left(g\right), \; \forall N \in \mathbb{N}, \tag{3.54}$$

(iii)

$$L_N\left(f + g\right) \leq L_N\left(f\right) + L_N\left(g\right), \; \forall f, g \in C_+\left([a, b]\right). \tag{3.55}$$

We call $\{L_N\}_{N \in \mathbb{N}}$ positive sublinear operators.

We need a Hölder's type inequality, see next:

Theorem 3.18 (see [2]) *Let $L : C_+\left([a, b]\right) \to C_+\left([a, b]\right)$, be a positive sublinear operator and $f, g \in C_+\left([a, b]\right)$, furthermore let $p, q > 1 : \frac{1}{p} + \frac{1}{q} = 1$. Assume that $L\left(\left(f\left(\cdot\right)\right)^p\right)\left(s_*\right), L\left(\left(g\left(\cdot\right)\right)^q\right)\left(s_*\right) > 0$ for some $s_* \in [a, b]$. Then*

$$L\left(f\left(\cdot\right) g\left(\cdot\right)\right)\left(s_*\right) \leq \left(L\left(\left(f\left(\cdot\right)\right)^p\right)\left(s_*\right)\right)^{\frac{1}{p}} \left(L\left(\left(g\left(\cdot\right)\right)^q\right)\left(s_*\right)\right)^{\frac{1}{q}}. \tag{3.56}$$

We make

Remark 3.19 By [5], p. 17, we get: let $f, g \in C_+\left([a, b]\right)$, then

$$\left|L_N\left(f\right)\left(x\right) - L_N\left(g\right)\left(x\right)\right| \leq L_N\left(\left|f - g\right|\right)\left(x\right), \; \forall x \in [a, b] \subseteq [0, \infty). \tag{3.57}$$

Furthermore, we also have that

$$\left|L_N\left(f\right)\left(x\right) - f\left(x\right)\right| \leq L_N\left(\left|f\left(\cdot\right) - f\left(x\right)\right|\right)\left(x\right) + \left|f\left(x\right)\right| \left|L_N\left(e_0\right)\left(x\right) - 1\right|, \tag{3.58}$$

$\forall\, x \in [a, b] \subseteq [0, \infty); e_0\left(t\right) = 1$.
From now on we assume that $L_N\left(1\right) = 1$. Hence it holds

$$\left|L_N\left(f\right)\left(x\right) - f\left(x\right)\right| \leq L_N\left(\left|f\left(\cdot\right) - f\left(x\right)\right|\right)\left(x\right), \; \forall x \in [a, b] \subseteq [0, \infty). \tag{3.59}$$

Next we use Corollary 3.16.
Here $D_\alpha f\left(x\right) = 0$ for a specific $x \in [a, b] \subseteq [0, \infty)$. We also assume that $L_N\left(\left|\cdot - x\right|^{\alpha+1}\right)\left(x\right), L_N\left(\left(\cdot - x\right)^{2(\alpha+1)}\right)\left(x\right) > 0$. By (3.52) we have

$$\left|f\left(\cdot\right) - f\left(x\right)\right| \leq \frac{\omega_1^\alpha\left(D_\alpha f, \delta\right)}{\alpha}\left[\left|\cdot - x\right|^\alpha + \frac{\left|\cdot - x\right|^{2\alpha}}{2\delta}\right], \; \delta > 0, \tag{3.60}$$

true over $[a, b] \subseteq [0, \infty)$.

By (3.59) we get

$$|L_N(f)(x) - f(x)| \leq \frac{\omega_1^\alpha(D_\alpha f, \delta)}{\alpha} \left[L_N(|\cdot - x|^\alpha)(x) + \frac{L_N(|\cdot - x|^{2\alpha})(x)}{2\delta} \right]$$

(3.61)

$$\overset{\text{(by (3.56))}}{\leq} \frac{\omega_1^\alpha(D_\alpha f, \delta)}{\alpha} \left[(L_N(|\cdot - x|^{\alpha+1})(x))^{\frac{\alpha}{\alpha+1}} + \frac{(L_N((\cdot - x)^{2(\alpha+1)})(x))^{\frac{\alpha}{\alpha+1}}}{2\delta} \right]$$

(3.62)

(choose $\qquad \delta := \left((L_N((\cdot - x)^{2(\alpha+1)})(x))^{\frac{\alpha}{\alpha+1}} \right)^{\frac{1}{2}} > 0,$ $\qquad$ hence

$\delta^2 = (L_N((\cdot - x)^{2(\alpha+1)})(x))^{\frac{\alpha}{\alpha+1}})$

$$= \frac{\omega_1^\alpha \left(D_\alpha f, (L_N((\cdot - x)^{2(\alpha+1)})(x))^{\frac{\alpha}{2(\alpha+1)}} \right)}{\alpha}$$

$$\left[(L_N(|\cdot - x|^{\alpha+1})(x))^{\frac{\alpha}{\alpha+1}} + \frac{1}{2} (L_N((\cdot - x)^{2(\alpha+1)})(x))^{\frac{\alpha}{2(\alpha+1)}} \right].$$

(3.63)

We have proved:

Theorem 3.20 *Let $\alpha \in (0, 1]$, $[a, b] \subseteq [0, \infty)$. Suppose f is α-conformable fractional differentiable on $[a, b]$. $D_\alpha f$ is continuous on $[a, b]$. Let an $x \in [a, b]$ such that $D_\alpha f(x) = 0$, and $L_N : C_+([a, b])$ into itself, positive sublinear operators. Assume that $L_N(1) = 1$ and $L_N(|\cdot - x|^{\alpha+1})(x)$, $L_N((\cdot - x)^{2(\alpha+1)})(x) > 0$, $\forall N \in \mathbb{N}$.*
Then

$$|L_N(f)(x) - f(x)| \leq \frac{\omega_1^\alpha \left(D_\alpha f, (L_N((\cdot - x)^{2(\alpha+1)})(x))^{\frac{\alpha}{2(\alpha+1)}} \right)}{\alpha} \cdot$$

$$\left[(L_N(|\cdot - x|^{\alpha+1})(x))^{\frac{\alpha}{\alpha+1}} + \frac{1}{2} (L_N((\cdot - x)^{2(\alpha+1)})(x))^{\frac{\alpha}{2(\alpha+1)}} \right], \forall N \in \mathbb{N}. \quad (3.64)$$

We make

Remark 3.21 By Theorem 3.18, we get that

$$L_N(|\cdot - x|^{\alpha+1})(x) \leq (L_N((\cdot - x)^{2(\alpha+1)})(x))^{\frac{1}{2}}. \quad (3.65)$$

As $N \to +\infty$, by (3.64) and (3.65), and $L_N((\cdot - x)^{2(\alpha+1)})(x) \to 0$, we obtain that $L_N(f)(x) \to f(x)$.

We continue with

Remark 3.22 In the assumptions of Corollary 3.15 and (3.51) we can write over $[a, b] \subseteq [0, \infty)$, that

$$|f(\cdot) - f(x)| \leq \frac{\omega_1^\alpha \left(D_\alpha^n f, \delta\right)}{\alpha^n n!} \left[|\cdot - x|^{n\alpha} + \frac{|\cdot - x|^{(n+1)\alpha}}{(n+1)\delta}\right], \quad \delta > 0. \quad (3.66)$$

By (3.59) we get

$$|L_N(f)(x) - f(x)| \leq \frac{\omega_1^\alpha \left(D_\alpha^n f, \delta\right)}{\alpha^n n!}.$$

$$\left[L_N\left(|\cdot - x|^{n\alpha}\right)(x) + \frac{1}{(n+1)\delta} L_N\left(|\cdot - x|^{(n+1)\alpha}\right)(x)\right]$$

$$\overset{\text{(by (3.56))}}{\leq} \frac{\omega_1^\alpha \left(D_\alpha^n f, \delta\right)}{\alpha^n n!}. \quad (3.67)$$

$$\left[\left(L_N\left(|\cdot - x|^{n(\alpha+1)}\right)(x)\right)^{\frac{\alpha}{\alpha+1}} + \frac{1}{(n+1)\delta}\left(L_N\left(|\cdot - x|^{(n+1)(\alpha+1)}\right)(x)\right)^{\frac{\alpha}{\alpha+1}}\right]$$

[(here is assumed $L_N(1) = 1$, and $L_N\left(|\cdot - x|^{n(\alpha+1)}\right)(x)$, $L_N\left(|\cdot - x|^{(n+1)(\alpha+1)}\right)$ $(x) > 0$, $\forall\ N \in \mathbb{N}$), (we take $\delta := \left(L_N\left(|\cdot - x|^{(n+1)(\alpha+1)}\right)(x)\right)^{\frac{\alpha}{(n+1)(\alpha+1)}} > 0$, then $\delta^{n+1} = \left(L_N\left(|\cdot - x|^{(n+1)(\alpha+1)}\right)(x)\right)^{\frac{\alpha}{\alpha+1}}$)]

$$= \frac{\omega_1^\alpha \left(D_\alpha^n f, \left(L_N\left(|\cdot - x|^{(n+1)(\alpha+1)}\right)(x)\right)^{\frac{\alpha}{(n+1)(\alpha+1)}}\right)}{\alpha^n n!}.$$

$$\left[\left(L_N\left(|\cdot - x|^{n(\alpha+1)}\right)(x)\right)^{\frac{\alpha}{\alpha+1}} + \frac{1}{(n+1)}\left(L_N\left(|\cdot - x|^{(n+1)(\alpha+1)}\right)(x)\right)^{\frac{n\alpha}{(n+1)(\alpha+1)}}\right]. \quad (3.68)$$

We have proved

Theorem 3.23 *Let* $\alpha \in (0, 1], n \in \mathbb{N}$. *Suppose* f *is* n *times conformable* α-*fractional differentiable on* $[a, b] \subseteq [0, \infty)$, *and* $D_\alpha^n f$ *is continuous on* $[a, b]$. *For a fixed* $x \in [a, b]$ *we have* $D_\alpha^k f(x) = 0, k = 1, ..., n$. *Let positive sublinear operators* $\{L_N\}_{N \in \mathbb{N}}$ *from* $C_+([a, b])$ *into itself, such that* $L_N(1) = 1$, *and* $L_N\left(|\cdot - x|^{n(\alpha+1)}\right)(x)$, $L_N\left(|\cdot - x|^{(n+1)(\alpha+1)}\right)(x) > 0, \forall\ N \in \mathbb{N}$. *Then*

$$|L_N(f)(x) - f(x)| \leq \frac{\omega_1^\alpha \left(D_\alpha^n f, \left(L_N\left(|\cdot - x|^{(n+1)(\alpha+1)}\right)(x)\right)^{\frac{\alpha}{(n+1)(\alpha+1)}}\right)}{\alpha^n n!}. \quad (3.69)$$

$$\left[\left(L_N \left(|\cdot - x|^{n(\alpha+1)} \right) (x) \right)^{\frac{\alpha}{\alpha+1}} + \frac{1}{(n+1)} \left(L_N \left(|\cdot - x|^{(n+1)(\alpha+1)} \right) (x) \right)^{\frac{n\alpha}{(n+1)(\alpha+1)}} \right],$$

$\forall N \in \mathbb{N}$.

We make

Remark 3.24 By Theorem 3.18, we get that

$$L_N \left(|\cdot - x|^{n(\alpha+1)} \right) (x) \le \left(L_N \left(|\cdot - x|^{(n+1)(\alpha+1)} \right) (x) \right)^{\frac{n}{n+1}}. \tag{3.70}$$

As $N \to +\infty$, by (3.69), (3.70), and $L_N \left(|\cdot - x|^{(n+1)(\alpha+1)} \right) (x) \to 0$, we derive that $L_N (f) (x) \to f (x)$.

3.4 Applications

Here we apply Theorems 3.20 and 3.23 to well known Max-product operators.
We make

Remark 3.25 The Max-product Bernstein operators $B_N^{(M)} (f) (x)$ are defined by (3.5), see also [5], p. 10; here $f : [0, 1] \to \mathbb{R}_+$ is a continuous function.
We have $B_N^{(M)} (1) = 1$, and

$$B_N^{(M)} \left(|\cdot - x| \right) (x) \le \frac{6}{\sqrt{N+1}}, \ \forall x \in [0, 1], \forall N \in \mathbb{N},$$

see [5], p. 31.
$B_N^{(M)}$ are positive sublinear operators and thus they possess the monotonicity property, also since $|\cdot - x| \le 1$, then $|\cdot - x|^{\beta} \le 1, \forall x \in [0, 1], \forall \beta > 0$.
Therefore it holds

$$B_N^{(M)} \left(|\cdot - x|^{1+\beta} \right) (x) \le \frac{6}{\sqrt{N+1}}, \ \forall x \in [0, 1], \forall N \in \mathbb{N}, \forall \beta > 0. \tag{3.71}$$

Furthermore, clearly it holds that

$$B_N^{(M)} \left(|\cdot - x|^{1+\beta} \right) (x) > 0, \forall N \in \mathbb{N}, \forall \beta \ge 0 \text{ and any } x \in (0, 1). \tag{3.72}$$

The operator $B_N^{(M)}$ maps $C_+ ([0, 1])$ into itself.

We have the following results:

Theorem 3.26 Let $\alpha \in (0, 1]$, f is α-conformable fractional differentiable on $[0, 1]$, $D_\alpha f$ is continuous on $[0, 1]$. Let $x \in (0, 1)$ such that $D_\alpha f (x) = 0$. Then

$$\left| B_N^{(M)} (f) (x) - f (x) \right| \leq \frac{\omega_1^\alpha \left(D_\alpha f, \left(\frac{6}{\sqrt{N+1}} \right)^{\frac{\alpha}{2(\alpha+1)}} \right)}{\alpha}. \tag{3.73}$$

$$\left[\left(\frac{6}{\sqrt{N+1}} \right)^{\frac{\alpha}{\alpha+1}} + \frac{1}{2} \left(\frac{6}{\sqrt{N+1}} \right)^{\frac{\alpha}{2(\alpha+1)}} \right], \forall N \in \mathbb{N}.$$

Proof By Theorem 3.20. ∎

Theorem 3.27 *Let* $\alpha \in (0, 1]$, f *is* n *times conformable* α *-fractional differentiable on* $[0, 1]$, *and* $D_\alpha^n f$ *is continuous on* $[0, 1]$. *For a fixed* $x \in (0, 1)$ *we have* $D_\alpha^k f (x) = 0$, $k = 1, ..., n \in \mathbb{N}$. *Then*

$$\left| B_N^{(M)} (f) (x) - f (x) \right| \leq \frac{\omega_1^\alpha \left(D_\alpha^n f, \left(\frac{6}{\sqrt{N+1}} \right)^{\frac{\alpha}{(n+1)(\alpha+1)}} \right)}{\alpha^n n!}. \tag{3.74}$$

$$\left[\left(\frac{6}{\sqrt{N+1}} \right)^{\frac{\alpha}{\alpha+1}} + \frac{1}{(n+1)} \left(\frac{6}{\sqrt{N+1}} \right)^{\frac{n\alpha}{(n+1)(\alpha+1)}} \right], \forall N \in \mathbb{N}.$$

Proof By Theorem 3.23. ∎

Note: By (3.73) and/or (3.74), as $N \to +\infty$, we get $B_N^{(M)} (f) (x) \to f (x)$. We continue with

Remark 3.28 The truncated Favard-Szász-Mirakjan operators are given by

$$T_N^{(M)} (f) (x) = \frac{\bigvee_{k=0}^N s_{N,k} (x) f \left(\frac{k}{N} \right)}{\bigvee_{k=0}^N s_{N,k} (x)}, \ x \in [0, 1], N \in \mathbb{N}, f \in C_+ ([0, 1]), \tag{3.75}$$

$s_{N,k} (x) = \frac{(Nx)^k}{k!}$, see also [5], p. 11.

By [5], pp. 178–179, we get that

$$T_N^{(M)} (|\cdot - x|) (x) \leq \frac{3}{\sqrt{N}}, \forall x \in [0, 1], \forall N \in \mathbb{N}. \tag{3.76}$$

Clearly it holds

$$T_N^{(M)} \left(|\cdot - x|^{1+\beta} \right) (x) \leq \frac{3}{\sqrt{N}}, \forall x \in [0, 1], \forall N \in \mathbb{N}, \forall \beta > 0. \tag{3.77}$$

The operators $T_N^{(M)}$ are positive sublinear operators mapping $C_+ ([0, 1])$ into itself, with $T_N^{(M)} (1) = 1$.

Furthermore it holds

$$T_N^{(M)} \left(|\cdot - x|^\lambda \right) (x) = \frac{\bigvee_{k=0}^{N} \frac{(Nx)^k}{k!} \left| \frac{k}{N} - x \right|^\lambda}{\bigvee_{k=0}^{N} \frac{(Nx)^k}{k!}} > 0, \ \forall x \in (0, 1], \forall \lambda \geq 1, \forall N \in \mathbb{N}.$$

(3.78)

We give the following results:

Theorem 3.29 *Let* $\alpha \in (0, 1]$, f *is* α-*conformable fractional differentiable on* $[0, 1]$. $D_\alpha f$ *is continuous on* $[0, 1]$. *Let* $x \in (0, 1]$ *such that* $D_\alpha f(x) = 0$. *Then*

$$\left| T_N^{(M)}(f)(x) - f(x) \right| \leq \frac{\omega_1^\alpha \left(D_\alpha f, \left(\frac{3}{\sqrt{N}} \right)^{\frac{\alpha}{2(\alpha+1)}} \right)}{\alpha}.$$

(3.79)

$$\left[\left(\frac{3}{\sqrt{N}} \right)^{\frac{\alpha}{\alpha+1}} + \frac{1}{2} \left(\frac{3}{\sqrt{N}} \right)^{\frac{\alpha}{2(\alpha+1)}} \right], \ \forall N \in \mathbb{N}.$$

Proof By Theorem 3.20. ∎

Theorem 3.30 *Let* $\alpha \in (0, 1]$, f *is* n *times conformable* α-*fractional differentiable on* $[0, 1]$, *and* $D_\alpha^n f$ *is continuous on* $[0, 1]$. *For a fixed* $x \in (0, 1]$ *we have* $D_\alpha^k f(x) = 0$, $k = 1, ..., n \in \mathbb{N}$. *Then*

$$\left| T_N^{(M)}(f)(x) - f(x) \right| \leq \frac{\omega_1^\alpha \left(D_\alpha^n f, \left(\frac{3}{\sqrt{N}} \right)^{\frac{\alpha}{(n+1)(\alpha+1)}} \right)}{\alpha^n n!}.$$

(3.80)

$$\left[\left(\frac{3}{\sqrt{N}} \right)^{\frac{\alpha}{\alpha+1}} + \frac{1}{(n+1)} \left(\frac{3}{\sqrt{N}} \right)^{\frac{n\alpha}{(n+1)(\alpha+1)}} \right], \ \forall N \in \mathbb{N}.$$

Proof By Theorem 3.23. ∎

Note: By (3.79) and/or (3.80), as $N \to +\infty$, we get $T_N^{(M)}(f)(x) \to f(x)$. We continue with

Remark 3.31 Next we study the truncated Max-product Baskakov operators (see [5], p. 11)

$$U_N^{(M)}(f)(x) = \frac{\bigvee_{k=0}^{N} b_{N,k}(x) f\left(\frac{k}{N} \right)}{\bigvee_{k=0}^{N} b_{N,k}(x)}, \ x \in [0, 1], \ f \in C_+([0, 1]), \ N \in \mathbb{N},$$

(3.81)

where

$$b_{N,k}(x) = \binom{N+k-1}{k} \frac{x^k}{(1+x)^{N+k}}.$$

(3.82)

From [5], pp. 217–218, we get ($x \in [0, 1]$)

$$\left(U_N^{(M)} \left(|\cdot - x| \right) \right)(x) \le \frac{2\sqrt{3}\left(\sqrt{2} + 2 \right)}{\sqrt{N+1}}, \; N \ge 2, N \in \mathbb{N}. \tag{3.83}$$

Let $\lambda \ge 1$, clearly then it holds

$$\left(U_N^{(M)} \left(|\cdot - x|^{\lambda} \right) \right)(x) \le \frac{2\sqrt{3}\left(\sqrt{2} + 2 \right)}{\sqrt{N+1}}, \; \forall N \ge 2, N \in \mathbb{N}. \tag{3.84}$$

Also it holds $U_N^{(M)}(1) = 1$, and $U_N^{(M)}$ are positive sublinear operators from $C_+([0, 1])$ into itself. Furthermore it holds

$$U_N^{(M)} \left(|\cdot - x|^{\lambda} \right)(x) > 0, \; \forall x \in (0, 1], \forall \lambda \ge 1, \forall N \in \mathbb{N}. \tag{3.85}$$

We give

Theorem 3.32 *Let* $\alpha \in (0, 1]$, *f is* α-conformable fractional differentiable on $[0, 1]$. $D_\alpha f$ is continuous on $[0, 1]$. Let $x \in (0, 1]$ such that $D_\alpha f(x) = 0$. Then

$$\left| U_N^{(M)}(f)(x) - f(x) \right| \le \frac{\omega_1^\alpha \left(D_\alpha f, \left(\frac{2\sqrt{3}(\sqrt{2}+2)}{\sqrt{N+1}} \right)^{\frac{\alpha}{2(\alpha+1)}} \right)}{\alpha} \cdot \tag{3.86}$$

$$\left[\left(\frac{2\sqrt{3}\left(\sqrt{2}+2 \right)}{\sqrt{N+1}} \right)^{\frac{\alpha}{\alpha+1}} + \frac{1}{2} \left(\frac{2\sqrt{3}\left(\sqrt{2}+2 \right)}{\sqrt{N+1}} \right)^{\frac{\alpha}{2(\alpha+1)}} \right], \; \forall N \ge 2, N \in \mathbb{N}.$$

Proof By Theorem 3.20. ∎

Theorem 3.33 *Let* $\alpha \in (0, 1]$, *f is n times conformable* α -*fractional differentiable on* $[0, 1]$, *and* $D_\alpha^n f$ *is continuous on* $[0, 1]$. *For a fixed* $x \in (0, 1]$ *we have* $D_\alpha^k f(x) = 0$, $k = 1, ..., n \in \mathbb{N}$. *Then*

$$\left| U_N^{(M)}(f)(x) - f(x) \right| \le \frac{\omega_1^\alpha \left(D_\alpha^n f, \left(\frac{2\sqrt{3}(\sqrt{2}+2)}{\sqrt{N+1}} \right)^{\frac{\alpha}{(n+1)(\alpha+1)}} \right)}{\alpha^n n!} \cdot \tag{3.87}$$

$$\left[\left(\frac{2\sqrt{3}\left(\sqrt{2}+2 \right)}{\sqrt{N+1}} \right)^{\frac{\alpha}{\alpha+1}} + \frac{1}{(n+1)} \left(\frac{2\sqrt{3}\left(\sqrt{2}+2 \right)}{\sqrt{N+1}} \right)^{\frac{n\alpha}{(n+1)(\alpha+1)}} \right],$$

$\forall N \ge 2, N \in \mathbb{N}$.

Proof By Theorem 3.23. ∎

Note: By (3.86) and/or (3.87), as $N \to +\infty$, we get that $U_N^{(M)}(f)(x) \to f(x)$. We continue with

Remark 3.34 Here we study the Max-product Meyer-Köning and Zeller operators (see [5], p. 11) defined by

$$Z_N^{(M)}(f)(x) = \frac{\bigvee_{k=0}^{\infty} s_{N,k}(x) f\left(\frac{k}{N+k}\right)}{\bigvee_{k=0}^{\infty} s_{N,k}(x)}, \quad \forall N \in \mathbb{N}, f \in C_+([0,1]), \quad (3.88)$$

$$s_{N,k}(x) = \binom{N+k}{k} x^k, x \in [0,1].$$

By [5], p. 253, we get that

$$Z_N^{(M)}(|\cdot - x|)(x) \le \frac{8\left(1+\sqrt{5}\right)}{3} \frac{\sqrt{x}(1-x)}{\sqrt{N}}, \quad \forall x \in [0,1], \forall N \ge 4, N \in \mathbb{N}. \quad (3.89)$$

As before we get that (for $\lambda \ge 1$)

$$Z_N^{(M)}(|\cdot - x|^\lambda)(x) \le \frac{8\left(1+\sqrt{5}\right)}{3} \frac{\sqrt{x}(1-x)}{\sqrt{N}} := \rho(x), \quad (3.90)$$

$\forall x \in [0,1], N \ge 4, N \in \mathbb{N}$.

Also it holds $Z_N^{(M)}(1) = 1$, and $Z_N^{(M)}$ are positive sublinear operators from $C_+([0,1])$ into itself. Also it holds

$$Z_N^{(M)}(|\cdot - x|^\lambda)(x) > 0, \quad \forall x \in (0,1), \forall \lambda \ge 1, \forall N \in \mathbb{N}. \quad (3.91)$$

We give

Theorem 3.35 *Let* $\alpha \in (0,1]$, *f is α-conformable fractional differentiable on* $[0,1]$. $D_\alpha f$ *is continuous on* $[0,1]$. *Let* $x \in (0,1)$ *such that* $D_\alpha f(x) = 0$. *Then*

$$\left| Z_N^{(M)}(f)(x) - f(x) \right| \le \frac{\omega_1^\alpha \left(D_\alpha f, (\rho(x))^{\frac{\alpha}{2(\alpha+1)}}\right)}{\alpha} \cdot \quad (3.92)$$

$$\left[(\rho(x))^{\frac{\alpha}{\alpha+1}} + \frac{1}{2}(\rho(x))^{\frac{\alpha}{2(\alpha+1)}} \right], \quad \forall N \ge 4, N \in \mathbb{N}.$$

Proof By Theorem 3.20. ∎

Theorem 3.36 *Let* $\alpha \in (0,1]$, *f is n times conformable α-fractional differentiable on* $[0,1]$, *and* $D_\alpha^n f$ *is continuous on* $[0,1]$. *For a fixed* $x \in (0,1)$ *we have* $D_\alpha^k f(x) = 0$, $k = 1, ..., n \in \mathbb{N}$. *Then*

$$\left| Z_N^{(M)} (f)(x) - f(x) \right| \le \frac{\omega_1^{\alpha} \left(D_{\alpha}^n f, (\rho(x))^{\frac{\alpha}{(n+1)(\alpha+1)}} \right)}{\alpha^n n!}. \tag{3.93}$$

$$\left[(\rho(x))^{\frac{\alpha}{\alpha+1}} + \frac{1}{(n+1)} (\rho(x))^{\frac{n\alpha}{(n+1)(\alpha+1)}} \right], \forall N \ge 4, N \in \mathbb{N}.$$

Proof By Theorem 3.23. ∎

Note: By (3.92) and/or (3.93), as $N \to +\infty$, we get that $Z_N^{(M)} (f)(x) \to f(x)$. We continue with

Remark 3.37 Here we deal with the Max-product truncated sampling operators (see [5], p. 13) defined by

$$W_N^{(M)} (f)(x) = \frac{\bigvee_{k=0}^{N} \frac{\sin(Nx - k\pi)}{Nx - k\pi} f\left(\frac{k\pi}{N}\right)}{\bigvee_{k=0}^{N} \frac{\sin(Nx - k\pi)}{Nx - k\pi}}, \tag{3.94}$$

and

$$K_N^{(M)} (f)(x) = \frac{\bigvee_{k=0}^{N} \frac{\sin^2(Nx - k\pi)}{(Nx - k\pi)^2} f\left(\frac{k\pi}{N}\right)}{\bigvee_{k=0}^{N} \frac{\sin^2(Nx - k\pi)}{(Nx - k\pi)^2}}, \tag{3.95}$$

$\forall x \in [0, \pi]$, $f : [0, \pi] \to \mathbb{R}_+$ a continuous function.

Following [5], p. 343, and making the convention $\frac{\sin(0)}{0} = 1$ and denoting $s_{N,k}(x) = \frac{\sin(Nx - k\pi)}{Nx - k\pi}$, we get that $s_{N,k}\left(\frac{k\pi}{N}\right) = 1$, and $s_{N,k}\left(\frac{j\pi}{N}\right) = 0$, if $k \ne j$, furthermore $W_N^{(M)}(f)\left(\frac{j\pi}{N}\right) = f\left(\frac{j\pi}{N}\right)$, for all $j \in \{0, ..., N\}$.

Clearly $W_N^{(M)}(f)$ is a well-defined function for all $x \in [0, \pi]$, and it is continuous on $[0, \pi]$, also $W_N^{(M)}(1) = 1$.

By [5], p. 344, $W_N^{(M)}$ are positive sublinear operators.

Call $I_N^+(x) = \{k \in \{0, 1, ..., N\} ; s_{N,k}(x) > 0\}$, and set $x_{N,k} := \frac{k\pi}{N}$, $k \in \{0, 1, ..., N\}$.

We see that

$$W_N^{(M)} (f)(x) = \frac{\bigvee_{k \in I_N^+(x)} s_{N,k}(x) f(x_{N,k})}{\bigvee_{k \in I_N^+(x)} s_{N,k}(x)}. \tag{3.96}$$

By [5], p. 346, we have

$$W_N^{(M)} (|\cdot - x|)(x) \le \frac{\pi}{2N}, \forall N \in \mathbb{N}, \forall x \in [0, \pi]. \tag{3.97}$$

Notice also $|x_{N,k} - x| \le \pi, \forall x \in [0, \pi]$.

Therefore ($\lambda \geq 1$) it holds

$$W_N^{(M)}\left(|\cdot - x|^\lambda\right)(x) \leq \frac{\pi^{\lambda-1}\pi}{2N} = \frac{\pi^\lambda}{2N}, \ \forall x \in [0, \pi], \forall N \in \mathbb{N}. \tag{3.98}$$

If $x \in \left(\frac{j\pi}{N}, \frac{(j+1)\pi}{N}\right)$, with $j \in \{0, 1, ..., N\}$, we obtain $nx - j\pi \in (0, \pi)$ and thus $s_{N,j}(x) = \frac{\sin(Nx - j\pi)}{Nx - j\pi} > 0$, see [5], pp. 343–344.

Consequently it holds ($\lambda \geq 1$)

$$W_N^{(M)}\left(|\cdot - x|^\lambda\right)(x) = \frac{\bigvee_{k \in I_N^+(x)} s_{N,k}(x) |x_{N,k} - x|^\lambda}{\bigvee_{k \in I_N^+(x)} s_{N,k}(x)} > 0, \ \forall x \in [0, \pi], \tag{3.99}$$

such that $x \neq x_{N,k}$, for any $k \in \{0, 1, ..., N\}$.

We give

Theorem 3.38 *Let* $\alpha \in (0, 1]$, f *is* α-*conformable fractional differentiable on* $[0, \pi]$. $D_\alpha f$ *is continuous on* $[0, \pi]$. *Let* $x \in [0, \pi]$ *be such that* $x \neq \frac{k\pi}{N}$, $k \in \{0, 1, ..., N\}$, $\forall N \in \mathbb{N}$, *and* $D_\alpha f(x) = 0$. *Then*

$$\left|W_N^{(M)}(f)(x) - f(x)\right| \leq \frac{\omega_1^\alpha\left(D_\alpha f, \left(\frac{\pi^{2(\alpha+1)}}{2N}\right)^{\frac{\alpha}{2(\alpha+1)}}\right)}{\alpha}\cdot$$

$$\left[\left(\frac{\pi^{\alpha+1}}{2N}\right)^{\frac{\alpha}{\alpha+1}} + \frac{1}{2}\left(\frac{\pi^{2(\alpha+1)}}{2N}\right)^{\frac{\alpha}{\alpha+1}}\right] =$$

$$\frac{\omega_1^\alpha\left(D_\alpha f, \frac{\pi^\alpha}{(2N)^{\frac{\alpha}{2(\alpha+1)}}}\right)}{\alpha}\left[\frac{\pi^\alpha}{(2N)^{\frac{\alpha}{(\alpha+1)}}} + \frac{\pi^\alpha}{2(2N)^{\frac{\alpha}{2(\alpha+1)}}}\right], \ \forall N \in \mathbb{N}. \tag{3.100}$$

Proof By Theorem 3.20. ∎

Theorem 3.39 *Let* $\alpha \in (0, 1]$, $n \in \mathbb{N}$. *Suppose* f *is* n *times conformable* α-*fractional differentiable on* $[0, \pi]$, *and* $D_\alpha^n f$ *is continuous on* $[0, \pi]$. *For a fixed* $x \in [0, \pi] : x \neq \frac{k\pi}{N}$, $k \in \{0, 1, ..., N\}$, $\forall N \in \mathbb{N}$, *we have* $D_\alpha^k f(x) = 0$, $k = 1, ..., n$. *Then*

$$\left|W_N^{(M)}(f)(x) - f(x)\right| \leq \frac{\omega_1^\alpha\left(D_\alpha^n f, \frac{\pi^\alpha}{(2N)^{\frac{\alpha}{(n+1)(\alpha+1)}}}\right)}{\alpha^n n!}\cdot$$

$$\left[\frac{\pi^{n\alpha}}{(2N)^{\frac{\alpha}{(\alpha+1)}}} + \frac{\pi^{n\alpha}}{(n+1)(2N)^{\frac{n\alpha}{(n+1)(\alpha+1)}}}\right], \ \forall N \in \mathbb{N}. \tag{3.101}$$

Proof By Theorem 3.23. ∎

Note: (i) if $x = \frac{j\pi}{N}$, $j \in \{0, ..., N\}$, then the left hand sides of (3.100) and (3.101) are zero, so these inequalities are trivially valid.

(ii) from (3.100) and/or (3.101), as $N \to +\infty$, we get that $W_N^{(M)}(f)(x) \to f(x)$.

We make

Remark 3.40 Here we continue with the Max-product truncated sampling operators (see [5], p. 13) defined by

$$K_N^{(M)}(f)(x) = \frac{\bigvee_{k=0}^{N} \frac{\sin^2(Nx - k\pi)}{(Nx - k\pi)^2} f\left(\frac{k\pi}{N}\right)}{\bigvee_{k=0}^{N} \frac{\sin^2(Nx - k\pi)}{(Nx - k\pi)^2}}, \tag{3.102}$$

$\forall x \in [0, \pi]$, $f : [0, \pi] \to \mathbb{R}_+$ a continuous function.

Following [5], p. 350, and making the convention $\frac{\sin(0)}{0} = 1$ and denoting $s_{N,k}(x) = \frac{\sin^2(Nx - k\pi)}{(Nx - k\pi)^2}$, we get that $s_{N,k}\left(\frac{k\pi}{N}\right) = 1$, and $s_{N,k}\left(\frac{j\pi}{N}\right) = 0$, if $k \neq j$, furthermore $K_N^{(M)}(f)\left(\frac{j\pi}{N}\right) = f\left(\frac{j\pi}{N}\right)$, for all $j \in \{0, ..., N\}$.

Since $s_{N,j}\left(\frac{j\pi}{N}\right) = 1$ it follows that $\bigvee_{k=0}^{N} s_{N,k}\left(\frac{j\pi}{N}\right) \geq 1 > 0$, for all $j \in \{0, 1, ..., N\}$. Hence $K_N^{(M)}(f)$ is well-defined function for all $x \in [0, \pi]$, and it is continuous on $[0, \pi]$, also $K_N^{(M)}(1) = 1$. By [5], p. 350, $K_N^{(M)}$ are positive sublinear operators.

Denote $x_{N,k} := \frac{k\pi}{N}$, $k \in \{0, 1, ..., N\}$.

By [5], p. 352, we have

$$K_N^{(M)}(|\cdot - x|)(x) \leq \frac{\pi}{2N}, \forall N \in \mathbb{N}, \forall x \in [0, \pi]. \tag{3.103}$$

Notice also $|x_{N,k} - x| \leq \pi$, $\forall x \in [0, \pi]$.

Therefore ($\lambda \geq 1$) it holds

$$K_N^{(M)}(|\cdot - x|^\lambda)(x) \leq \frac{\pi^{\lambda-1}\pi}{2N} = \frac{\pi^\lambda}{2N}, \forall x \in [0, \pi], \forall N \in \mathbb{N}. \tag{3.104}$$

If $x \in \left(\frac{j\pi}{N}, \frac{(j+1)\pi}{N}\right)$, with $j \in \{0, 1, ..., N\}$, we obtain $nx - j\pi \in (0, \pi)$ and thus $s_{N,j}(x) = \frac{\sin^2(Nx - j\pi)}{(Nx - j\pi)^2} > 0$, see [5], pp. 350.

Consequently it holds ($\lambda \geq 1$)

$$K_N^{(M)}(|\cdot - x|^\lambda)(x) = \frac{\bigvee_{k=0}^{N} s_{N,k}(x) |x_{N,k} - x|^\lambda}{\bigvee_{k=0}^{N} s_{N,k}(x)} > 0, \forall x \in [0, \pi], \tag{3.105}$$

such that $x \neq x_{N,k}$, for any $k \in \{0, 1, ..., N\}$.

We give

Theorem 3.41 *Let* $\alpha \in (0, 1]$, f *is* α-*conformable fractional differentiable on* $[0, \pi]$. $D_\alpha f$ *is continuous on* $[0, \pi]$. *Let* $x \in [0, \pi]$ *be such that* $x \neq \frac{k\pi}{N}$, $k \in \{0, 1, ..., N\}$, $\forall N \in \mathbb{N}$, *and* $D_\alpha f(x) = 0$. *Then*

$$\left| K_N^{(M)}(f)(x) - f(x) \right| \leq \frac{\omega_1^\alpha \left(D_\alpha f, \left(\frac{\pi^{2(\alpha+1)}}{2N} \right)^{\frac{\alpha}{2(\alpha+1)}} \right)}{\alpha} \cdot$$

$$\left[\left(\frac{\pi^{\alpha+1}}{2N} \right)^{\frac{\alpha}{\alpha+1}} + \frac{1}{2} \left(\frac{\pi^{2(\alpha+1)}}{2N} \right)^{\frac{\alpha}{2(\alpha+1)}} \right] =$$

$$\frac{\omega_1^\alpha \left(D_\alpha f, \frac{\pi^\alpha}{(2N)^{\frac{\alpha}{2(\alpha+1)}}} \right)}{\alpha} \left[\frac{\pi^\alpha}{(2N)^{\frac{\alpha}{(\alpha+1)}}} + \frac{\pi^\alpha}{2(2N)^{\frac{\alpha}{2(\alpha+1)}}} \right], \ \forall N \in \mathbb{N}. \quad (3.106)$$

Proof By Theorem 3.20. ■

Theorem 3.42 *Let* $\alpha \in (0, 1]$, $n \in \mathbb{N}$. *Suppose* f *is* n *times conformable* α-*fractional differentiable on* $[0, \pi]$, *and* $D_\alpha^n f$ *is continuous on* $[0, \pi]$. *For a fixed* $x \in [0, \pi] : x \neq \frac{k\pi}{N}$, $k \in \{0, 1, ..., N\}$, $\forall N \in \mathbb{N}$, *we have* $D_\alpha^k f(x) = 0$, $k = 1, ..., n$. *Then*

$$\left| K_N^{(M)}(f)(x) - f(x) \right| \leq \frac{\omega_1^\alpha \left(D_\alpha^n f, \frac{\pi^\alpha}{(2N)^{\frac{\alpha}{(n+1)(\alpha+1)}}} \right)}{\alpha^n n!} \cdot$$

$$\left[\frac{\pi^{n\alpha}}{(2N)^{\frac{\alpha}{(\alpha+1)}}} + \frac{\pi^{n\alpha}}{(n+1)(2N)^{\frac{n\alpha}{(n+1)(\alpha+1)}}} \right], \ \forall N \in \mathbb{N}. \quad (3.107)$$

Proof By Theorem 3.23. ■

Note: (i) if $x = \frac{j\pi}{N}$, $j \in \{0, ..., N\}$, then the left hand sides of (3.106) and (3.107) are zero, so these inequalities are trivially valid.

(ii) from (3.106) and/or (3.107), as $N \to +\infty$, we get that $K_N^{(M)}(f)(x) \to f(x)$.

References

1. M. Abu Hammad, R. Khalil, Abel's formula and Wronskian for conformable fractional differential equations. Int. J. Differ. Equ. Appl. **13**(3), 177–183 (2014)
2. G. Anastassiou, *Approximation by Sublinear Operators* (2017, submitted)
3. G. Anastassiou, *Conformable Fractional Approximation by Max-Product Operators* (2017, submitted)

4. D. Anderson, Taylor's formula and integral inequalities for conformable fractional derivatives, in *Contributions in Mathematics and Engineering. Honor of Constantin Carathéodory* (Springer, Berlin, 2016), pp. 25–43
5. B. Bede, L. Coroianu, S. Gal, *Approximation by Max-Product Type Operators* (Springer, New York, 2016)
6. O. Iyiola, E. Nwaeze, Some new results on the new conformable fractional calculus with application using D'Alambert approach. Progr. Fract. Differ. Appl. **2**(2), 115–122 (2016)
7. R. Khalil, M. Al Horani, A. Yousef, M. Sababheh, A new definition of fractional derivative. J. Comput. Appl. Math. **264**, 65–70 (2014)
8. G.G. Lorentz, *Bernstein Polynomials*, 2nd edn. (Chelsea Publishing Company, New York, 1986)
9. T. Popoviciu, Sur l'approximation de fonctions convexes d'order superieur. Mathematica (Cluj) **10**, 49–54 (1935)

Chapter 4
Caputo Fractional Approximation Using Positive Sublinear Operators

Here we consider the approximation of functions by sublinear positive operators with applications to a big variety of Max-Product operators under Caputo fractional differentiability. Our study is based on our general fractional results about positive sublinear operators. We produce Jackson type inequalities under simple initial conditions. So our approach is quantitative by producing inequalities with their right hand sides involving the modulus of continuity of fractional derivative of the function under approximation. It follows [4].

4.1 Introduction

The main motivation here is the monograph by B. Bede, L. Coroianu and S. Gal [6], 2016.

Let $N \in \mathbb{N}$, the well-known Bernstein polynomials [12] are positive linear operators, defined by the formula

$$B_N (f) (x) = \sum_{k=0}^{N} \binom{N}{k} x^k (1 - x)^{N-k} f \left(\frac{k}{N} \right), \quad x \in [0, 1], \ f \in C ([0, 1]).$$

(4.1)

T. Popoviciu in [13], 1935, proved for $f \in C ([0, 1])$ that

$$|B_N (f) (x) - f (x)| \leq \frac{5}{4} \omega_1 \left(f, \frac{1}{\sqrt{N}} \right), \quad \forall \, x \in [0, 1],$$

(4.2)

© Springer International Publishing AG, part of Springer Nature 2018
G. A. Anastassiou, *Nonlinearity: Ordinary and Fractional Approximations by Sublinear and Max-Product Operators*, Studies in Systems, Decision and Control 147, https://doi.org/10.1007/978-3-319-89509-3_4

where

$$\omega_1(f, \delta) = \sup_{\substack{x, y \in [a,b]: \\ |x-y| \leq \delta}} |f(x) - f(y)|, \; \delta > 0, \tag{4.3}$$

is the first modulus of continuity, here $[a, b] = [0, 1]$.
G. G. Lorentz in [12], 1986, p. 21, proved for $f \in C^1([0, 1])$ that

$$|B_N(f)(x) - f(x)| \leq \frac{3}{4\sqrt{N}} \omega_1\left(f', \frac{1}{\sqrt{N}}\right), \; \forall \, x \in [0, 1], \tag{4.4}$$

In [6], p. 10, the authors introduced the basic Max-product Bernstein operators,

$$B_N^{(M)}(f)(x) = \frac{\bigvee_{k=0}^{N} p_{N,k}(x) f\left(\frac{k}{N}\right)}{\bigvee_{k=0}^{N} p_{N,k}(x)}, \; N \in \mathbb{N}, \tag{4.5}$$

where $\bigvee$ stands for maximum, and $p_{N,k}(x) = \binom{N}{k} x^k (1-x)^{N-k}$ and $f : [0, 1] \rightarrow \mathbb{R}_+ = [0, \infty)$.

These are nonlinear and piecewise rational operators.

The authors in [6] studied similar such nonlinear operators such as: the Max-product Favard–Szász–Mirakjan operators and their truncated version, the Max-product Baskakov operators and their truncated version, also many other similar specific operators. The study in [6] is based on presented there general theory of sublinear operators. These Max-product operators tend to converge faster to the on hand function.

So we mention from [6], p. 30, that for $f : [0, 1] \rightarrow \mathbb{R}_+$ continuous, we have the estimate

$$\left|B_N^{(M)}(f)(x) - f(x)\right| \leq 12\omega_1\left(f, \frac{1}{\sqrt{N+1}}\right), \; \text{for all } N \in \mathbb{N}, \; x \in [0, 1], \tag{4.6}$$

Also from [6], p. 36, we mention that for $f : [0, 1] \rightarrow \mathbb{R}_+$ being concave function we get that

$$\left|B_N^{(M)}(f)(x) - f(x)\right| \leq 2\omega_1\left(f, \frac{1}{N}\right), \; \text{for all } x \in [0, 1], \tag{4.7}$$

a much faster convergence.

In this chapter we expand the study in [6] by considering Caputo fractional smoothness of functions. So our inequalities are with respect to $\omega_1(D^\alpha f, \delta), \delta > 0$, where $D^\alpha f$ with $\alpha > 0$ is the Caputo fractional derivative.

4.2 Main Results

We need

Definition 4.1 Let $\nu \geq 0$, $n = \lceil \nu \rceil$ ($\lceil \cdot \rceil$ is the ceiling of the number), $f \in AC^n$ ($[a, b]$) (space of functions f with $f^{(n-1)} \in AC$ ($[a, b]$), absolutely continuous functions). We call left Caputo fractional derivative (see [8], p. 49, [11, 14]) the function

$$D_{*a}^{\nu} f(x) = \frac{1}{\Gamma(n-\nu)} \int_a^x (x-t)^{n-\nu-1} f^{(n)}(t)\, dt, \; \forall\, x \in [a, b], \qquad (4.8)$$

where Γ is the gamma function $\Gamma(v) = \int_0^\infty e^{-t} t^{v-1} dt$, $v > 0$.
    We set $D_{*a}^0 f(x) = f(x)$, $\forall\, x \in [a, b]$.

Lemma 4.2 ([2]) *Let* $\nu > 0$, $\nu \notin \mathbb{N}$, $n = \lceil \nu \rceil$, $f \in C^{n-1}$ ($[a, b]$) *and* $f^{(n)} \in L_\infty$ ($[a, b]$). *Then* $D_{*a}^{\nu} f(a) = 0$.

We need

Definition 4.3 (*see also* [1, 9, 11]) Let $f \in AC^m$ ($[a, b]$), $m = \lceil \alpha \rceil$, $\alpha > 0$. We right Caputo fractional derivative of order $\alpha > 0$ is given by

$$D_{b-}^{\alpha} f(x) = \frac{(-1)^m}{\Gamma(m-\alpha)} \int_x^b (\zeta - x)^{m-\alpha-1} f^{(m)}(\zeta)\, d\zeta, \; \forall\, x \in [a, b]. \qquad (4.9)$$

We set $D_{b-}^0 f(x) = f(x)$.

Lemma 4.4 ([2]) *Let* $f \in C^{m-1}$ ($[a, b]$), $f^{(m)} \in L_\infty$ ($[a, b]$), $m = \lceil \alpha \rceil$, $\alpha > 0$, $\alpha \notin \mathbb{N}$. *Then* $D_{b-}^{\alpha} f(b) = 0$.

Convention 4.5 *We assume that*

$$D_{*x_0}^{a} f(x) = 0, \text{ for } x < x_0, \qquad (4.10)$$

and

$$D_{x_0-}^{\alpha} f(x) = 0, \text{ for } x > x_0, \qquad (4.11)$$

for all $x, x_0 \in [a, b]$.

We mention

Proposition 4.6 ([2]) *Let* $f \in C^n$ ($[a, b]$), $n = \lceil \nu \rceil$, $\nu > 0$. *Then* $D_{*a}^{\nu} f(x)$ *is continuous in* $x \in [a, b]$.

Proposition 4.7 ([2]) *Let* $f \in C^m$ ($[a, b]$), $m = \lceil \alpha \rceil$, $\alpha > 0$. *Then* $D_{b-}^{\alpha} f(x)$ *is continuous in* $x \in [a, b]$.

The modulus of continuity $\omega_1 (f, \delta)$ is defined the same way for bounded functions, see (4.3), and it is finite.

We make

Remark 4.8 ([2]) Let $f \in C^{n-1} ([a, b])$, $f^{(n)} \in L_\infty ([a, b])$, $n = \lceil \nu \rceil, \nu > 0, \nu \notin \mathbb{N}$. Then

$$\omega_1 \left(D_{*a}^\nu f, \delta \right) \leq \frac{2 \left\| f^{(n)} \right\|_\infty}{\Gamma (n - \nu + 1)} (b - a)^{n - \nu} . \tag{4.12}$$

Similarly, let $f \in C^{m-1} ([a, b])$, $f^{(m)} \in L_\infty ([a, b])$, $m = \lceil \alpha \rceil, \alpha > 0, \alpha \notin \mathbb{N}$, then

$$\omega_1 \left(D_{b-}^\alpha f, \delta \right) \leq \frac{2 \left\| f^{(m)} \right\|_\infty}{\Gamma (m - \alpha + 1)} (b - a)^{m - \alpha} . \tag{4.13}$$

That is $\omega_1 \left(D_{*a}^\nu f, \delta \right), \omega_1 \left(D_{b-}^\alpha f, \delta \right)$ are finite.

Clearly, above $D_{*a}^\nu f$ and $D_{b-}^\alpha f$ are bounded, from

$$\left| D_{*a}^\nu f (x) \right| \leq \frac{\left\| f^{(n)} \right\|_\infty}{\Gamma (n - \nu + 1)} (b - a)^{n - \nu} , \ \forall \ x \in [a, b], \tag{4.14}$$

see [2].

We need

Definition 4.9 Let $D_{x_0}^\alpha f$ denote any of $D_{x_0-}^\alpha f$, $D_{*x_0}^\alpha f$, and $\delta > 0$. We set

$$\omega_1 \left(D_{x_0}^\alpha f, \delta \right) := \max \left\{ \omega_1 \left(D_{x_0-}^\alpha f, \delta \right)_{[a, x_0]} , \omega_1 \left(D_{*x_0}^\alpha f, \delta \right)_{[x_0, b]} \right\}, \tag{4.15}$$

where $x_0 \in [a, b]$. Here the moduli of continuity are considered over $[a, x_0]$ and $[x_0, b]$, respectively.

We need

Theorem 4.10 Let $\alpha > 0$, $\alpha \notin \mathbb{N}$, $m = \lceil \alpha \rceil$, $x_0 \in [a, b] \subset \mathbb{R}$, $f \in AC^m ([a, b], \mathbb{R}_+)$ (i.e. $f^{(m-1)} \in AC ([a, b])$, absolutely continuous functions on $[a, b]$), and $f^{(m)} \in L_\infty ([a, b])$. Furthermore we assume that $f^{(k)} (x_0) = 0$, $k = 1, ..., m - 1$. Then

$$|f (x) - f (x_0)| \leq \frac{\omega_1 \left(D_{x_0}^\alpha f, \delta \right)}{\Gamma (\alpha + 1)} \left[|x - x_0|^\alpha + \frac{|x - x_0|^{\alpha+1}}{(\alpha + 1) \delta} \right], \ \delta > 0, \tag{4.16}$$

for all $a \leq x \leq b$.

If $0 < \alpha < 1$, then we do not need initial conditions.

Proof From [8], p. 54, we get by left Caputo Taylor formula that

$$f(x) = \sum_{k=0}^{m-1} \frac{f^{(k)}(x_0)}{k!}(x-x_0)^k + \frac{1}{\Gamma(\alpha)}\int_{x_0}^{x}(x-z)^{\alpha-1}D_{*x_0}^{\alpha}f(z)\,dz, \quad (4.17)$$

for all $x_0 \le x \le b$.

Also from [1], using the right Caputo fractional Taylor formula we get

$$f(x) = \sum_{k=0}^{m-1} \frac{f^{(k)}(x_0)}{k!}(x-x_0)^k + \frac{1}{\Gamma(\alpha)}\int_{x}^{x_0}(z-x)^{\alpha-1}D_{x_0-}^{\alpha}f(z)\,dz, \quad (4.18)$$

for all $a \le x \le x_0$.

By the assumption $f^{(k)}(x_0) = 0, k = 1, ..., m-1$, we get

$$f(x) - f(x_0) = \frac{1}{\Gamma(\alpha)}\int_{x_0}^{x}(x-z)^{\alpha-1}D_{*x_0}^{\alpha}f(z)\,dz, \quad (4.19)$$

for all $x_0 \le x \le b$.

And it holds

$$f(x) - f(x_0) = \frac{1}{\Gamma(\alpha)}\int_{x}^{x_0}(z-x)^{\alpha-1}D_{x_0-}^{\alpha}f(z)\,dz, \quad (4.20)$$

for all $a \le x \le x_0$.

Notice that when $0 < \alpha < 1$, then $m = 1$, and (4.19) and (4.20) are valid without initial conditions.

Since $D_{x_0-}^{\alpha}f(x_0) = D_{*x_0}^{\alpha}f(x_0) = 0$, we get

$$f(x) - f(x_0) = \frac{1}{\Gamma(\alpha)}\int_{x_0}^{x}(x-z)^{\alpha-1}\left((D_{*x_0}^{\alpha}f)(z) - D_{*x_0}^{\alpha}f(x_0)\right)dz, \quad (4.21)$$

$x_0 \le x \le b$,

and

$$f(x) - f(x_0) = \frac{1}{\Gamma(\alpha)}\int_{x}^{x_0}(z-x)^{\alpha-1}\left(D_{x_0-}^{\alpha}f(z) - D_{x_0-}^{\alpha}f(x_0)\right)dz, \quad (4.22)$$

$a \le x \le x_0$.

We have that ($x_0 \le x \le b$)

$$|f(x) - f(x_0)| \le \frac{1}{\Gamma(\alpha)}\int_{x_0}^{x}(x-z)^{\alpha-1}\left|(D_{*x_0}^{\alpha}f)(z) - D_{*x_0}^{\alpha}f(x_0)\right|dz \underset{(\delta_1>0)}{\le}$$

$$\frac{1}{\Gamma(\alpha)}\int_{x_0}^{x}(x-z)^{\alpha-1}\omega_1\left(D_{*x_0}^{\alpha}f, \frac{\delta_1|z-x_0|}{\delta_1}\right)_{[x_0,b]}dz \le \quad (4.23)$$

$$\frac{\omega_1 \left(D_{*x_0}^\alpha f, \delta_1\right)_{[x_0,b]}}{\Gamma(\alpha)} \int_{x_0}^x (x-z)^{\alpha-1} \left(1 + \frac{(z-x_0)}{\delta_1}\right) dz =$$

$$\frac{\omega_1 \left(D_{*x_0}^\alpha f, \delta_1\right)_{[x_0,b]}}{\Gamma(\alpha)} \left[\frac{(x-x_0)^\alpha}{\alpha} + \frac{1}{\delta_1} \int_{x_0}^x (x-z)^{\alpha-1} (z-x_0)^{2-1} dz\right] =$$

$$\frac{\omega_1 \left(D_{*x_0}^\alpha f, \delta_1\right)_{[x_0,b]}}{\Gamma(\alpha)} \left[\frac{(x-x_0)^\alpha}{\alpha} + \frac{1}{\delta_1} \frac{\Gamma(\alpha)\Gamma(2)}{\Gamma(\alpha+2)} (x-x_0)^{\alpha+1}\right] = \qquad (4.24)$$

$$\frac{\omega_1 \left(D_{*x_0}^\alpha f, \delta_1\right)_{[x_0,b]}}{\Gamma(\alpha)} \left[\frac{(x-x_0)^\alpha}{\alpha} + \frac{1}{\delta_1} \frac{1}{(\alpha+1)\alpha} (x-x_0)^{\alpha+1}\right] =$$

$$\frac{\omega_1 \left(D_{*x_0}^\alpha f, \delta_1\right)_{[x_0,b]}}{\Gamma(\alpha+1)} \left[(x-x_0)^\alpha + \frac{(x-x_0)^{\alpha+1}}{(\alpha+1)\delta_1}\right].$$

We have proved that

$$|f(x) - f(x_0)| \le \frac{\omega_1 \left(D_{*x_0}^\alpha f, \delta_1\right)_{[x_0,b]}}{\Gamma(\alpha+1)} \left[(x-x_0)^\alpha + \frac{(x-x_0)^{\alpha+1}}{(\alpha+1)\delta_1}\right], \qquad (4.25)$$

$\delta_1 > 0$, and $x_0 \le x \le b$.

Similarly acting, we get $(a \le x \le x_0)$

$$|f(x) - f(x_0)| \le \frac{1}{\Gamma(\alpha)} \int_x^{x_0} (z-x)^{\alpha-1} \left|D_{x_0-}^\alpha f(z) - D_{x_0-}^\alpha f(x_0)\right| dz \le$$

$$\frac{1}{\Gamma(\alpha)} \int_x^{x_0} (z-x)^{\alpha-1} \omega_1 \left(D_{x_0-}^\alpha f, |z-x_0|\right)_{[a,x_0]} dz =$$

$(\delta_2 > 0)$

$$\frac{1}{\Gamma(\alpha)} \int_x^{x_0} (z-x)^{\alpha-1} \omega_1 \left(D_{x_0-}^\alpha f, \frac{\delta_2(x_0-z)}{\delta_2}\right)_{[a,x_0]} dz \le \qquad (4.26)$$

$$\frac{\omega_1 \left(D_{x_0-}^\alpha f, \delta_2\right)_{[a,x_0]}}{\Gamma(\alpha)} \left[\int_x^{x_0} (z-x)^{\alpha-1} \left(1 + \frac{x_0-z}{\delta_2}\right) dz\right] =$$

$$\frac{\omega_1 \left(D_{x_0-}^\alpha f, \delta_2\right)_{[a,x_0]}}{\Gamma(\alpha)} \left[\frac{(x_0-x)^\alpha}{\alpha} + \frac{1}{\delta_2} \int_x^{x_0} (x_0-z)^{2-1} (z-x)^{\alpha-1} dz\right] =$$

$$\frac{\omega_1 \left(D_{x_0-}^\alpha f, \delta_2\right)_{[a,x_0]}}{\Gamma(\alpha)} \left[\frac{(x_0-x)^\alpha}{\alpha} + \frac{1}{\delta_2} \frac{\Gamma(\alpha)\Gamma(2)}{\Gamma(\alpha+2)} (x_0-x)^{\alpha+1}\right] =$$

$$\frac{\omega_1 \left(D_{x_0-}^{\alpha} f, \delta_2 \right)_{[a,x_0]}}{\Gamma(\alpha)} \left[\frac{(x_0 - x)^{\alpha}}{\alpha} + \frac{1}{\delta_2} \frac{(x_0 - x)^{\alpha+1}}{(\alpha+1)\alpha} \right] = \qquad (4.27)$$

$$\frac{\omega_1 \left(D_{x_0-}^{\alpha} f, \delta_2 \right)_{[a,x_0]}}{\Gamma(\alpha+1)} \left[(x_0 - x)^{\alpha} + \frac{(x_0 - x)^{\alpha+1}}{(\alpha+1)\delta_2} \right].$$

We have proved that

$$|f(x) - f(x_0)| \le \frac{\omega_1 \left(D_{x_0-}^{\alpha} f, \delta_2 \right)_{[a,x_0]}}{\Gamma(\alpha+1)} \left[(x_0 - x)^{\alpha} + \frac{(x_0 - x)^{\alpha+1}}{(\alpha+1)\delta_2} \right], \qquad (4.28)$$

$\delta_2 > 0$, and $(a \le x \le x_0)$. Choosing $\delta = \delta_1 = \delta_2 > 0$, by (4.25) and (4.28), we get (4.16). ∎

We need

Definition 4.11 Here $C_+([a,b]) := \{f : [a,b] \to \mathbb{R}_+, \text{continuous functions}\}$. Let $L_N : C_+([a,b]) \to C_+([a,b])$, operators, $\forall N \in \mathbb{N}$, such that
(i)
$$L_N(\alpha f) = \alpha L_N(f), \quad \forall \alpha \ge 0, \forall f \in C_+([a,b]), \qquad (4.29)$$

(ii) if $f, g \in C_+([a,b]) : f \le g$, then

$$L_N(f) \le L_N(g), \quad \forall N \in \mathbb{N}, \qquad (4.30)$$

(iii)
$$L_N(f + g) \le L_N(f) + L_N(g), \quad \forall f, g \in C_+([a,b]). \qquad (4.31)$$

We call $\{L_N\}_{N \in \mathbb{N}}$ positive sublinear operators.

We need a Hölder's type inequality, see next:

Theorem 4.12 (see [3]) *Let* $L : C_+([a,b]) \to C_+([a,b])$, *be a positive sublinear operator and* $f, g \in C_+([a,b])$, *furthermore let* $p, q > 1 : \frac{1}{p} + \frac{1}{q} = 1$. *Assume that* $L\left((f(\cdot))^p\right)(s_*), L\left((g(\cdot))^q\right)(s_*) > 0$ *for some* $s_* \in [a,b]$. *Then*

$$L(f(\cdot)g(\cdot))(s_*) \le \left(L\left((f(\cdot))^p\right)(s_*) \right)^{\frac{1}{p}} \left(L\left((g(\cdot))^q\right)(s_*) \right)^{\frac{1}{q}}. \qquad (4.32)$$

We make

Remark 4.13 By [6], p. 17, we get: let $f, g \in C_+([a,b])$, then

$$|L_N(f)(x) - L_N(g)(x)| \le L_N(|f - g|)(x), \quad \forall x \in [a,b]. \qquad (4.33)$$

Furthermore, we also have that

$$|L_N(f)(x) - f(x)| \le L_N(|f(\cdot) - f(x)|)(x) + |f(x)| |L_N(e_0)(x) - 1|,$$
(4.34)

$\forall\, x \in [a,b];\ e_0(t) = 1.$

From now on we assume that $L_N(1) = 1$. Hence it holds

$$|L_N(f)(x) - f(x)| \le L_N(|f(\cdot) - f(x)|)(x), \ \forall\, x \in [a,b].$$
(4.35)

Using Theorem 4.10 and (4.16) with (4.35) we get:

$$|L_N(f)(x_0) - f(x_0)| \le \frac{\omega_1\left(D_{x_0}^\alpha f, \delta\right)}{\Gamma(\alpha + 1)}.$$
(4.36)

$$\left[L_N(|\cdot - x_0|^\alpha)(x_0) + \frac{L_N\left(|\cdot - x_0|^{\alpha+1}\right)(x_0)}{(\alpha + 1)\delta} \right], \ \delta > 0.$$

We have proved

Theorem 4.14 *Let* $\alpha > 0$, $\alpha \notin \mathbb{N}$, $m = \lceil \alpha \rceil$, $x_0 \in [a,b] \subset \mathbb{R}$, $f \in AC^m([a,b],$ $\mathbb{R}_+)$, *and* $f^{(m)} \in L_\infty([a,b])$. *Furthermore we assume that* $f^{(k)}(x_0) = 0$, $k = 1, ..., m - 1$. *Let* $L_N : C_+([a,b]) \to C_+([a,b])$, $\forall\, N \in \mathbb{N}$, *be positive sublinear operators, such that* $L_N(1) = 1$, $\forall\, N \in \mathbb{N}$. *Then*

$$|L_N(f)(x_0) - f(x_0)| \le \frac{\omega_1\left(D_{x_0}^\alpha f, \delta\right)}{\Gamma(\alpha + 1)}.$$

$$\left[L_N(|\cdot - x_0|^\alpha)(x_0) + \frac{L_N\left(|\cdot - x_0|^{\alpha+1}\right)(x_0)}{(\alpha + 1)\delta} \right],$$
(4.37)

$\delta > 0$, $\forall\, N \in \mathbb{N}$.

In particular (4.37) is true for $\alpha > 1$, $\alpha \notin \mathbb{N}$.

Corollary 4.15 *Let* $0 < \alpha < 1$, $x_0 \in [a,b] \subset \mathbb{R}$, $f \in AC([a,b], \mathbb{R}_+)$, *and* $f' \in L_\infty([a,b])$. *Let* $L_N : C_+([a,b]) \to C_+([a,b])$, $\forall\, N \in \mathbb{N}$, *be positive sublinear operators, such that* $L_N(1) = 1$, $\forall\, N \in \mathbb{N}$. *Then (4.37) is valid.*

We give

Theorem 4.16 *Let* $0 < \alpha < 1$, $x_0 \in [a,b] \subset \mathbb{R}$, $f \in AC([a,b], \mathbb{R}_+)$, *and* $f' \in L_\infty([a,b])$. *Let* L_N *from* $C_+([a,b])$ *into itself be positive sublinear operators, such that* $L_N(1) = 1$, $\forall\, N \in \mathbb{N}$. *Assume that* $L_N\left(|\cdot - x_0|^{\alpha+1}\right)(x_0) > 0$, $\forall\, N \in \mathbb{N}$. *Then*

$$|L_N(f)(x_0) - f(x_0)| \le$$

$$\frac{(\alpha+2)\,\omega_1\left(D_{x_0}^{\alpha}f,\left(L_N\left(|\cdot-x_0|^{\alpha+1}\right)(x_0)\right)^{\frac{1}{\alpha+1}}\right)}{\Gamma(\alpha+2)}\left(L_N\left(|\cdot-x_0|^{\alpha+1}\right)(x_0)\right)^{\frac{\alpha}{\alpha+1}}.$$

$$(4.38)$$

Proof By Theorem 4.12, see (4.32), we get

$$L_N\left(|\cdot-x_0|^{\alpha}\right)(x_0)\le\left(L_N\left(|\cdot-x_0|^{\alpha+1}\right)(x_0)\right)^{\frac{\alpha}{\alpha+1}}.\qquad(4.39)$$

Choose

$$\delta:=\left(L_N\left(|\cdot-x_0|^{\alpha+1}\right)(x_0)\right)^{\frac{1}{\alpha+1}}>0,\qquad(4.40)$$

i.e. $\delta^{\alpha+1}=L_N\left(|\cdot-x_0|^{\alpha+1}\right)(x_0)$.

By (4.37) we obtain

$$|L_N(f)(x_0)-f(x_0)|\le\frac{1}{\Gamma(\alpha+1)}\omega_1\left(D_{x_0}^{\alpha}f,\left(L_N\left(|\cdot-x_0|^{\alpha+1}\right)(x_0)\right)^{\frac{1}{\alpha+1}}\right).$$

$$\left[\left(L_N\left(|\cdot-x_0|^{\alpha+1}\right)(x_0)\right)^{\frac{\alpha}{\alpha+1}}+\frac{1}{(\alpha+1)}\left(L_N\left(|\cdot-x_0|^{\alpha+1}\right)(x_0)\right)^{\frac{\alpha}{\alpha+1}}\right]=$$

$$\frac{\omega_1\left(D_{x_0}^{\alpha}f,\left(L_N\left(|\cdot-x_0|^{\alpha+1}\right)(x_0)\right)^{\frac{1}{\alpha+1}}\right)}{\Gamma(\alpha+1)}.$$

$$\left(L_N\left(|\cdot-x_0|^{\alpha+1}\right)(x_0)\right)^{\frac{\alpha}{\alpha+1}}\left[1+\frac{1}{\alpha+1}\right]=\qquad(4.41)$$

$$\frac{\omega_1\left(D_{x_0}^{\alpha}f,\left(L_N\left(|\cdot-x_0|^{\alpha+1}\right)(x_0)\right)^{\frac{1}{\alpha+1}}\right)}{\Gamma(\alpha+1)}\left(L_N\left(|\cdot-x_0|^{\alpha+1}\right)(x_0)\right)^{\frac{\alpha}{\alpha+1}}\left(\frac{\alpha+2}{\alpha+1}\right)=$$

$$\frac{(\alpha+2)\,\omega_1\left(D_{x_0}^{\alpha}f,\left(L_N\left(|\cdot-x_0|^{\alpha+1}\right)(x_0)\right)^{\frac{1}{\alpha+1}}\right)}{\Gamma(\alpha+2)}\left(L_N\left(|\cdot-x_0|^{\alpha+1}\right)(x_0)\right)^{\frac{\alpha}{\alpha+1}},$$

proving (4.38). ∎

4.3 Applications

(I) Case $0<\alpha<1$.

Here we apply Theorem 4.16 to well known Max-product operators.
We make

Remark 4.17 The Max-product Bernstein operators $B_N^{(M)}(f)(x)$ are defined by (4.5), see also [6], p. 10; here $f : [0, 1] \to \mathbb{R}_+$ is a continuous function.

We have $B_N^{(M)}(1) = 1$, and

$$B_N^{(M)}(|\cdot - x|)(x) \le \frac{6}{\sqrt{N+1}}, \forall x \in [0, 1], \forall N \in \mathbb{N}, \tag{4.42}$$

see [6], p. 31.

$B_N^{(M)}$ are positive sublinear operators and thus they possess the monotonicity property, also since $|\cdot - x| \le 1$, then $|\cdot - x|^\beta \le 1, \forall x \in [0, 1], \forall \beta > 0$.

Therefore it holds

$$B_N^{(M)}\left(|\cdot - x|^{1+\beta}\right)(x) \le \frac{6}{\sqrt{N+1}}, \forall x \in [0, 1], \forall N \in \mathbb{N}, \forall \beta > 0. \tag{4.43}$$

Furthermore, clearly it holds that

$$B_N^{(M)}\left(|\cdot - x|^{1+\beta}\right)(x) > 0, \forall N \in \mathbb{N}, \forall \beta \ge 0 \text{ and any } x \in (0, 1). \tag{4.44}$$

The operator $B_N^{(M)}$ maps $C_+([0, 1])$ into itself.

We present

Theorem 4.18 *Let* $0 < \alpha < 1$, *any* $x \in (0, 1)$, $f \in AC([0, 1], \mathbb{R}_+)$, *and* $f' \in L_\infty$ $([0, 1])$. *Then*

$$\left| B_N^{(M)}(f)(x) - f(x) \right| \le \frac{(\alpha + 2)\,\omega_1\left(D_x^\alpha f, \left(\frac{6}{\sqrt{N+1}}\right)^{\frac{1}{\alpha+1}}\right)}{\Gamma(\alpha + 2)} \left(\frac{6}{\sqrt{N+1}}\right)^{\frac{\alpha}{\alpha+1}}, \tag{4.45}$$

$\forall N \in \mathbb{N}$.

As $N \to +\infty$, *we get* $B_N^{(M)}(f)(x) \to f(x)$, *any* $x \in (0, 1)$.

Proof By Theorem 4.16 ∎

We continue with

Remark 4.19 The truncated Favard–Szász–Mirakjan operators are given by

$$T_N^{(M)}(f)(x) = \frac{\bigvee_{k=0}^N s_{N,k}(x)\, f\left(\frac{k}{N}\right)}{\bigvee_{k=0}^N s_{N,k}(x)}, \quad x \in [0, 1], \ N \in \mathbb{N}, \ f \in C_+([0, 1]), \tag{4.46}$$

$s_{N,k}(x) = \frac{(Nx)^k}{k!}$, see also [6], p. 11.

By [6], pp. 178–179, we get that

$$T_N^{(M)}(|\cdot - x|)(x) \le \frac{3}{\sqrt{N}}, \forall x \in [0, 1], \forall N \in \mathbb{N}. \tag{4.47}$$

Clearly it holds

$$T_N^{(M)}\left(|\cdot - x|^{1+\beta}\right)(x) \le \frac{3}{\sqrt{N}}, \ \forall \, x \in [0, 1], \ \forall \, N \in \mathbb{N}, \forall \, \beta > 0. \qquad (4.48)$$

The operators $T_N^{(M)}$ are positive sublinear operators mapping $C_+ ([0, 1])$ into itself, with $T_N^{(M)} (1) = 1$.

Furthermore it holds

$$T_N^{(M)}\left(|\cdot - x|^{\lambda}\right)(x) = \frac{\bigvee_{k=0}^{N} \frac{(Nx)^k}{k!} \left|\frac{k}{N} - x\right|^{\lambda}}{\bigvee_{k=0}^{N} \frac{(Nx)^k}{k!}} > 0, \ \forall \, x \in (0, 1], \ \forall \, \lambda \ge 1, \forall \, N \in \mathbb{N}.$$

$$(4.49)$$

We give

Theorem 4.20 *Let* $0 < \alpha < 1$, *any* $x \in (0, 1]$, $f \in AC ([0, 1], \mathbb{R}_+)$, *and* $f' \in L_\infty$ *([0, 1]). Then*

$$\left|T_N^{(M)} (f)(x) - f(x)\right| \le \frac{(\alpha + 2) \, \omega_1 \left(D_x^\alpha f, \left(\frac{3}{\sqrt{N}}\right)^{\frac{1}{\alpha+1}}\right)}{\Gamma(\alpha + 2)} \left(\frac{3}{\sqrt{N}}\right)^{\frac{\alpha}{\alpha+1}}, \ \forall \, N \in \mathbb{N}.$$

$$(4.50)$$

As $N \to +\infty$, *we get* $T_N^{(M)} (f)(x) \to f(x)$, *for any* $x \in (0, 1]$.

Proof By Theorem 4.16. ∎

We make

Remark 4.21 Next we study the truncated Max-product Baskakov operators (see [6], p. 11)

$$U_N^{(M)} (f)(x) = \frac{\bigvee_{k=0}^{N} b_{N,k} (x) \, f\left(\frac{k}{N}\right)}{\bigvee_{k=0}^{N} b_{N,k} (x)}, \ x \in [0, 1], \ f \in C_+ ([0, 1]), \ N \in \mathbb{N},$$

$$(4.51)$$

where

$$b_{N,k} (x) = \binom{N + k - 1}{k} \frac{x^k}{(1 + x)^{N+k}}. \qquad (4.52)$$

From [6], pp. 217–218, we get $(x \in [0, 1])$

$$\left(U_N^{(M)} (|\cdot - x|)\right)(x) \le \frac{2\sqrt{3}\left(\sqrt{2} + 2\right)}{\sqrt{N + 1}}, \ N \ge 2, N \in \mathbb{N}. \qquad (4.53)$$

Let $\lambda \ge 1$, clearly then it holds

$$\left(U_N^{(M)}\left(|\cdot - x|^\lambda\right)\right)(x) \le \frac{2\sqrt{3}\left(\sqrt{2}+2\right)}{\sqrt{N+1}}, \ \forall \ N \ge 2, N \in \mathbb{N}. \tag{4.54}$$

Also it holds $U_N^{(M)}(1) = 1$, and $U_N^{(M)}$ are positive sublinear operators from $C_+([0, 1])$ into itself. Furthermore it holds

$$U_N^{(M)}\left(|\cdot - x|^\lambda\right)(x) > 0, \ \forall \ x \in (0, 1], \ \forall \ \lambda \ge 1, \forall \ N \in \mathbb{N}. \tag{4.55}$$

We give

Theorem 4.22 *Let* $0 < \alpha < 1$, *any* $x \in (0, 1]$, $f \in AC([0, 1], \mathbb{R}_+)$, *and* $f' \in L_\infty$ *$([0, 1])$. Then*

$$\left|U_N^{(M)}(f)(x) - f(x)\right| \le \frac{(\alpha+2)\,\omega_1\left(D_x^\alpha f, \left(\frac{2\sqrt{3}(\sqrt{2}+2)}{\sqrt{N+1}}\right)^{\frac{1}{\alpha+1}}\right)}{\Gamma(\alpha+2)} \cdot$$

$$\left(\frac{2\sqrt{3}\left(\sqrt{2}+2\right)}{\sqrt{N+1}}\right)^{\frac{\alpha}{\alpha+1}}, \ \forall \ N \ge 2, N \in \mathbb{N}. \tag{4.56}$$

As $N \to +\infty$, *we get* $U_N^{(M)}(f)(x) \to f(x)$, *for any* $x \in (0, 1]$.

Proof By Theorem 4.16. ∎

We continue with

Remark 4.23 Here we study the Max-product Meyer–Köning and Zeller operators (see [6], p. 11) defined by

$$Z_N^{(M)}(f)(x) = \frac{\bigvee_{k=0}^\infty s_{N,k}(x)\, f\left(\frac{k}{N+k}\right)}{\bigvee_{k=0}^\infty s_{N,k}(x)}, \ \forall \ N \in \mathbb{N}, f \in C_+([0, 1]), \tag{4.57}$$

$$s_{N,k}(x) = \binom{N+k}{k} x^k, x \in [0, 1].$$

By [6], p. 253, we get that

$$Z_N^{(M)}(|\cdot - x|)(x) \le \frac{8\left(1 + \sqrt{5}\right)}{3} \frac{\sqrt{x}\,(1-x)}{\sqrt{N}}, \ \forall \ x \in [0, 1], \ \forall \ N \ge 4, N \in \mathbb{N}. \tag{4.58}$$

As before we get that (for $\lambda \ge 1$)

$$Z_N^{(M)}\left(|\cdot - x|^\lambda\right)(x) \le \frac{8\left(1 + \sqrt{5}\right)}{3} \frac{\sqrt{x}\,(1-x)}{\sqrt{N}} := \rho(x), \tag{4.59}$$

$\forall\, x \in [0, 1], N \geq 4, N \in \mathbb{N}.$

Also it holds $Z_N^{(M)}(1) = 1$, and $Z_N^{(M)}$ are positive sublinear operators from $C_+([0, 1])$ into itself. Also it holds

$$Z_N^{(M)}\left(|\cdot - x|^\lambda\right)(x) > 0, \ \forall\, x \in (0, 1), \ \forall\, \lambda \geq 1, \forall\, N \in \mathbb{N}. \qquad (4.60)$$

We give

Theorem 4.24 *Let* $0 < \alpha < 1$, *any* $x \in (0, 1)$, $f \in AC([0, 1], \mathbb{R}_+)$, *and* $f' \in L_\infty$ $([0, 1])$. *Then*

$$\left| Z_N^{(M)}(f)(x) - f(x) \right| \leq \frac{(\alpha + 2)\, \omega_1\left(D_x^\alpha f, (\rho(x))^{\frac{1}{\alpha+1}} \right)}{\Gamma(\alpha + 2)} (\rho(x))^{\frac{\alpha}{\alpha+1}} \qquad (4.61)$$

$\forall\, N \geq 4, N \in \mathbb{N}.$

As $N \to +\infty$, *we get* $Z_N^{(M)}(f)(x) \to f(x)$, *for any* $x \in (0, 1)$.

Proof By Theorem 4.16. ∎

We continue with

Remark 4.25 Here we deal with the Max-product truncated sampling operators (see [6], p. 13) defined by

$$W_N^{(M)}(f)(x) = \frac{\bigvee_{k=0}^{N} \frac{\sin(Nx - k\pi)}{Nx - k\pi} f\left(\frac{k\pi}{N}\right)}{\bigvee_{k=0}^{N} \frac{\sin(Nx - k\pi)}{Nx - k\pi}}, \qquad (4.62)$$

and

$$K_N^{(M)}(f)(x) = \frac{\bigvee_{k=0}^{N} \frac{\sin^2(Nx - k\pi)}{(Nx - k\pi)^2} f\left(\frac{k\pi}{N}\right)}{\bigvee_{k=0}^{N} \frac{\sin^2(Nx - k\pi)}{(Nx - k\pi)^2}}, \qquad (4.63)$$

$\forall\, x \in [0, \pi]$, $f : [0, \pi] \to \mathbb{R}_+$ a continuous function.

Following [6], p. 343, and making the convention $\frac{\sin(0)}{0} = 1$ and denoting $s_{N,k}$ $(x) = \frac{\sin(Nx - k\pi)}{Nx - k\pi}$, we get that $s_{N,k}\left(\frac{k\pi}{N}\right) = 1$, and $s_{N,k}\left(\frac{j\pi}{N}\right) = 0$, if $k \neq j$, furthermore $W_N^{(M)}(f)\left(\frac{j\pi}{N}\right) = f\left(\frac{j\pi}{N}\right)$, for all $j \in \{0, ..., N\}$.

Clearly $W_N^{(M)}(f)$ is a well-defined function for all $x \in [0, \pi]$, and it is continuous on $[0, \pi]$, also $W_N^{(M)}(1) = 1$.

By [6], p. 344, $W_N^{(M)}$ are positive sublinear operators.

Call $I_N^+(x) = \left\{ k \in \{0, 1, ..., N\}; s_{N,k}(x) > 0 \right\}$, and set $x_{N,k} := \frac{k\pi}{N}$, $k \in \{0, 1, ..., N\}$.

We see that

$$W_N^{(M)}(f)(x) = \frac{\bigvee_{k \in I_N^+(x)} s_{N,k}(x)\, f\left(x_{N,k}\right)}{\bigvee_{k \in I_N^+(x)} s_{N,k}(x)}. \qquad (4.64)$$

By [6], p. 346, we have

$$W_N^{(M)} \left(|\cdot - x| \right)(x) \le \frac{\pi}{2N}, \ \forall \, N \in \mathbb{N}, \ \forall \, x \in [0, \pi]. \tag{4.65}$$

Notice also $\left| x_{N,k} - x \right| \le \pi, \forall \, x \in [0, \pi]$.

Therefore $(\lambda \ge 1)$ it holds

$$W_N^{(M)} \left(|\cdot - x|^\lambda \right)(x) \le \frac{\pi^{\lambda-1}\pi}{2N} = \frac{\pi^\lambda}{2N}, \ \forall \, x \in [0, \pi], \forall \, N \in \mathbb{N}. \tag{4.66}$$

If $x \in \left(\frac{j\pi}{N}, \frac{(j+1)\pi}{N} \right)$, with $j \in \{0, 1, ..., N\}$, we obtain $nx - j\pi \in (0, \pi)$ and thus $s_{N,j}(x) = \frac{\sin(Nx-j\pi)}{Nx-j\pi} > 0$, see [6], pp. 343–344.

Consequently it holds $(\lambda \ge 1)$

$$W_N^{(M)} \left(|\cdot - x|^\lambda \right)(x) = \frac{\bigvee_{k \in I_N^+(x)} s_{N,k}(x) \left| x_{N,k} - x \right|^\lambda}{\bigvee_{k \in I_N^+(x)} s_{N,k}(x)} > 0, \ \forall \, x \in [0, \pi], \tag{4.67}$$

such that $x \ne x_{N,k}$, for any $k \in \{0, 1, ..., N\}$.

We give

Theorem 4.26 *Let* $0 < \alpha < 1$, *any* $x \in [0, \pi]$ *be such that* $x \ne \frac{k\pi}{N}, k \in \{0, 1, ..., N\}$, $\forall \, N \in \mathbb{N}$; $f \in AC([0, \pi], \mathbb{R}_+)$, *and* $f' \in L_\infty([0, \pi])$. *Then*

$$\left| W_N^{(M)}(f)(x) - f(x) \right| \le \frac{(\alpha + 2) \, \omega_1 \left(D_x^\alpha f, \left(\frac{\pi^{\alpha+1}}{2N} \right)^{\frac{1}{\alpha+1}} \right)}{\Gamma(\alpha + 2)} \left(\frac{\pi^{\alpha+1}}{2N} \right)^{\frac{\alpha}{\alpha+1}}, \ \forall \, N \in \mathbb{N}. \tag{4.68}$$

As $N \to +\infty$, *we get* $W_N^{(M)}(f)(x) \to f(x)$.

Proof By Theorem 4.16. ∎

We make

Remark 4.27 Here we continue with the Max-product truncated sampling operators (see [6], p. 13) defined by

$$K_N^{(M)}(f)(x) = \frac{\bigvee_{k=0}^N \frac{\sin^2(Nx-k\pi)}{(Nx-k\pi)^2} f \left(\frac{k\pi}{N} \right)}{\bigvee_{k=0}^N \frac{\sin^2(Nx-k\pi)}{(Nx-k\pi)^2}}, \tag{4.69}$$

$\forall \, x \in [0, \pi]$, $f : [0, \pi] \to \mathbb{R}_+$ a continuous function.

Following [6], p. 350, and making the convention $\frac{\sin(0)}{0} = 1$ and denoting $s_{N,k}$ $(x) = \frac{\sin^2(Nx-k\pi)}{(Nx-k\pi)^2}$, we get that $s_{N,k} \left(\frac{k\pi}{N} \right) = 1$, and $s_{N,k} \left(\frac{j\pi}{N} \right) = 0$, if $k \ne j$, furthermore $K_N^{(M)}(f) \left(\frac{j\pi}{N} \right) = f \left(\frac{j\pi}{N} \right)$, for all $j \in \{0, ..., N\}$.

Since $s_{N,j}\left(\frac{j\pi}{N}\right) = 1$ it follows that $\bigvee_{k=0}^{N} s_{N,k}\left(\frac{j\pi}{N}\right) \geq 1 > 0$, for all $j \in \{0, 1, ..., N\}$. Hence $K_N^{(M)}(f)$ is well-defined function for all $x \in [0, \pi]$, and it is continuous on $[0, \pi]$, also $K_N^{(M)}(1) = 1$. By [6], p. 350, $K_N^{(M)}$ are positive sublinear operators.

Denote $x_{N,k} := \frac{k\pi}{N}, k \in \{0, 1, ..., N\}$.

By [6], p. 352, we have

$$K_N^{(M)}(|\cdot - x|)(x) \leq \frac{\pi}{2N}, \ \forall N \in \mathbb{N}, \ \forall x \in [0, \pi]. \tag{4.70}$$

Notice also $|x_{N,k} - x| \leq \pi, \forall x \in [0, \pi]$.

Therefore $(\lambda \geq 1)$ it holds

$$K_N^{(M)}(|\cdot - x|^\lambda)(x) \leq \frac{\pi^{\lambda-1}\pi}{2N} = \frac{\pi^\lambda}{2N}, \ \forall x \in [0, \pi], \forall N \in \mathbb{N}. \tag{4.71}$$

If $x \in \left(\frac{j\pi}{N}, \frac{(j+1)\pi}{N}\right)$, with $j \in \{0, 1, ..., N\}$, we obtain $nx - j\pi \in (0, \pi)$ and thus $s_{N,j}(x) = \frac{\sin^2(Nx-j\pi)}{(Nx-j\pi)^2} > 0$, see [6], p. 350.

Consequently it holds $(\lambda \geq 1)$

$$K_N^{(M)}(|\cdot - x|^\lambda)(x) = \frac{\bigvee_{k=0}^{N} s_{N,k}(x)|x_{N,k} - x|^\lambda}{\bigvee_{k=0}^{N} s_{N,k}(x)} > 0, \ \forall x \in [0, \pi], \tag{4.72}$$

such that $x \neq x_{N,k}$, for any $k \in \{0, 1, ..., N\}$.

We give

Theorem 4.28 *Let $0 < \alpha < 1$, $x \in [0, \pi]$ be such that $x \neq \frac{k\pi}{N}$, $k \in \{0, 1, ..., N\}$, $\forall N \in \mathbb{N}$; $f \in AC([0, \pi], \mathbb{R}_+)$, and $f' \in L_\infty([0, \pi])$. Then*

$$\left|K_N^{(M)}(f)(x) - f(x)\right| \leq \frac{(\alpha + 2)\omega_1\left(D_x^\alpha f, \left(\frac{\pi^{\alpha+1}}{2N}\right)^{\frac{1}{\alpha+1}}\right)}{\Gamma(\alpha + 2)}\left(\frac{\pi^{\alpha+1}}{2N}\right)^{\frac{\alpha}{\alpha+1}}, \ \forall N \in \mathbb{N}. \tag{4.73}$$

As $N \to +\infty$, we get $K_N^{(M)}(f)(x) \to f(x)$.

Proof By Theorem 4.16. ∎

When $\alpha = \frac{1}{2}$ we get:

Corollary 4.29 *Let $f \in AC([0, 1], \mathbb{R}_+)$, $f' \in L_\infty([0, 1])$. Then*

$$\left|B_N^{(M)}(f)(x) - f(x)\right| \leq \frac{10\sqrt[3]{6}\omega_1\left(D_x^{\frac{1}{2}}f, \frac{\sqrt[3]{36}}{\sqrt[3]{N+1}}\right)}{3\sqrt{\pi}\sqrt[6]{N+1}}, \ \forall x \in (0, 1), \forall N \in \mathbb{N}. \tag{4.74}$$

Proof By Theorem 4.18. ∎

Due to lack of space we avoid to give other applications when $\alpha = \frac{1}{2}$ from the other Max-product operators.

(II) Case $\alpha > 1$, $\alpha \notin \mathbb{N}$.

Here we apply Theorem 4.14 to well known Max-product operators.

We present

Theorem 4.30 *Let $\alpha > 1$, $\alpha \notin \mathbb{N}$, $m = \lceil \alpha \rceil$, $x \in [0, 1]$, $f \in AC^m([0, 1], \mathbb{R}_+)$, and $f^{(m)} \in L_\infty([0, 1])$. Furthermore we assume that $f^{(k)}(x) = 0$, $k = 1, ..., m - 1$. Then*

$$\left| B_N^{(M)}(f)(x) - f(x) \right| \leq \frac{\omega_1\left(D_x^\alpha f, \left(\frac{6}{\sqrt{N+1}} \right)^{\frac{1}{\alpha+1}} \right)}{\Gamma(\alpha+1)} \cdot \tag{4.75}$$

$$\left[\frac{6}{\sqrt{N+1}} + \frac{1}{(\alpha+1)} \left(\frac{6}{\sqrt{N+1}} \right)^{\frac{\alpha}{\alpha+1}} \right], \forall N \in \mathbb{N}.$$

We get $\lim\limits_{N \to +\infty} B_N^{(M)}(f)(x) = f(x)$.

Proof Applying (4.37) for $B_N^{(M)}$ and using (4.43), we get

$$\left| B_N^{(M)}(f)(x) - f(x) \right| \leq \frac{\omega_1\left(D_x^\alpha f, \delta \right)}{\Gamma(\alpha+1)} \left[\frac{6}{\sqrt{N+1}} + \frac{\frac{6}{\sqrt{N+1}}}{(\alpha+1)\delta} \right]. \tag{4.76}$$

Choose $\delta = \left(\frac{6}{\sqrt{N+1}} \right)^{\frac{1}{\alpha+1}}$, then $\delta^{\alpha+1} = \frac{6}{\sqrt{N+1}}$, and apply it to (4.76). Clearly we derive (4.75). ∎

We continue with

Theorem 4.31 *Same assumptions as in Theorem 4.30. Then*

$$\left| T_N^{(M)}(f)(x) - f(x) \right| \leq \frac{\omega_1\left(D_x^\alpha f, \left(\frac{3}{\sqrt{N}} \right)^{\frac{1}{\alpha+1}} \right)}{\Gamma(\alpha+1)} \cdot \tag{4.77}$$

$$\left[\frac{3}{\sqrt{N}} + \frac{1}{(\alpha+1)} \left(\frac{3}{\sqrt{N}} \right)^{\frac{\alpha}{\alpha+1}} \right], \forall N \in \mathbb{N}.$$

We get $\lim\limits_{N \to +\infty} T_N^{(M)}(f)(x) = f(x)$.

Proof Use of Theorem 4.14, similar to the proof of Theorem 4.30. ∎

We give

Theorem 4.32 *Same assumptions as in Theorem 4.30. Then*

$$\left| U_N^{(M)}(f)(x) - f(x) \right| \leq \frac{\omega_1\left(D_x^\alpha f, \left(\frac{2\sqrt{3}(\sqrt{2}+2)}{\sqrt{N+1}} \right)^{\frac{1}{\alpha+1}} \right)}{\Gamma(\alpha+1)}. \tag{4.78}$$

$$\left[\frac{2\sqrt{3}\left(\sqrt{2}+2\right)}{\sqrt{N+1}} + \frac{1}{(\alpha+1)} \left(\frac{2\sqrt{3}\left(\sqrt{2}+2\right)}{\sqrt{N+1}} \right)^{\frac{\alpha}{\alpha+1}} \right], \ \forall N \in \mathbb{N}, \ N \geq 2.$$

We get $\lim_{N \to +\infty} U_N^{(M)}(f)(x) = f(x).$

Proof Use of Theorem 4.14, similar to the proof of Theorem 4.30. ∎

We give

Theorem 4.33 *Same assumptions as in Theorem 4.30. Then*

$$\left| Z_N^{(M)}(f)(x) - f(x) \right| \leq \frac{\omega_1\left(D_x^\alpha f, (\rho(x))^{\frac{1}{\alpha+1}} \right)}{\Gamma(\alpha+1)}. \tag{4.79}$$

$$\left[\rho(x) + \frac{1}{(\alpha+1)} (\rho(x))^{\frac{\alpha}{\alpha+1}} \right], \ \forall N \in \mathbb{N}, \ N \geq 4.$$

We get $\lim_{N \to +\infty} Z_N^{(M)}(f)(x) = f(x),$ *where* $\rho(x)$ *is as in (4.59).*

Proof Use of Theorem 4.14, similar to the proof of Theorem 4.30. ∎

We continue with

Theorem 4.34 *Let* $\alpha > 1$, $\alpha \notin \mathbb{N}$, $m = \lceil \alpha \rceil$, $x \in [0, \pi] \subset \mathbb{R}$, $f \in AC^m([0, \pi],$ $\mathbb{R}_+)$, *and* $f^{(m)} \in L_\infty([0, \pi])$. *Furthermore we assume that* $f^{(k)}(x) = 0$, $k = 1, ...,$ $m - 1$. *Then*

$$\left| W_N^{(M)}(f)(x) - f(x) \right| \leq \frac{\omega_1\left(D_x^\alpha f, \left(\frac{\pi^{\alpha+1}}{2N} \right)^{\frac{1}{\alpha+1}} \right)}{\Gamma(\alpha+1)}. \tag{4.80}$$

$$\left[\frac{\pi^\alpha}{2N} + \frac{1}{(\alpha+1)} \left(\frac{\pi^{\alpha+1}}{2N} \right)^{\frac{\alpha}{\alpha+1}} \right], \ \forall N \in \mathbb{N}.$$

We have that $\lim_{N \to +\infty} W_N^{(M)}(f)(x) = f(x).$

Proof Applying (4.37) for $W_N^{(M)}$ and using (4.66), we get

$$\left| W_N^{(M)}(f)(x) - f(x) \right| \le \frac{\omega_1\left(D_x^\alpha f, \delta\right)}{\Gamma(\alpha+1)} \left[\frac{\pi^\alpha}{2N} + \frac{\frac{\pi^{\alpha+1}}{2N}}{(\alpha+1)\delta} \right]. \tag{4.81}$$

Choose $\delta = \left(\frac{\pi^{\alpha+1}}{2N}\right)^{\frac{1}{\alpha+1}}$, i.e. $\delta^{\alpha+1} = \frac{\pi^{\alpha+1}}{2N}$, and $\delta^\alpha = \left(\frac{\pi^{\alpha+1}}{2N}\right)^{\frac{\alpha}{\alpha+1}}$. We use the last into (4.81) and we obtain (4.80). ∎

We also have

Theorem 4.35 *Let* $\alpha > 1$, $\alpha \notin \mathbb{N}$, $m = \lceil\alpha\rceil$, $x \in [0,\pi] \subset \mathbb{R}$, $f \in AC^m([0,\pi]$, $\mathbb{R}_+)$, *and* $f^{(m)} \in L_\infty([0,\pi])$. *Furthermore we assume that* $f^{(k)}(x) = 0$, $k = 1, ...,$ $m-1$. *Then*

$$\left| K_N^{(M)}(f)(x) - f(x) \right| \le \frac{\omega_1\left(D_x^\alpha f, \left(\frac{\pi^{\alpha+1}}{2N}\right)^{\frac{1}{\alpha+1}}\right)}{\Gamma(\alpha+1)}. \tag{4.82}$$

$$\left[\frac{\pi^\alpha}{2N} + \frac{1}{(\alpha+1)} \left(\frac{\pi^{\alpha+1}}{2N}\right)^{\frac{\alpha}{\alpha+1}} \right], \ \forall N \in \mathbb{N}.$$

We have that $\lim_{N \to +\infty} K_N^{(M)}(f)(x) = f(x)$.

Proof As in Theorem 4.34. ∎

We make

Remark 4.36 We mention the interpolation Hermite–Fejér polynomials on Chebyshev knots of the first kind (see [6], p. 4): Let $f : [-1,1] \to \mathbb{R}$ and based on the knots $x_{N,k} = \cos\left(\frac{(2(N-k)+1)}{2(N+1)}\pi\right) \in (-1,1)$, $k \in \{0, ..., N\}$, $-1 < x_{N,0} < x_{N,1} < ... < x_{N,N} < 1$, which are the roots of the first kind Chebyshev polynomial $T_{N+1}(x) = \cos((N+1)\arccos x)$, we define (see Fejér [10])

$$H_{2N+1}(f)(x) = \sum_{k=0}^{N} h_{N,k}(x) f\left(x_{N,k}\right), \tag{4.83}$$

where

$$h_{N,k}(x) = \left(1 - x \cdot x_{N,k}\right) \left(\frac{T_{N+1}(x)}{(N+1)\left(x - x_{N,k}\right)}\right)^2, \tag{4.84}$$

the fundamental interpolation polynomials.

The Max-product interpolation Hermite–Fejér operators on Chebyshev knots of the first kind (see p. 12 of [6]) are defined by

$$H_{2N+1}^{(M)}(f)(x) = \frac{\bigvee_{k=0}^{N} h_{N,k}(x) f(x_{N,k})}{\bigvee_{k=0}^{N} h_{N,k}(x)}, \ \forall \ N \in \mathbb{N}, \tag{4.85}$$

where $f : [-1, 1] \to \mathbb{R}_+$ is continuous.

Call

$$E_N(x) := H_{2N+1}^{(M)}(|\cdot - x|)(x) = \frac{\bigvee_{k=0}^{N} h_{N,k}(x) |x_{N,k} - x|}{\bigvee_{k=0}^{N} h_{N,k}(x)}, \ x \in [-1, 1]. \tag{4.86}$$

Then by [6], p. 287 we obtain that

$$E_N(x) \le \frac{2\pi}{N+1}, \ \forall \ x \in [-1, 1], \ N \in \mathbb{N}. \tag{4.87}$$

For $m > 1$, we get

$$H_{2N+1}^{(M)}(|\cdot - x|^m)(x) = \frac{\bigvee_{k=0}^{N} h_{N,k}(x) |x_{N,k} - x|^m}{\bigvee_{k=0}^{N} h_{N,k}(x)} =$$

$$\frac{\bigvee_{k=0}^{N} h_{N,k}(x) |x_{N,k} - x| |x_{N,k} - x|^{m-1}}{\bigvee_{k=0}^{N} h_{N,k}(x)} \le 2^{m-1} \frac{\bigvee_{k=0}^{N} h_{N,k}(x) |x_{N,k} - x|}{\bigvee_{k=0}^{N} h_{N,k}(x)} \tag{4.88}$$

$$\le \frac{2^m \pi}{N+1}, \ \forall \ x \in [-1, 1], \ N \in \mathbb{N}.$$

Hence it holds

$$H_{2N+1}^{(M)}(|\cdot - x|^m)(x) \le \frac{2^m \pi}{N+1}, \ \forall \ x \in [-1, 1], \ m > 1, \forall \ N \in \mathbb{N}. \tag{4.89}$$

Furthermore we have

$$H_{2N+1}^{(M)}(1)(x) = 1, \ \forall \ x \in [-1, 1], \tag{4.90}$$

and $H_{2N+1}^{(M)}$ maps continuous functions to continuous functions over $[-1, 1]$ and for any $x \in \mathbb{R}$ we have $\bigvee_{k=0}^{N} h_{N,k}(x) > 0$.

We also have $h_{N,k}(x_{N,k}) = 1$, and $h_{N,k}(x_{N,j}) = 0$, if $k \ne j$, furthermore it holds $H_{2N+1}^{(M)}(f)(x_{N,j}) = f(x_{N,j})$, for all $j \in \{0, ..., N\}$, see [6], p. 282.

$H_{2N+1}^{(M)}$ are positive sublinear operators, [6], p. 282.

We give

Theorem 4.37 *Let $\alpha > 1$, $\alpha \notin \mathbb{N}$, $m = \lceil \alpha \rceil$, $x \in [-1, 1]$, $f \in AC^m([-1, 1], \mathbb{R}_+)$, and $f^{(m)} \in L_\infty([-1, 1])$. Furthermore we assume that $f^{(k)}(x) = 0$, $k = 1, ..., m - 1$. Then*

$$\left| H_{2N+1}^{(M)} (f) (x) - f (x) \right| \leq \frac{\omega_1 \left(D_x^\alpha f, \left(\frac{2^{\alpha+1}\pi}{N+1} \right)^{\frac{1}{\alpha+1}} \right)}{\Gamma (\alpha + 1)}. \tag{4.91}$$

$$\left[\frac{2^\alpha \pi}{N + 1} + \frac{1}{(\alpha + 1)} \left(\frac{2^{\alpha+1}\pi}{N + 1} \right)^{\frac{\alpha}{\alpha+1}} \right], \; \forall \, N \in \mathbb{N}.$$

Furthermore it holds $\lim_{N \to +\infty} H_{2N+1}^{(M)} (f) (x) = f (x)$.

Proof By Theorem 4.14, choose $\delta := \left(\frac{2^{\alpha+1}\pi}{N+1} \right)^{\frac{1}{\alpha+1}}$, use (4.37), (4.89). ∎

We continue with

Remark 4.38 Here we deal with Lagrange interpolation polynomials on Chebyshev knots of second kind plus the endpoints ± 1 (see [6], p. 5). These polynomials are linear operators attached to $f : [-1, 1] \to \mathbb{R}$ and to the knots $x_{N,k} = \cos \left(\left(\frac{N-k}{N-1} \right) \pi \right) \in [-1, 1]$, $k = 1, ..., N$, $N \in \mathbb{N}$, which are the roots of $\omega_N (x) = \sin (N - 1) t \sin t$, $x = \cos t$. Notice that $x_{N,1} = -1$ and $x_{N,N} = 1$. Their formula is given by ([6], p. 377)

$$L_N (f) (x) = \sum_{k=1}^{N} l_{N,k} (x) f \left(x_{N,k} \right), \tag{4.92}$$

where

$$l_{N,k} (x) = \frac{(-1)^{k-1} \omega_N (x)}{(1 + \delta_{k,1} + \delta_{k,N}) (N - 1) \left(x - x_{N,k} \right)}, \tag{4.93}$$

$N \geq 2$, $k = 1, ..., N$, and $\omega_N (x) = \prod_{k=1}^{N} \left(x - x_{N,k} \right)$ and $\delta_{i,j}$ denotes the Kronecher's symbol, that is $\delta_{i,j} = 1$, if $i = j$, and $\delta_{i,j} = 0$, if $i \neq j$.

The Max-product Lagrange interpolation operators on Chebyshev knots of second kind, plus the endpoints ± 1, are defined by ([6], p. 12)

$$L_N^{(M)} (f) (x) = \frac{\bigvee_{k=1}^{N} l_{N,k} (x) f \left(x_{N,k} \right)}{\bigvee_{k=1}^{N} l_{N,k} (x)}, \; x \in [-1, 1], \tag{4.94}$$

where $f : [-1, 1] \to \mathbb{R}_+$ continuous.

First we see that $L_N^{(M)} (f) (x)$ is well defined and continuous for any $x \in [-1, 1]$. Following [6], p. 289, because $\sum_{k=1}^{N} l_{N,k} (x) = 1$, $\forall \, x \in \mathbb{R}$, for any x there exists $k \in \{1, ..., N\} : l_{N,k} (x) > 0$, hence $\bigvee_{k=1}^{N} l_{N,k} (x) > 0$. We have that $l_{N,k} \left(x_{N,k} \right) = 1$, and $l_{N,k} \left(x_{N,j} \right) = 0$, if $k \neq j$. Furthermore it holds $L_N^{(M)} (f) \left(x_{N,j} \right) = f \left(x_{N,j} \right)$, all $j \in \{1, ..., N\}$, and $L_N^{(M)} (1) = 1$.

Call $I_N^+ (x) = \left\{ k \in \{1, ..., N\} ; l_{N,k} (x) > 0 \right\}$, then $I_N^+ (x) \neq \emptyset$.

So for $f \in C_+ ([-1, 1])$ we get

$$L_N^{(M)}(f)(x) = \frac{\bigvee_{k \in I_N^+(x)} l_{N,k}(x) f(x_{N,k})}{\bigvee_{k \in I_N^+(x)} l_{N,k}(x)} \geq 0. \tag{4.95}$$

Notice here that $|x_{N,k} - x| \leq 2, \forall x \in [-1, 1]$.

By [6], p. 297, we get that

$$L_N^{(M)}(|\cdot - x|)(x) = \frac{\bigvee_{k=1}^N l_{N,k}(x)|x_{N,k} - x|}{\bigvee_{k=1}^N l_{N,k}(x)} =$$

$$\frac{\bigvee_{k \in I_N^+(x)} l_{N,k}(x)|x_{N,k} - x|}{\bigvee_{k \in I_N^+(x)} l_{N,k}(x)} \leq \frac{\pi^2}{6(N-1)}, \tag{4.96}$$

$N \geq 3, \forall x \in (-1, 1)$, N is odd.

We get that $(m > 1)$

$$L_N^{(M)}(|\cdot - x|^m)(x) = \frac{\bigvee_{k \in I_N^+(x)} l_{N,k}(x)|x_{N,k} - x|^m}{\bigvee_{k \in I_N^+(x)} l_{N,k}(x)} \leq \frac{2^{m-1}\pi^2}{6(N-1)}, \tag{4.97}$$

$N \geq 3$ odd, $\forall x \in (-1, 1)$.

$L_N^{(M)}$ are positive sublinear operators, [6], p. 290.

We give

Theorem 4.39 *Same assumptions as in Theorem 4.37. Then*

$$\left| L_N^{(M)}(f)(x) - f(x) \right| \leq \frac{\omega_1\left(D_x^\alpha f, \left(\frac{2^\alpha \pi^2}{6(N-1)}\right)^{\frac{1}{\alpha+1}}\right)}{\Gamma(\alpha+1)} \cdot \tag{4.98}$$

$$\left[\frac{2^{\alpha-1}\pi^2}{6(N-1)} + \frac{1}{(\alpha+1)}\left(\frac{2^\alpha \pi^2}{6(N-1)}\right)^{\frac{\alpha}{\alpha+1}} \right], \forall N \in \mathbb{N} : N \geq 3, odd.$$

It holds $\lim_{N \to +\infty} L_N^{(M)}(f)(x) = f(x)$.

Proof By Theorem 4.14, choose $\delta := \left(\frac{2^\alpha \pi^2}{6(N-1)}\right)^{\frac{1}{\alpha+1}}$, use of (4.37) and (4.97). At ± 1 the left hand side of (4.98) is zero, thus (4.98) is trivially true. ∎

We make

Remark 4.40 Let $f \in C_+([-1, 1])$, $N \geq 4$, $N \in \mathbb{N}$, N even.

By [6], p. 298, we get

$$L_N^{(M)}(|\cdot - x|)(x) \leq \frac{4\pi^2}{3(N-1)} = \frac{2^2\pi^2}{3(N-1)}, \forall x \in (-1, 1). \tag{4.99}$$

Hence $(m > 1)$

$$L_N^{(M)} \left(|\cdot - x|^m \right) (x) \le \frac{2^{m+1} \pi^2}{3 (N - 1)}, \; \forall \, x \in (-1, 1). \tag{4.100}$$

We present

Theorem 4.41 *Same assumptions as in Theorem 4.37. Then*

$$\left| L_N^{(M)} (f) (x) - f (x) \right| \le \frac{\omega_1 \left(D_x^\alpha f, \left(\frac{2^{\alpha+2} \pi^2}{3(N-1)} \right)^{\frac{1}{\alpha+1}} \right)}{\Gamma (\alpha + 1)} \cdot \tag{4.101}$$

$$\left[\frac{2^{\alpha+1} \pi^2}{3 (N - 1)} + \frac{1}{(\alpha + 1)} \left(\frac{2^{\alpha+2} \pi^2}{3 (N - 1)} \right)^{\frac{\alpha}{\alpha+1}} \right], \; \forall \, N \in \mathbb{N}, \; N \ge 4, \; N \text{ is even.}$$

It holds $\lim_{N \to +\infty} L_N^{(M)} (f) (x) = f (x).$

Proof By Theorem 4.14, choose $\delta := \left(\frac{2^{\alpha+2} \pi^2}{3(N-1)} \right)^{\frac{1}{\alpha+1}}$, use of (4.37) and (4.100). At ± 1, (4.101) is trivially true. ∎

We need

Definition 4.42 ([7], p. 41) Let $I \subset \mathbb{R}$ be an interval of finite or infinite length, and $f : I \to \mathbb{R}$ a bounded or uniformly continuous function. We define the first modulus of continuity

$$\omega_1 (f, \delta)_I = \sup_{\substack{x, y \in I \\ |x-y| \le \delta}} |f (x) - f (y)|, \; \delta > 0. \tag{4.102}$$

Clearly, it holds $\omega_1 (f, \delta)_I < +\infty$.
We also have

$$\omega_1 (f, r\delta)_I \le (r + 1) \omega_1 (f, \delta)_I, \; \text{any } r \ge 0. \tag{4.103}$$

Convention 4.43 *Let a real number $m > 1$, from now on we assume that $D_{x_0-}^m f$ is either bounded or uniformly continuous function on $(-\infty, x_0]$, similarly from now on we assume that $D_{*x_0}^m f$ is either bounded or uniformly continuous function on $[x_0, +\infty)$.*

We need

Definition 4.44 Let $D_{x_0}^m f$ (real number $m > 1$) denote any of $D_{x_0-}^m f$, $D_{*x_0}^m f$ and $\delta > 0$. We set

$$\omega_1 \left(D_{x_0}^m f, \delta \right)_{\mathbb{R}} := \max \left\{ \omega_1 \left(D_{x_0-}^m f, \delta \right)_{(-\infty, x_0]}, \omega_1 \left(D_{*x_0}^m f, \delta \right)_{[x_0, +\infty)} \right\}, \tag{4.104}$$

where $x_0 \in \mathbb{R}$. Notice that $\omega_1 \left(D_{x_0}^m f, \delta \right)_{\mathbb{R}} < +\infty$.

We will use

Theorem 4.45 *Let the real number $m > 0$, $m \notin \mathbb{N}$, $\lambda = \lceil m \rceil$, $x_0 \in \mathbb{R}$, $f \in AC^{\lambda}$ $([a, b], \mathbb{R}_+)$ (i.e. $f^{(\lambda-1)} \in AC[a, b]$, absolutely continuous functions on $[a, b]$), $\forall [a, b] \subset \mathbb{R}$, and $f^{(\lambda)} \in L_{\infty}(\mathbb{R})$. Furthermore we assume that $f^{(k)}(x_0) = 0$, $k = 1, ..., \lambda - 1$. The Convention 4.43 is imposed here. Then*

$$|f(x) - f(x_0)| \le \frac{\omega_1 \left(D_{x_0}^m f, \delta \right)_{\mathbb{R}}}{\Gamma(m+1)} \left[|x - x_0|^m + \frac{|x - x_0|^{m+1}}{(m+1)\delta} \right], \quad \delta > 0, \quad (4.105)$$

for all $x \in \mathbb{R}$.

If $0 < m < 1$, then we do not need initial conditions.

Proof Similar to Theorem 4.10. ∎

We continue with

Remark 4.46 Let $b : \mathbb{R} \to \mathbb{R}_+$ be a centered (it takes a global maximum at 0) bell-shaped function, with compact support $[-T, T]$, $T > 0$ (that is $b(x) > 0$ for all $x \in (-T, T)$) and $I = \int_{-T}^{T} b(x) \, dx > 0$.

The Cardaliaguet–Euvrard neural network operators are defined by (see [5])

$$C_{N,\alpha}(f)(x) = \sum_{k=-N^2}^{N^2} \frac{f\left(\frac{k}{n}\right)}{I N^{1-\alpha}} b \left(N^{1-\alpha} \left(x - \frac{k}{N} \right) \right), \quad (4.106)$$

$0 < \alpha < 1$, $N \in \mathbb{N}$ and typically here $f : \mathbb{R} \to \mathbb{R}$ is continuous and bounded or uniformly continuous on $\mathbb{R}$.

$CB(\mathbb{R})$ denotes the continuous and bounded function on $\mathbb{R}$, and

$$CB_+(\mathbb{R}) = \{f : \mathbb{R} \to [0, \infty); \ f \in CB(\mathbb{R})\}.$$

The corresponding max-product Cardaliaguet–Euvrard neural network operators will be given by

$$C_{N,\alpha}^{(M)}(f)(x) = \frac{\bigvee_{k=-N^2}^{N^2} b \left(N^{1-\alpha} \left(x - \frac{k}{N} \right) \right) f \left(\frac{k}{N} \right)}{\bigvee_{k=-N^2}^{N^2} b \left(N^{1-\alpha} \left(x - \frac{k}{N} \right) \right)}, \quad (4.107)$$

$x \in \mathbb{R}$, typically here $f \in CB_+(\mathbb{R})$, see also [5].

Next we follow [5].

For any $x \in \mathbb{R}$, denoting

$$J_{T,N}(x) = \left\{ k \in \mathbb{Z}; \ -N^2 \le k \le N^2, N^{1-\alpha} \left(x - \frac{k}{N} \right) \in (-T, T) \right\},$$

we can write

$$C_{N,\alpha}^{(M)}(f)(x) = \frac{\bigvee_{k \in J_{T,N}(x)} b\left(N^{1-\alpha}\left(x - \frac{k}{N}\right)\right) f\left(\frac{k}{N}\right)}{\bigvee_{k \in J_{T,N}(x)} b\left(N^{1-\alpha}\left(x - \frac{k}{N}\right)\right)}, \tag{4.108}$$

$x \in \mathbb{R}$, $N > \max\left\{T + |x|, T^{-\frac{1}{\alpha}}\right\}$, where $J_{T,N}(x) \neq \emptyset$. Indeed, we have $\bigvee_{k \in J_{T,N}(x)} b\left(N^{1-\alpha}\left(x - \frac{k}{N}\right)\right) > 0, \forall x \in \mathbb{R}$ and $N > \max\left\{T + |x|, T^{-\frac{1}{\alpha}}\right\}$.

We have that $C_{N,\alpha}^{(M)}(1)(x) = 1, \forall x \in \mathbb{R}$ and $N > \max\left\{T + |x|, T^{-\frac{1}{\alpha}}\right\}$.

See in [5] there: Lemma 2.1, Corollary 2.2 and Remarks.
We need

Theorem 4.47 ([5]) *Let $b(x)$ be a centered bell-shaped function, continuous and with compact support $[-T, T]$, $T > 0$, $0 < \alpha < 1$ and $C_{N,\alpha}^{(M)}$ be defined as in (4.107).*

(i) If $|f(x)| \leq c$ for all $x \in \mathbb{R}$ then $\left|C_{N,\alpha}^{(M)}(f)(x)\right| \leq c$, for all $x \in \mathbb{R}$ and $N > \max\left\{T + |x|, T^{-\frac{1}{\alpha}}\right\}$ and $C_{N,\alpha}^{(M)}(f)(x)$ is continuous at any point $x \in \mathbb{R}$, for all $N > \max\left\{T + |x|, T^{-\frac{1}{\alpha}}\right\}$;

(ii) If $f, g \in CB_{+}(\mathbb{R})$ satisfy $f(x) \leq g(x)$ for all $x \in \mathbb{R}$, then $C_{N,\alpha}^{(M)}(f)(x) \leq C_{N,\alpha}^{(M)}(g)(x)$ for all $x \in \mathbb{R}$ and $N > \max\left\{T + |x|, T^{-\frac{1}{\alpha}}\right\}$;

(iii) $C_{N,\alpha}^{(M)}(f + g)(x) \leq C_{N,\alpha}^{(M)}(f)(x) + C_{N,\alpha}^{(M)}(g)(x)$ for all $f, g \in CB_{+}(\mathbb{R})$, $x \in \mathbb{R}$ and $N > \max\left\{T + |x|, T^{-\frac{1}{\alpha}}\right\}$;

(iv) For all $f, g \in CB_{+}(\mathbb{R})$, $x \in \mathbb{R}$ and $N > \max\left\{T + |x|, T^{-\frac{1}{\alpha}}\right\}$, we have

$$\left|C_{N,\alpha}^{(M)}(f)(x) - C_{N,\alpha}^{(M)}(g)(x)\right| \leq C_{N,\alpha}^{(M)}(|f - g|)(x);$$

(v) $C_{N,\alpha}^{(M)}$ is positive homogeneous, that is $C_{N,\alpha}^{(M)}(\lambda f)(x) = \lambda C_{N,\alpha}^{(M)}(f)(x)$ for all $\lambda \geq 0$, $x \in \mathbb{R}$, $N > \max\left\{T + |x|, T^{-\frac{1}{\alpha}}\right\}$ and $f \in CB_{+}(\mathbb{R})$.

We make

Remark 4.48 We have that

$$E_{N,\alpha}(x) := C_{N,\alpha}^{(M)}(|\cdot - x|)(x) = \frac{\bigvee_{k \in J_{T,N}(x)} b\left(N^{1-\alpha}\left(x - \frac{k}{N}\right)\right)\left|x - \frac{k}{N}\right|}{\bigvee_{k \in J_{T,N}(x)} b\left(N^{1-\alpha}\left(x - \frac{k}{N}\right)\right)}, \tag{4.109}$$

$\forall x \in \mathbb{R}$, and $N > \max\left\{T + |x|, T^{-\frac{1}{\alpha}}\right\}$.

We mention from [5] the following:

Theorem 4.49 ([5]) *Let $b(x)$ be a centered bell-shaped function, continuous and with compact support $[-T, T]$, $T > 0$ and $0 < \alpha < 1$. In addition, suppose that the following requirements are fulfilled:*

(i) There exist $0 < m_1 \le M_1 < \infty$ such that $m_1(T - x) \le b(x) \le M_1(T - x)$, $\forall x \in [0, T]$;

(ii) There exist $0 < m_2 \le M_2 < \infty$ such that $m_2(x + T) \le b(x) \le M_2(x + T)$, $\forall x \in [-T, 0]$.

Then for all $f \in CB_+(\mathbb{R})$, $x \in \mathbb{R}$ and for all $N \in \mathbb{N}$ satisfying $N > \max\left\{T + |x|, \left(\frac{2}{T}\right)^{\frac{1}{\alpha}}\right\}$, we have the estimate

$$\left| C_{N,\alpha}^{(M)}(f)(x) - f(x) \right| \le c\omega_1\left(f, N^{\alpha-1}\right)_{\mathbb{R}}, \tag{4.110}$$

where

$$c := 2\left(\max\left\{\frac{TM_2}{2m_2}, \frac{TM_1}{2m_1}\right\} + 1\right),$$

and

$$\omega_1(f, \delta)_{\mathbb{R}} := \sup_{\substack{x, y \in \mathbb{R}: \\ |x-y| \le \delta}} |f(x) - f(y)|.$$

We make

Remark 4.50 In [5], was proved that

$$E_{N,\alpha}(x) \le \max\left\{\frac{TM_2}{2m_2}, \frac{TM_1}{2m_1}\right\} N^{\alpha-1}, \ \forall N > \max\left\{T + |x|, \left(\frac{2}{T}\right)^{\frac{1}{\alpha}}\right\}. \tag{4.111}$$

That is

$$C_{N,\alpha}^{(M)}(|\cdot - x|)(x) \le \max\left\{\frac{TM_2}{2m_2}, \frac{TM_1}{2m_1}\right\} N^{\alpha-1}, \ \forall N > \max\left\{T + |x|, \left(\frac{2}{T}\right)^{\frac{1}{\alpha}}\right\}. \tag{4.112}$$

From (4.109) we have that $\left|x - \frac{k}{N}\right| \le \frac{T}{N^{1-\alpha}}$.

Hence $(m > 1)$ $(\forall x \in \mathbb{R}$ and $N > \max\left\{T + |x|, \left(\frac{2}{T}\right)^{\frac{1}{\alpha}}\right\})$

$$C_{N,\alpha}^{(M)}(|\cdot - x|^m)(x) = \frac{\bigvee_{k \in J_{T,N}(x)} b\left(N^{1-\alpha}\left(x - \frac{k}{N}\right)\right) \left|x - \frac{k}{N}\right|^m}{\bigvee_{k \in J_{T,N}(x)} b\left(N^{1-\alpha}\left(x - \frac{k}{N}\right)\right)} \le \tag{4.113}$$

$$\left(\frac{T}{N^{1-\alpha}}\right)^{m-1} \max\left\{\frac{TM_2}{2m_2}, \frac{TM_1}{2m_1}\right\} N^{\alpha-1}, \ \forall N > \max\left\{T + |x|, \left(\frac{2}{T}\right)^{\frac{1}{\alpha}}\right\}.$$

Then $(m > 1)$ it holds

$$C_{N,\alpha}^{(M)} \left(|\cdot - x|^m \right)(x) \le$$

$$T^{m-1} \max \left\{ \frac{TM_2}{2m_2}, \frac{TM_1}{2m_1} \right\} \frac{1}{N^{m(1-\alpha)}}, \ \forall \ N > \max \left\{ T + |x|, \left(\frac{2}{T} \right)^{\frac{1}{\alpha}} \right\}. \quad (4.114)$$

Call

$$\theta := \max \left\{ \frac{TM_2}{2m_2}, \frac{TM_1}{2m_1} \right\} > 0. \quad (4.115)$$

Consequently ($m > 1$) we derive

$$C_{N,\alpha}^{(M)} \left(|\cdot - x|^m \right)(x) \le \frac{\theta T^{m-1}}{N^{m(1-\alpha)}}, \ \forall \ N > \max \left\{ T + |x|, \left(\frac{2}{T} \right)^{\frac{1}{\alpha}} \right\}. \quad (4.116)$$

We need

Theorem 4.51 *All here as in Theorem 4.45, where $x = x_0 \in \mathbb{R}$ is fixed. Let b be a centered bell-shaped function, continuous and with compact support $[-T, T]$, $T > 0, 0 < \alpha < 1$ and $C_{N,\alpha}^{(M)}$ be defined as in (4.107). Then*

$$\left| C_{N,\alpha}^{(M)}(f)(x) - f(x) \right| \le$$

$$\frac{\omega_1 \left(D_x^m f, \delta \right)_{\mathbb{R}}}{\Gamma(m+1)} \left[C_{N,\alpha}^{(M)} \left(|\cdot - x|^m \right)(x) + \frac{C_{N,\alpha}^{(M)} \left(|\cdot - x|^{m+1} \right)(x)}{(m+1)\delta} \right], \quad (4.117)$$

$$\forall \ N \in \mathbb{N} : N > \max \left\{ T + |x|, T^{-\frac{1}{\alpha}} \right\}.$$

Proof By Theorem 4.45 and (4.105) we get

$$|f(\cdot) - f(x)| \le \frac{\omega_1 \left(D_x^m f, \delta \right)_{\mathbb{R}}}{\Gamma(m+1)} \left[|\cdot - x|^m + \frac{|\cdot - x|^{m+1}}{(m+1)\delta} \right], \ \delta > 0, \quad (4.118)$$

true over $\mathbb{R}$.

As in Theorem 4.47 and using similar reasoning and $C_{N,\alpha}^{(M)}(1) = 1$, we get

$$\left| C_{N,\alpha}^{(M)}(f)(x) - f(x) \right| \le C_{N,\alpha}^{(M)} \left(|f(\cdot) - f(x)| \right)(x) \overset{(4.118)}{\le}$$

$$\frac{\omega_1 \left(D_x^m f, \delta \right)_{\mathbb{R}}}{\Gamma(m+1)} \left[C_{N,\alpha}^{(M)} \left(|\cdot - x|^m \right)(x) + \frac{C_{N,\alpha}^{(M)} \left(|\cdot - x|^{m+1} \right)(x)}{(m+1)\delta} \right], \quad (4.119)$$

$$\forall \ N \in \mathbb{N} : N > \max \left\{ T + |x|, T^{-\frac{1}{\alpha}} \right\}. \quad \blacksquare$$

We continue with

Theorem 4.52 *Here all as in Theorem 4.45, where $x = x_0 \in \mathbb{R}$ is fixed and $m > 1$. Also the same assumptions as in Theorem 4.49. Then*

$$\left| C_{N,\alpha}^{(M)} (f)(x) - f(x) \right| \le \frac{1}{\Gamma(m+1)} \omega_1 \left(D_x^m f, \left(\frac{\theta T^m}{N^{(m+1)(1-\alpha)}} \right)^{\frac{1}{m+1}} \right)_{\mathbb{R}} \cdot$$

$$\left[\frac{\theta T^{m-1}}{N^{m(1-\alpha)}} + \frac{1}{(m+1)} \left(\frac{\theta T^m}{N^{(m+1)(1-\alpha)}} \right)^{\frac{m}{m+1}} \right], \qquad (4.120)$$

$\forall\, N \in \mathbb{N} : N > \max \left\{ T + |x|, \left(\frac{2}{T} \right)^{\frac{1}{\alpha}} \right\}.$

We have that $\lim\limits_{N \to +\infty} C_{N,\alpha}^{(M)}(f)(x) = f(x).$

Proof We apply Theorem 4.51. In (4.117) we choose

$$\delta := \left(\frac{\theta T^m}{N^{(m+1)(1-\alpha)}} \right)^{\frac{1}{m+1}},$$

thus $\delta^{m+1} = \frac{\theta T^m}{N^{(m+1)(1-\alpha)}}$, and

$$\delta^m = \left(\frac{\theta T^m}{N^{(m+1)(1-\alpha)}} \right)^{\frac{m}{m+1}}. \qquad (4.121)$$

Therefore we have

$$\left| C_{N,\alpha}^{(M)}(f)(x) - f(x) \right| \overset{(4.116)}{\le} \frac{1}{\Gamma(m+1)} \omega_1 \left(D_x^m f, \left(\frac{\theta T^m}{N^{(m+1)(1-\alpha)}} \right)^{\frac{1}{m+1}} \right)_{\mathbb{R}} \cdot \quad (4.122)$$

$$\left[\frac{\theta T^{m-1}}{N^{m(1-\alpha)}} + \frac{1}{(m+1)\,\delta} \frac{\theta T^m}{N^{(m+1)(1-\alpha)}} \right] =$$

$$\frac{1}{\Gamma(m+1)} \omega_1 \left(D_x^m f, \left(\frac{\theta T^m}{N^{(m+1)(1-\alpha)}} \right)^{\frac{1}{m+1}} \right)_{\mathbb{R}} \left[\frac{\theta T^{m-1}}{N^{m(1-\alpha)}} + \frac{1}{(m+1)\,\delta} \delta^{m+1} \right] \overset{(4.121)}{=}$$

$$\frac{1}{\Gamma(m+1)} \omega_1 \left(D_x^m f, \left(\frac{\theta T^m}{N^{(m+1)(1-\alpha)}} \right)^{\frac{1}{m+1}} \right)_{\mathbb{R}} \cdot$$

$$\left[\frac{\theta T^{m-1}}{N^{m(1-\alpha)}} + \frac{1}{(m+1)} \left(\frac{\theta T^m}{N^{(m+1)(1-\alpha)}} \right)^{\frac{m}{m+1}} \right], \qquad (4.123)$$

$\forall N \in \mathbb{N}: N > \max \left\{ T + |x|, \left(\frac{2}{T}\right)^{\frac{1}{\alpha}} \right\}$, proving the inequality (4.120). ∎

We finish with (case of $\alpha = 1.5$)

Corollary 4.53 *Let* $x \in [0, 1]$, $f \in AC^2 ([0, 1], \mathbb{R}_+)$ *and* $f^{(2)} \in L_\infty ([0, 1])$. *Assume that* $f'(x) = 0$. *Then*

$$\left| B_N^{(M)} (f)(x) - f(x) \right| \leq \frac{4\omega_1 \left(D_x^{1.5} f, \left(\frac{6}{\sqrt{N+1}}\right)^{\frac{2}{5}} \right)}{3\sqrt{\pi}}$$

$$\left[\frac{6}{\sqrt{N+1}} + \frac{2}{5} \left(\frac{6}{\sqrt{N+1}}\right)^{\frac{3}{5}} \right], \forall N \in \mathbb{N}. \tag{4.124}$$

Proof By Theorem 4.30, apply (4.75). ∎

Due to lack of space we do not give other example applications.

References

1. G. Anastassiou, On right fractional calculus. Chaos, Solitons Fractals **42**, 365–376 (2009)
2. G. Anastassiou, Fractional Korovkin theory. Chaos, Solitons Fractals **42**, 2080–2094 (2009)
3. G. Anastassiou, *Approximation by Sublinear Operators* (2017, submitted)
4. G. Anastassiou, *Caputo Fractional Approximation by Sublinear Operators* (2017, submitted)
5. G. Anastassiou, L. Coroianu, S. Gal, Approximation by a nonlinear Cardaliaguet-Euvrard neural network operator of max-product kind. J. Comput. Anal. Appl. **12**(2), 396–406 (2010)
6. B. Bede, L. Coroianu, S. Gal, *Approximation by Max-Product Type Operators* (Springer, Heidelberg, 2016)
7. R.A. DeVore, G.G. Lorentz, *Constructive Approximation* (Springer, Berlin, 1993)
8. K. Diethelm, *The Analysis of Fractional Differential Equations* (Springer, Heidelberg, 2010)
9. A.M.A. El-Sayed, M. Gaber, On the finite Caputo and finite Riesz derivatives. Electron. J. Theor. Phys. 3(12), 81–95 (2006)
10. L. Fejér, *Über Interpolation*, Göttingen Nachrichten, (1916), pp. 66–91
11. G.S. Frederico, D.F.M. Torres, Fractional optimal control in the sense of Caputo and the fractional Noether's theorem. Int. Math. Forum 3(10), 479–493 (2008)
12. G.G. Lorentz, *Bernstein Polynomials*, 2nd edn. (Chelsea Publishing Company, New York, 1986)
13. T. Popoviciu, Sur l'approximation de fonctions convexes d'order superieur. Mathematica (Cluj) **10**, 49–54 (1935)
14. S.G. Samko, A.A. Kilbas and O.I. Marichev, *Fractional Integrals and Derivatives, Theory and Applications*, (Gordon and Breach, Amsterdam, 1993) [English translation from the Russian, Integrals and Derivatives of Fractional Order and Some of Their Applications (Nauka i Tekhnika, Minsk, 1987)]

Chapter 5
Canavati Fractional Approximations Using Max-Product Operators

Here we study the approximation of functions by sublinear positive operators with applications to a large variety of Max-Product operators under Canavati fractional differentiability. Our approach is based on our general fractional results about positive sublinear operators. We derive Jackson type inequalities under simple initial conditions. So our way is quantitative by producing inequalities with their right hand sides involving the modulus of continuity of Canavati fractional derivative of the function under approximation. It follows [3].

5.1 Introduction

The inspiring motivation here is the monograph by B. Bede, L. Coroianu and S. Gal [6], 2016.

Let $N \in \mathbb{N}$, the well-known Bernstein polynomials [10] are positive linear operators, defined by the formula

$$B_N(f)(x) = \sum_{k=0}^{N} \binom{N}{k} x^k (1-x)^{N-k} f\left(\frac{k}{N}\right), x \in [0, 1], f \in C([0, 1]).$$

(5.1)

T. Popoviciu in [11], 1935, proved for $f \in C([0, 1])$ that

$$|B_N(f)(x) - f(x)| \le \frac{5}{4}\omega_1\left(f, \frac{1}{\sqrt{N}}\right), \forall x \in [0, 1],$$

(5.2)

where

$$\omega_1(f, \delta) = \sup_{\substack{x,y\in[a,b]:\\|x-y|\le\delta}} |f(x) - f(y)|, \delta > 0,$$

(5.3)

© Springer International Publishing AG, part of Springer Nature 2018
G. A. Anastassiou, *Nonlinearity: Ordinary and Fractional Approximations by Sublinear and Max-Product Operators*, Studies in Systems, Decision and Control 147, https://doi.org/10.1007/978-3-319-89509-3_5

is the first modulus of continuity, here $[a, b] = [0, 1]$.

G. G. Lorentz in [10], 1986, p. 21, proved for $f \in C^1 ([0, 1])$ that

$$|B_N (f) (x) - f (x)| \leq \frac{3}{4\sqrt{N}} \omega_1 \left(f', \frac{1}{\sqrt{N}} \right), \forall x \in [0, 1], \qquad (5.4)$$

In [6], p. 10, the authors introduced the basic Max-product Bernstein operators,

$$B_N^{(M)} (f) (x) = \frac{\bigvee_{k=0}^{N} p_{N,k} (x) f \left(\frac{k}{N} \right)}{\bigvee_{k=0}^{N} p_{N,k} (x)}, N \in \mathbb{N}, \qquad (5.5)$$

where $\bigvee$ stands for maximum, and $p_{N,k} (x) = \binom{N}{k} x^k (1 - x)^{N-k}$ and $f :$
$[0, 1] \rightarrow \mathbb{R}_+ = [0, \infty)$.

These are nonlinear and piecewise rational operators.

The authors in [6] studied similar such nonlinear operators such as: the Max-product Favard-Szász-Mirakjan operators and their truncated version, the Max-product Baskakov operators and their truncated version, also many other similar specific operators. The study in [6] is based on presented there general theory of sublinear operators. These Max-product operators tend to converge faster to the on hand function.

So we mention from [6], p. 30, that for $f : [0, 1] \rightarrow \mathbb{R}_+$ continuous, we have the estimate

$$\left| B_N^{(M)} (f) (x) - f (x) \right| \leq 12\omega_1 \left(f, \frac{1}{\sqrt{N + 1}} \right), \text{ for all } N \in \mathbb{N}, x \in [0, 1], \quad (5.6)$$

Also from [6], p. 36, we mention that for $f : [0, 1] \rightarrow \mathbb{R}_+$ being concave function we get that

$$\left| B_N^{(M)} (f) (x) - f (x) \right| \leq 2\omega_1 \left(f, \frac{1}{N} \right), \text{ for all } x \in [0, 1], \qquad (5.7)$$

a much faster convergence.

In this chapter we expand the study of [6] by considering Canavati fractional smoothness of functions. So our inequalities are with respect to $\omega_1 (D^\alpha f, \delta), \delta > 0$, where $D^\alpha f$ with $\alpha > 0$ is the Canavati fractional derivative.

5.2 Main Results

We make

Remark 5.1 (I) Here see [1], pp. 7-10.

Let $x, x_0 \in [a, b]$ such that $x \geq x_0$, $\nu > 0$, $\nu \notin \mathbb{N}$, such that $p = [\nu]$, $[\cdot]$ the integral part, $\alpha = \nu - p$ $(0 < \alpha < 1)$.

Let $f \in C^p([a, b])$ and define

$$\left(J_\nu^{x_0} f\right)(x) := \frac{1}{\Gamma(\nu)} \int_{x_0}^x (x - t)^{\nu - 1} f(t)\, dt\,, \quad x_0 \leq x \leq b. \tag{5.8}$$

the left generalized Riemann-Liouville fractional integral.

Here Γ stands for the gamma function.

Clearly here it holds $\left(J_\nu^{x_0} f\right)(x_0) = 0$. We define $\left(J_\nu^{x_0} f\right)(x) = 0$ for $x < x_0$. By [1], p. 388, $\left(J_\nu^{x_0} f\right)(x)$ is a continuous function in x, for a fixed x_0.

We define the subspace $C_{x_0+}^\nu([a, b])$ of $C^p([a, b])$:

$$C_{x_0+}^\nu([a, b]) := \left\{ f \in C^p([a, b]) : J_{1-\alpha}^{x_0} f^{(p)} \in C^1([x_0, b]) \right\}. \tag{5.9}$$

So let $f \in C_{x_0+}^\nu([a, b])$, we define the left generalized ν-fractional derivative of f over $[x_0, b]$ as

$$D_{x_0+}^\nu f = \left(J_{1-\alpha}^{x_0} f^{(p)}\right)', \tag{5.10}$$

that is

$$\left(D_{x_0+}^\nu f\right)(x) = \frac{1}{\Gamma(1 - \alpha)} \frac{d}{dx} \int_{x_0}^x (x - t)^{-\alpha} f^{(p)}(t)\, dt, \tag{5.11}$$

which exists for $f \in C_{x_0+}^\nu([a, b])$, for $a \leq x_0 \leq x \leq b$.

Canavati in [7] first introduced this kind of left fractional derivative over $[0, 1]$.

We mention the following left generalized fractional Taylor formula ($f \in C_{x_0+}^\nu([a, b])$, $\nu > 1$).

It holds

$$f(x) - f(x_0) = \sum_{k=1}^{p-1} \frac{f^{(k)}(x_0)}{k!} (x - x_0)^k + \frac{1}{\Gamma(\nu)} \int_{x_0}^x (x - t)^{\nu - 1} \left(D_{x_0+}^\nu f\right)(t)\, dt, \tag{5.12}$$

for $x, x_0 \in [a, b]$ with $x \geq x_0$.

(II) Here see [2], p. 333, and again [2], pp. 345–348.

Let $x, x_0 \in [a, b]$ such that $x \leq x_0$, $\nu > 0$, $\nu \notin \mathbb{N}$, such that $p = [\nu]$, $\alpha = \nu - p$ $(0 < \alpha < 1)$.

Let $f \in C^p([a, b])$ and define

$$\left(J_{x_0-}^\nu f\right)(x) := \frac{1}{\Gamma(\nu)} \int_x^{x_0} (z - x)^{\nu - 1} f(z)\, dz, \quad a \leq x \leq x_0. \tag{5.13}$$

the right generalized Riemann-Liouville fractional integral.

Define the subspace of functions

$$C_{x_0-}^\nu([a, b]) := \left\{ f \in C^p([a, b]) : J_{x_0-}^{1-\alpha} f^{(p)} \in C^1([a, x_0]) \right\}. \tag{5.14}$$

Define the right generalized ν-fractional derivative of f over $[a, x_0]$ as

$$D_{x_0-}^{\nu} f = (-1)^{p-1} \left(J_{x_0-}^{1-\alpha} f^{(p)} \right)'. \tag{5.15}$$

Notice that

$$J_{x_0-}^{1-\alpha} f^{(p)} (x) = \frac{1}{\Gamma (1-\alpha)} \int_x^{x_0} (z-x)^{-\alpha} f^{(p)} (z) \, dz, \tag{5.16}$$

exists for $f \in C_{x_0-}^{\nu} ([a, b])$, and

$$\left(D_{x_0-}^{\nu} f \right) (x) = \frac{(-1)^{p-1}}{\Gamma (1-\alpha)} \frac{d}{dx} \int_x^{x_0} (z-x)^{-\alpha} f^{(p)} (z) \, dz. \tag{5.17}$$

I.e.

$$\left(D_{x_0-}^{\nu} f \right) (x) = \frac{(-1)^{p-1}}{\Gamma (p-\nu+1)} \frac{d}{dx} \int_x^{x_0} (z-x)^{p-\nu} f^{(p)} (z) \, dz, \tag{5.18}$$

which exists for $f \in C_{x_0-}^{\nu} ([a, b])$, for $a \le x \le x_0 \le b$.

We mention the following right generalized fractional Taylor formula ($f \in C_{x_0-}^{\nu} ([a, b]), \nu > 1$).

It holds

$$f (x) - f (x_0) = \sum_{k=1}^{p-1} \frac{f^{(k)} (x_0)}{k!} (x - x_0)^k + \frac{1}{\Gamma (\nu)} \int_x^{x_0} (z-x)^{\nu-1} \left(D_{x_0-}^{\nu} f \right) (z) \, dz, \tag{5.19}$$

for $x, x_0 \in [a, b]$ with $x \le x_0$.

We need

Definition 5.2 Let $D_{x_0}^{\nu} f$ denote any of $D_{x_0-}^{\nu} f$, $D_{x_0+}^{\nu} f$, and $\delta > 0$. We set

$$\omega_1 \left(D_{x_0}^{\nu} f, \delta \right) := \max \left\{ \omega_1 \left(D_{x_0-}^{\nu} f, \delta \right)_{[a, x_0]}, \omega_1 \left(D_{x_0+}^{\nu} f, \delta \right)_{[x_0, b]} \right\}, \tag{5.20}$$

where $x_0 \in [a, b]$. Here the moduli of continuity are considered over $[a, x_0]$ and $[x_0, b]$, respectively.

We need

Theorem 5.3 Let $\nu > 1$, $\nu \notin \mathbb{N}$, $p = [\nu]$, $x_0 \in [a, b]$ and $f \in C_{x_0+}^{\nu} ([a, b]) \cap C_{x_0-}^{\nu} ([a, b])$. Assume that $f^{(k)} (x_0) = 0$, $k = 1, ..., p-1$, and $\left(D_{x_0+}^{\nu} f \right) (x_0) = \left(D_{x_0-}^{\nu} f \right) (x_0) = 0$. Then

$$|f (x) - f (x_0)| \le \frac{\omega_1 \left(D_{x_0}^{\nu} f, \delta \right)}{\Gamma (\nu + 1)} \left[|x - x_0|^{\nu} + \frac{|x - x_0|^{\nu+1}}{(\nu + 1) \delta} \right], \delta > 0, \tag{5.21}$$

for all $a \le x \le b$.

Proof We use (5.12) and (5.19), and the assumption $f^{(k)}(x_0) = 0, k = 1, ..., p - 1$ and $\left(D_{x_{0+}}^{\nu} f\right)(x_0) = \left(D_{x_{0-}}^{\nu} f\right)(x_0) = 0$. We have that

$$f(x) - f(x_0) = \frac{1}{\Gamma(\nu)} \int_{x_0}^{x} (x - z)^{\nu-1} \left(\left(D_{x_{0+}}^{\nu} f\right)(z) - \left(D_{x_{0+}}^{\nu} f\right)(x_0)\right) dz, \quad (5.22)$$

for all $x_0 \le x \le b$,

and

$$f(x) - f(x_0) = \frac{1}{\Gamma(\nu)} \int_{x}^{x_0} (z - x)^{\nu-1} \left(\left(D_{x_{0-}}^{\nu} f\right)(z) - \left(D_{x_{0-}}^{\nu} f\right)(x_0)\right) dz, \quad (5.23)$$

for all $a \le x \le x_0$.

We observe that ($x_0 \le x \le b$)

$$|f(x) - f(x_0)| \le \frac{1}{\Gamma(\nu)} \int_{x_0}^{x} (x - z)^{\nu-1} \left|\left(D_{x_{0+}}^{\nu} f\right)(z) - \left(D_{x_{0+}}^{\nu} f\right)(x_0)\right| dz \underset{(\delta_1 > 0)}{\le}$$

$$\frac{1}{\Gamma(\nu)} \int_{x_0}^{x} (x - z)^{\nu-1} \omega_1 \left(D_{x_{0+}}^{\nu} f, \frac{\delta_1 |z - x_0|}{\delta_1}\right)_{[x_0, b]} dz \le$$

$$\frac{\omega_1 \left(D_{x_{0+}}^{\nu} f, \delta_1\right)_{[x_0, b]}}{\Gamma(\nu)} \int_{x_0}^{x} (x - z)^{\nu-1} \left(1 + \frac{(z - x_0)}{\delta_1}\right) dz =$$

$$\frac{\omega_1 \left(D_{x_{0+}}^{\nu} f, \delta_1\right)_{[x_0, b]}}{\Gamma(\nu)} \left[\frac{(x - x_0)^{\nu}}{\nu} + \frac{1}{\delta_1} \int_{x_0}^{x} (x - z)^{\nu-1} (z - x_0)^{2-1} dz\right] = \quad (5.24)$$

$$\frac{\omega_1 \left(D_{x_{0+}}^{\nu} f, \delta_1\right)_{[x_0, b]}}{\Gamma(\nu)} \left[\frac{(x - x_0)^{\nu}}{\nu} + \frac{1}{\delta_1} \frac{\Gamma(\nu) \Gamma(2)}{\Gamma(\nu + 2)} (x - x_0)^{\nu+1}\right] =$$

$$\frac{\omega_1 \left(D_{x_{0+}}^{\nu} f, \delta_1\right)_{[x_0, b]}}{\Gamma(\nu)} \left[\frac{(x - x_0)^{\nu}}{\nu} + \frac{1}{\delta_1} \frac{1}{(\nu + 1)\nu} (x - x_0)^{\nu+1}\right] =$$

$$\frac{\omega_1 \left(D_{x_{0+}}^{\nu} f, \delta_1\right)_{[x_0, b]}}{\Gamma(\nu + 1)} \left[(x - x_0)^{\nu} + \frac{(x - x_0)^{\nu+1}}{(\nu + 1)\delta_1}\right].$$

We have proved

$$|f(x) - f(x_0)| \le \frac{\omega_1 \left(D_{x_{0+}}^{\nu} f, \delta_1\right)_{[x_0, b]}}{\Gamma(\nu + 1)} \left[(x - x_0)^{\nu} + \frac{(x - x_0)^{\nu+1}}{(\nu + 1)\delta_1}\right], \quad (5.25)$$

$\delta_1 > 0$, and $x_0 \leq x \leq b$.

Similarly acting, we get $(a \leq x \leq x_0)$

$$|f(x) - f(x_0)| \leq \frac{1}{\Gamma(\nu)} \int_x^{x_0} (z-x)^{\nu-1} \left| D_{x_0-}^{\nu} f(z) - D_{x_0-}^{\nu} f(x_0) \right| dz \leq$$

$$\frac{1}{\Gamma(\nu)} \int_x^{x_0} (z-x)^{\nu-1} \omega_1 \left(D_{x_0-}^{\nu} f, |z-x_0| \right)_{[a,x_0]} dz =$$

$(\delta_2 > 0)$

$$\frac{1}{\Gamma(\nu)} \int_x^{x_0} (z-x)^{\nu-1} \omega_1 \left(D_{x_0-}^{\nu} f, \frac{\delta_2 |x_0-z|}{\delta_2} \right)_{[a,x_0]} dz \leq \qquad (5.26)$$

$$\frac{\omega_1 \left(D_{x_0-}^{\nu} f, \delta_2 \right)_{[a,x_0]}}{\Gamma(\nu)} \left[\int_x^{x_0} (z-x)^{\nu-1} \left(1 + \frac{x_0-z}{\delta_2} \right) dz \right] =$$

$$\frac{\omega_1 \left(D_{x_0-}^{\nu} f, \delta_2 \right)_{[a,x_0]}}{\Gamma(\nu)} \left[\frac{(x_0-x)^{\nu}}{\nu} + \frac{1}{\delta_2} \int_x^{x_0} (x_0-z)^{2-1} (z-x)^{\nu-1} dz \right] =$$

$$\frac{\omega_1 \left(D_{x_0-}^{\nu} f, \delta_2 \right)_{[a,x_0]}}{\Gamma(\nu)} \left[\frac{(x_0-x)^{\nu}}{\nu} + \frac{1}{\delta_2} \frac{\Gamma(\nu) \Gamma(2)}{\Gamma(\nu+2)} (x_0-x)^{\nu+1} \right] =$$

$$\frac{\omega_1 \left(D_{x_0-}^{\nu} f, \delta_2 \right)_{[a,x_0]}}{\Gamma(\nu)} \left[\frac{(x_0-x)^{\nu}}{\nu} + \frac{1}{\delta_2} \frac{(x_0-x)^{\nu+1}}{(\nu+1)\nu} \right] = \qquad (5.27)$$

$$\frac{\omega_1 \left(D_{x_0-}^{\nu} f, \delta_2 \right)_{[a,x_0]}}{\Gamma(\nu+1)} \left[(x_0-x)^{\nu} + \frac{(x_0-x)^{\nu+1}}{(\nu+1)\delta_2} \right].$$

We have proved

$$|f(x) - f(x_0)| \leq \frac{\omega_1 \left(D_{x_0-}^{\nu} f, \delta_2 \right)_{[a,x_0]}}{\Gamma(\nu+1)} \left[(x_0-x)^{\nu} + \frac{(x_0-x)^{\nu+1}}{(\nu+1)\delta_2} \right], \qquad (5.28)$$

$\delta_2 > 0$, and $(a \leq x \leq x_0)$. Choosing $\delta = \delta_1 = \delta_2 > 0$, by (5.25) and (5.28), we get (5.21). ∎

We need

Definition 5.4 Here $C_+([a,b]) := \{ f : [a,b] \to \mathbb{R}_+, \text{ continuous functions} \}$. Let $L_N : C_+([a,b]) \to C_+([a,b])$, operators, $\forall N \in \mathbb{N}$, such that
(i)
$$L_N(\alpha f) = \alpha L_N(f), \quad \forall \alpha \geq 0, \forall f \in C_+([a,b]), \qquad (5.29)$$

(ii) if $f, g \in C_+([a,b]) : f \leq g$, then

$$L_N (f) \leq L_N (g), \forall N \in \mathbb{N},\tag{5.30}$$

(iii)

$$L_N (f + g) \leq L_N (f) + L_N (g), \forall f, g \in C_+ ([a, b]).\tag{5.31}$$

We call $\{L_N\}_{N \in \mathbb{N}}$ positive sublinear operators.

We make

Remark 5.5 By [6], p. 17, we get: let $f, g \in C_+ ([a, b])$, then

$$|L_N (f) (x) - L_N (g) (x)| \leq L_N (|f - g|) (x), \forall x \in [a, b].\tag{5.32}$$

Furthermore, we also have that

$$|L_N (f) (x) - f (x)| \leq L_N (|f (\cdot) - f (x)|) (x) + |f (x)| |L_N (e_0) (x) - 1|,\tag{5.33}$$

$\forall x \in [a, b]; e_0 (t) = 1$.

From now on we assume that $L_N (1) = 1$. Hence it holds

$$|L_N (f) (x) - f (x)| \leq L_N (|f (\cdot) - f (x)|) (x), \forall x \in [a, b].\tag{5.34}$$

Using Theorem 5.3 and (5.21) with (5.34) we get:

$$|L_N (f) (x_0) - f (x_0)| \leq \frac{\omega_1 \left(D_{x_0}^\nu f, \delta \right)}{\Gamma (\nu + 1)}.\tag{5.35}$$

$$\left[L_N (|\cdot - x_0|^\nu) (x_0) + \frac{L_N \left(|\cdot - x_0|^{\nu+1} \right) (x_0)}{(\nu + 1) \delta} \right], \delta > 0.$$

We have proved

Theorem 5.6 *Let* $\nu > 1$, $\nu \notin \mathbb{N}$, $p = [\nu]$, $x_0 \in [a, b]$ *and* $f : [a, b] \to \mathbb{R}_+$, $f \in C_{x_0+}^\nu ([a, b]) \cap C_{x_0-}^\nu ([a, b])$. *Assume that* $f^{(k)} (x_0) = 0, k = 1, ..., p - 1$, *and* $\left(D_{x_0+}^\nu f \right) (x_0) = \left(D_{x_0-}^\nu f \right) (x_0) = 0$. *Let* $L_N : C_+ ([a, b]) \to C_+ ([a, b])$, $\forall N \in \mathbb{N}$, *be positive sublinear operators, such that* $L_N (1) = 1$, $\forall N \in \mathbb{N}$. *Then*

$$|L_N (f) (x_0) - f (x_0)| \leq \frac{\omega_1 \left(D_{x_0}^\nu f, \delta \right)}{\Gamma (\nu + 1)}.$$

$$\left[L_N (|\cdot - x_0|^\nu) (x_0) + \frac{L_N \left(|\cdot - x_0|^{\nu+1} \right) (x_0)}{(\nu + 1) \delta} \right],\tag{5.36}$$

$\delta > 0, \forall N \in \mathbb{N}$.

5.3 Applications

We give

Theorem 5.7 *Let $\nu > 1$, $\nu \notin \mathbb{N}$, $p = [\nu]$, $x \in [0, 1]$, $f : [0, 1] \to \mathbb{R}_+$ and $f \in C_{x+}^{\nu} ([0, 1]) \cap C_{x-}^{\nu} ([0, 1])$. Assume that $f^{(k)} (x) = 0$, $k = 1, ..., p - 1$, and $\left(D_{x+}^{\nu} f \right) (x) = \left(D_{x-}^{\nu} f \right) (x) = 0$. Then*

$$\left| B_N^{(M)} (f) (x) - f (x) \right| \leq \frac{\omega_1 \left(D_x^{\nu} f, \left(\frac{6}{\sqrt{N+1}} \right)^{\frac{1}{\nu+1}} \right)}{\Gamma (\nu + 1)}.$$

$$\left[\frac{6}{\sqrt{N+1}} + \frac{1}{(\nu+1)} \left(\frac{6}{\sqrt{N+1}} \right)^{\frac{\nu}{\nu+1}} \right], \tag{5.37}$$

$\forall N \in \mathbb{N}$.
We get $\displaystyle\lim_{N \to +\infty} B_N^{(M)} (f) (x) = f (x)$.

Proof By [4] we get that

$$B_N^{(M)} (|\cdot - x|^{\nu}) (x) \leq \frac{6}{\sqrt{N+1}}, \forall x \in [0, 1], \tag{5.38}$$

$\forall N \in \mathbb{N}$, $\forall \nu > 1$.
   Also $B_N^{(M)}$ maps $C_+ ([0, 1])$ into itself, $B_N^{(M)} (1) = 1$, and it is positive sublinear operator.
   We apply Theorem 5.6 and (5.36), we get

$$\left| B_N^{(M)} (f) (x) - f (x) \right| \leq \frac{\omega_1 \left(D_x^{\nu} f, \delta \right)}{\Gamma (\nu + 1)} \left[\frac{6}{\sqrt{N+1}} + \frac{\frac{6}{\sqrt{N+1}}}{(\nu+1) \delta} \right]. \tag{5.39}$$

Choose $\delta = \left(\frac{6}{\sqrt{N+1}} \right)^{\frac{1}{\nu+1}}$, then $\delta^{\nu+1} = \frac{6}{\sqrt{N+1}}$, and apply it to (5.39). Clearly we derive (5.37). ∎

   We continue with

Remark 5.8 The truncated Favard-Szász-Mirakjan operators are given by

$$T_N^{(M)} (f) (x) = \frac{\bigvee_{k=0}^{N} s_{N,k} (x) f \left(\frac{k}{N} \right)}{\bigvee_{k=0}^{N} s_{N,k} (x)}, x \in [0, 1], N \in \mathbb{N}, f \in C_+ ([0, 1]), \tag{5.40}$$

$s_{N,k} (x) = \frac{(Nx)^k}{k!}$, see also [6], p. 11.
   By [6], pp. 178–179, we get that

$$T_N^{(M)} \left(|\cdot - x| \right) (x) \le \frac{3}{\sqrt{N}}, \forall x \in [0, 1], \forall N \in \mathbb{N}. \tag{5.41}$$

Clearly it holds

$$T_N^{(M)} \left(|\cdot - x|^{1+\beta} \right) (x) \le \frac{3}{\sqrt{N}}, \forall x \in [0, 1], \forall N \in \mathbb{N}, \forall \beta > 0. \tag{5.42}$$

The operators $T_N^{(M)}$ are positive sublinear operators mapping $C_+ ([0, 1])$ into itself, with $T_N^{(M)} (1) = 1$.

We continue with

Theorem 5.9 *Same assumptions as in Theorem 5.7. Then*

$$\left| T_N^{(M)} (f) (x) - f (x) \right| \le \frac{\omega_1 \left(D_x^\nu f, \left(\frac{3}{\sqrt{N}} \right)^{\frac{1}{\nu+1}} \right)}{\Gamma (\nu + 1)} \cdot$$

$$\left[\frac{3}{\sqrt{N}} + \frac{1}{(\nu + 1)} \left(\frac{3}{\sqrt{N}} \right)^{\frac{\nu}{\nu+1}} \right], \forall N \in \mathbb{N}. \tag{5.43}$$

We get $\lim_{N \to +\infty} T_N^{(M)} (f) (x) = f (x)$.

Proof Use of Theorem 5.6, similar to the proof of Theorem 5.7. ∎

We make

Remark 5.10 Next we study the truncated Max-product Baskakov operators (see [6], p. 11)

$$U_N^{(M)} (f) (x) = \frac{\bigvee_{k=0}^N b_{N,k} (x) f \left(\frac{k}{N} \right)}{\bigvee_{k=0}^N b_{N,k} (x)}, x \in [0, 1], f \in C_+ ([0, 1]), N \in \mathbb{N}, \tag{5.44}$$

where

$$b_{N,k} (x) = \binom{N + k - 1}{k} \frac{x^k}{(1 + x)^{N+k}}. \tag{5.45}$$

From [6], pp. 217–218, we get ($x \in [0, 1]$)

$$\left(U_N^{(M)} (|\cdot - x|) \right) (x) \le \frac{2\sqrt{3} \left(\sqrt{2} + 2 \right)}{\sqrt{N + 1}}, N \ge 2, N \in \mathbb{N}. \tag{5.46}$$

Let $\lambda \ge 1$, clearly then it holds

$$\left(U_N^{(M)}\left(|\cdot - x|^\lambda\right)\right)(x) \le \frac{2\sqrt{3}\left(\sqrt{2}+2\right)}{\sqrt{N+1}}, \forall N \ge 2, N \in \mathbb{N}. \tag{5.47}$$

Also it holds $U_N^{(M)}(1) = 1$, and $U_N^{(M)}$ are positive sublinear operators from $C_+([0, 1])$ into itself.

We give

Theorem 5.11 *Same assumptions as in Theorem 5.7. Then*

$$\left|U_N^{(M)}(f)(x) - f(x)\right| \le \frac{\omega_1\left(D_x^\nu f, \left(\frac{2\sqrt{3}(\sqrt{2}+2)}{\sqrt{N+1}}\right)^{\frac{1}{\nu+1}}\right)}{\Gamma(\nu+1)}. \tag{5.48}$$

$$\left[\frac{2\sqrt{3}\left(\sqrt{2}+2\right)}{\sqrt{N+1}} + \frac{1}{(\nu+1)}\left(\frac{2\sqrt{3}\left(\sqrt{2}+2\right)}{\sqrt{N+1}}\right)^{\frac{\nu}{\nu+1}}\right], \forall N \ge 2, N \in \mathbb{N}.$$

We get $\lim_{N\to+\infty} U_N^{(M)}(f)(x) = f(x)$.

Proof Use of Theorem 5.6, similar to the proof of Theorem 5.7. ∎

We continue with

Remark 5.12 Here we study the Max-product Meyer-Köning and Zeller operators (see [6], p. 11) defined by

$$Z_N^{(M)}(f)(x) = \frac{\bigvee_{k=0}^\infty s_{N,k}(x) f\left(\frac{k}{N+k}\right)}{\bigvee_{k=0}^\infty s_{N,k}(x)}, \forall N \in \mathbb{N}, f \in C_+([0, 1]), \tag{5.49}$$

$$s_{N,k}(x) = \binom{N+k}{k} x^k, x \in [0, 1].$$

By [6], p. 253, we get that

$$Z_N^{(M)}(|\cdot - x|)(x) \le \frac{8\left(1+\sqrt{5}\right)}{3} \frac{\sqrt{x}(1-x)}{\sqrt{N}}, \forall x \in [0, 1], \forall N \ge 4, N \in \mathbb{N}. \tag{5.50}$$

We have that (for $\lambda \ge 1$)

$$Z_N^{(M)}\left(|\cdot - x|^\lambda\right)(x) \le \frac{8\left(1+\sqrt{5}\right)}{3} \frac{\sqrt{x}(1-x)}{\sqrt{N}} := \rho(x), \tag{5.51}$$

$\forall x \in [0, 1], N \ge 4, N \in \mathbb{N}.$

Also it holds $Z_N^{(M)}(1) = 1$, and $Z_N^{(M)}$ are positive sublinear operators from $C_+([0, 1])$ into itself.

We give

Theorem 5.13 *Same assumptions as in Theorem 5.7. Then*

$$\left| Z_N^{(M)}(f)(x) - f(x) \right| \le \frac{\omega_1 \left(D_x^\nu f, (\rho(x))^{\frac{1}{\nu+1}} \right)}{\Gamma(\nu+1)}. \tag{5.52}$$

$$\left[\rho(x) + \frac{1}{(\nu+1)} (\rho(x))^{\frac{\nu}{\nu+1}} \right], \forall N \in \mathbb{N}, N \ge 4.$$

We get $\lim_{N \to +\infty} Z_N^{(M)}(f)(x) = f(x)$, *where* $\rho(x)$ *is as in (5.51).*

Proof Use of Theorem 5.6, similar to the proof of Theorem 5.7. ■

We continue with

Remark 5.14 Here we deal with the Max-product truncated sampling operators (see [6], p. 13) defined by

$$W_N^{(M)}(f)(x) = \frac{\bigvee_{k=0}^N \frac{\sin(Nx-k\pi)}{Nx-k\pi} f\left(\frac{k\pi}{N}\right)}{\bigvee_{k=0}^N \frac{\sin(Nx-k\pi)}{Nx-k\pi}}, \tag{5.53}$$

and

$$K_N^{(M)}(f)(x) = \frac{\bigvee_{k=0}^N \frac{\sin^2(Nx-k\pi)}{(Nx-k\pi)^2} f\left(\frac{k\pi}{N}\right)}{\bigvee_{k=0}^N \frac{\sin^2(Nx-k\pi)}{(Nx-k\pi)^2}}, \tag{5.54}$$

$\forall x \in [0, \pi]$, $f : [0, \pi] \to \mathbb{R}_+$ a continuous function.

Following [6], p. 343, and making the convention $\frac{\sin(0)}{0} = 1$ and denoting $s_{N,k}(x) = \frac{\sin(Nx-k\pi)}{Nx-k\pi}$, we get that $s_{N,k}\left(\frac{k\pi}{N}\right) = 1$, and $s_{N,k}\left(\frac{j\pi}{N}\right) = 0$, if $k \ne j$, furthermore $W_N^{(M)}(f)\left(\frac{j\pi}{N}\right) = f\left(\frac{j\pi}{N}\right)$, for all $j \in \{0, ..., N\}$.

Clearly $W_N^{(M)}(f)$ is a well-defined function for all $x \in [0, \pi]$, and it is continuous on $[0, \pi]$, also $W_N^{(M)}(1) = 1$.

By [6], p. 344, $W_N^{(M)}$ are positive sublinear operators.

Call $I_N^+(x) = \left\{ k \in \{0, 1, ..., N\}; s_{N,k}(x) > 0 \right\}$, and set $x_{N,k} := \frac{k\pi}{N}$, $k \in \{0, 1, ..., N\}$.

We see that

$$W_N^{(M)}(f)(x) = \frac{\bigvee_{k \in I_N^+(x)} s_{N,k}(x) f(x_{N,k})}{\bigvee_{k \in I_N^+(x)} s_{N,k}(x)}. \tag{5.55}$$

By [6], p. 346, we have

$$W_N^{(M)}\left(|\cdot - x|\right)(x) \le \frac{\pi}{2N}, \forall N \in \mathbb{N}, \forall x \in [0, \pi].\tag{5.56}$$

Notice also $\left|x_{N,k} - x\right| \le \pi, \forall\, x \in [0, \pi]$.

Therefore $(\lambda \ge 1)$ it holds

$$W_N^{(M)}\left(|\cdot - x|^{\lambda}\right)(x) \le \frac{\pi^{\lambda-1}\pi}{2N} = \frac{\pi^{\lambda}}{2N}, \forall x \in [0, \pi], \forall N \in \mathbb{N}.\tag{5.57}$$

We continue with

Theorem 5.15 *Let $\nu > 1$, $\nu \notin \mathbb{N}$, $p = [\nu]$, $x \in [0, \pi]$, $f : [0, \pi] \to \mathbb{R}_+$ and $f \in C_{x+}^{\nu}\left([0, \pi]\right) \cap C_{x-}^{\nu}\left([0, \pi]\right)$. Assume that $f^{(k)}(x) = 0$, $k = 1, ..., p - 1$, and $\left(D_{x+}^{\nu} f\right)(x) = \left(D_{x-}^{\nu} f\right)(x) = 0$. Then*

$$\left|W_N^{(M)}(f)(x) - f(x)\right| \le \frac{\omega_1\left(D_x^{\nu} f, \left(\frac{\pi^{\nu+1}}{2N}\right)^{\frac{1}{\nu+1}}\right)}{\Gamma(\nu + 1)} \cdot$$

$$\left[\frac{\pi^{\nu}}{2N} + \frac{1}{(\nu + 1)}\left(\frac{\pi^{\nu+1}}{2N}\right)^{\frac{\nu}{\nu+1}}\right], \forall N \in \mathbb{N}.\tag{5.58}$$

We have that $\lim_{N \to +\infty} W_N^{(M)}(f)(x) = f(x)$.

Proof Applying (5.36) for $W_N^{(M)}$ and using (5.57), we get

$$\left|W_N^{(M)}(f)(x) - f(x)\right| \le \frac{\omega_1\left(D_x^{\nu} f, \delta\right)}{\Gamma(\nu + 1)}\left[\frac{\pi^{\nu}}{2N} + \frac{\frac{\pi^{\nu+1}}{2N}}{(\nu + 1)\delta}\right].\tag{5.59}$$

Choose $\delta = \left(\frac{\pi^{\nu+1}}{2N}\right)^{\frac{1}{\nu+1}}$, then $\delta^{\nu+1} = \frac{\pi^{\nu+1}}{2N}$, and $\delta^{\nu} = \left(\frac{\pi^{\nu+1}}{2N}\right)^{\frac{\nu}{\nu+1}}$. We use the last into (5.59) and we obtain (5.58). ∎

We make

Remark 5.16 Here we continue with the Max-product truncated sampling operators (see [6], p. 13) defined by

$$K_N^{(M)}(f)(x) = \frac{\bigvee_{k=0}^{N} \frac{\sin^2(Nx-k\pi)}{(Nx-k\pi)^2} f\left(\frac{k\pi}{N}\right)}{\bigvee_{k=0}^{N} \frac{\sin^2(Nx-k\pi)}{(Nx-k\pi)^2}},\tag{5.60}$$

$\forall\, x \in [0, \pi]$, $f : [0, \pi] \to \mathbb{R}_+$ a continuous function.

Following [6], p. 350, and making the convention $\frac{\sin(0)}{0} = 1$ and denoting $s_{N,k}(x) = \frac{\sin^2(Nx - k\pi)}{(Nx - k\pi)^2}$, we get that $s_{N,k}\left(\frac{k\pi}{N}\right) = 1$, and $s_{N,k}\left(\frac{j\pi}{N}\right) = 0$, if $k \neq j$, furthermore $K_N^{(M)}(f)\left(\frac{j\pi}{N}\right) = f\left(\frac{j\pi}{N}\right)$, for all $j \in \{0, ..., N\}$.

Since $s_{N,j}\left(\frac{j\pi}{N}\right) = 1$ it follows that $\bigvee_{k=0}^{N} s_{N,k}\left(\frac{j\pi}{N}\right) \geq 1 > 0$, for all $j \in \{0, 1, ..., N\}$. Hence $K_N^{(M)}(f)$ is well-defined function for all $x \in [0, \pi]$, and it is continuous on $[0, \pi]$, also $K_N^{(M)}(1) = 1$. By [6], p. 350, $K_N^{(M)}$ are positive sublinear operators.

Denote $x_{N,k} := \frac{k\pi}{N}$, $k \in \{0, 1, ..., N\}$.

By [6], p. 352, we have

$$K_N^{(M)}(|\cdot - x|)(x) \leq \frac{\pi}{2N}, \forall N \in \mathbb{N}, \forall x \in [0, \pi]. \tag{5.61}$$

Notice also $|x_{N,k} - x| \leq \pi, \forall x \in [0, \pi]$.

Therefore ($\lambda \geq 1$) it holds

$$K_N^{(M)}\left(|\cdot - x|^\lambda\right)(x) \leq \frac{\pi^{\lambda-1}\pi}{2N} = \frac{\pi^\lambda}{2N}, \forall x \in [0, \pi], \forall N \in \mathbb{N}. \tag{5.62}$$

We give

Theorem 5.17 *All as in Theorem 5.15. Then*

$$\left| K_N^{(M)}(f)(x) - f(x) \right| \leq \frac{\omega_1\left(D_x^\nu f, \left(\frac{\pi^{\nu+1}}{2N}\right)^{\frac{1}{\nu+1}}\right)}{\Gamma(\nu+1)} \cdot$$

$$\left[\frac{\pi^\nu}{2N} + \frac{1}{(\nu+1)}\left(\frac{\pi^{\nu+1}}{2N}\right)^{\frac{\nu}{\nu+1}} \right], \forall N \in \mathbb{N}. \tag{5.63}$$

We have that $\lim_{N \to +\infty} K_N^{(M)}(f)(x) = f(x)$.

Proof As in Theorem 5.15. ∎

We make

Remark 5.18 We mention the interpolation Hermite-Fejér polynomials on Chebyshev knots of the first kind (see [6], p. 4): Let $f : [-1, 1] \to \mathbb{R}$ and based on the knots $x_{N,k} = \cos\left(\frac{(2(N-k)+1)}{2(N+1)}\pi\right) \in (-1, 1)$, $k \in \{0, ..., N\}$, $-1 < x_{N,0} < x_{N,1} < ... < x_{N,N} < 1$, which are the roots of the first kind Chebyshev polynomial $T_{N+1}(x) = \cos((N+1)\arccos x)$, we define (see Fejér [9])

$$H_{2N+1}(f)(x) = \sum_{k=0}^{N} h_{N,k}(x) f\left(x_{N,k}\right), \tag{5.64}$$

where

$$h_{N,k}(x) = \left(1 - x \cdot x_{N,k}\right) \left(\frac{T_{N+1}(x)}{(N+1)(x - x_{N,k})}\right)^2, \tag{5.65}$$

the fundamental interpolation polynomials.

The Max-product interpolation Hermite-Fejér operators on Chebyshev knots of the first kind (see p. 12 of [6]) are defined by

$$H_{2N+1}^{(M)}(f)(x) = \frac{\bigvee_{k=0}^{N} h_{N,k}(x) f\left(x_{N,k}\right)}{\bigvee_{k=0}^{N} h_{N,k}(x)}, \forall N \in \mathbb{N}, \tag{5.66}$$

where $f : [-1, 1] \to \mathbb{R}_+$ is continuous.

Call

$$E_N(x) := H_{2N+1}^{(M)}(|\cdot - x|)(x) = \frac{\bigvee_{k=0}^{N} h_{N,k}(x) \left|x_{N,k} - x\right|}{\bigvee_{k=0}^{N} h_{N,k}(x)}, x \in [-1, 1]. \tag{5.67}$$

Then by [6], p. 287 we obtain that

$$E_N(x) \le \frac{2\pi}{N+1}, \forall x \in [-1, 1], N \in \mathbb{N}. \tag{5.68}$$

For $m > 1$, we get

$$H_{2N+1}^{(M)}(|\cdot - x|^m)(x) = \frac{\bigvee_{k=0}^{N} h_{N,k}(x) \left|x_{N,k} - x\right|^m}{\bigvee_{k=0}^{N} h_{N,k}(x)} =$$

$$\frac{\bigvee_{k=0}^{N} h_{N,k}(x) \left|x_{N,k} - x\right| \left|x_{N,k} - x\right|^{m-1}}{\bigvee_{k=0}^{N} h_{N,k}(x)} \le 2^{m-1} \frac{\bigvee_{k=0}^{N} h_{N,k}(x) \left|x_{N,k} - x\right|}{\bigvee_{k=0}^{N} h_{N,k}(x)} \tag{5.69}$$

$$\le \frac{2^m \pi}{N+1}, \forall x \in [-1, 1], N \in \mathbb{N}.$$

Hence it holds

$$H_{2N+1}^{(M)}(|\cdot - x|^m)(x) \le \frac{2^m \pi}{N+1}, \forall x \in [-1, 1], m > 1, \forall N \in \mathbb{N}. \tag{5.70}$$

Furthermore we have

$$H_{2N+1}^{(M)}(1)(x) = 1, \forall x \in [-1, 1], \tag{5.71}$$

and $H_{2N+1}^{(M)}$ maps continuous functions to continuous functions over $[-1, 1]$ and for any $x \in \mathbb{R}$ we have $\bigvee_{k=0}^{N} h_{N,k}(x) > 0$.

We also have $h_{N,k}(x_{N,k}) = 1$, and $h_{N,k}(x_{N,j}) = 0$, if $k \neq j$, furthermore it holds $H_{2N+1}^{(M)}(f)(x_{N,j}) = f(x_{N,j})$, for all $j \in \{0, ..., N\}$, see [6], p. 282.

$H_{2N+1}^{(M)}$ are positive sublinear operators, [6], p. 282.

We give

Theorem 5.19 *Let* $\nu > 1$, $\nu \notin \mathbb{N}$, $p = [\nu]$, $x \in [-1, 1]$, $f : [-1, 1] \to \mathbb{R}_+$ *and* $f \in C_{x+}^{\nu}([-1, 1]) \cap C_{x-}^{\nu}([-1, 1])$. *Assume that* $f^{(k)}(x) = 0$, $k = 1, ..., p-1$, *and* $\left(D_{x+}^{\nu} f\right)(x) = \left(D_{x-}^{\nu} f\right)(x) = 0$. *Then*

$$\left| H_{2N+1}^{(M)}(f)(x) - f(x) \right| \leq \frac{\omega_1\left(D_x^{\nu} f, \left(\frac{2^{\nu+1}\pi}{N+1}\right)^{\frac{1}{\nu+1}}\right)}{\Gamma(\nu+1)}.$$

(5.72)

$$\left[\frac{2^{\nu}\pi}{N+1} + \frac{1}{(\nu+1)}\left(\frac{2^{\nu+1}\pi}{N+1}\right)^{\frac{\nu}{\nu+1}}\right], \forall N \in \mathbb{N}.$$

Furthermore it holds $\lim_{N \to +\infty} H_{2N+1}^{(M)}(f)(x) = f(x)$.

Proof Use of Theorem 5.6, (5.36) and (5.70). Choose $\delta := \left(\frac{2^{\nu+1}\pi}{N+1}\right)^{\frac{1}{\nu+1}}$, etc. ■

We continue with

Remark 5.20 Here we deal with Lagrange interpolation polynomials on Chebyshev knots of second kind plus the endpoints ± 1 (see [6], p. 5). These polynomials are linear operators attached to $f : [-1, 1] \to \mathbb{R}$ and to the knots $x_{N,k} = \cos\left(\left(\frac{N-k}{N-1}\right)\pi\right) \in [-1, 1]$, $k = 1, ..., N$, $N \in \mathbb{N}$, which are the roots of $\omega_N(x) = \sin(N-1)t \sin t$, $x = \cos t$. Notice that $x_{N,1} = -1$ and $x_{N,N} = 1$. Their formula is given by ([6], p. 377)

$$L_N(f)(x) = \sum_{k=1}^{N} l_{N,k}(x) f(x_{N,k}),$$

(5.73)

where

$$l_{N,k}(x) = \frac{(-1)^{k-1} \omega_N(x)}{(1 + \delta_{k,1} + \delta_{k,N})(N-1)(x - x_{N,k})},$$

(5.74)

$N \geq 2$, $k = 1, ..., N$, and $\omega_N(x) = \prod_{k=1}^{N}(x - x_{N,k})$ and $\delta_{i,j}$ denotes the Kronecker's symbol, that is $\delta_{i,j} = 1$, if $i = j$, and $\delta_{i,j} = 0$, if $i \neq j$.

The Max-product Lagrange interpolation operators on Chebyshev knots of second kind, plus the endpoints ± 1, are defined by ([6], p. 12)

$$L_N^{(M)}(f)(x) = \frac{\bigvee_{k=1}^{N} l_{N,k}(x) f(x_{N,k})}{\bigvee_{k=1}^{N} l_{N,k}(x)}, x \in [-1, 1], \tag{5.75}$$

where $f : [-1, 1] \to \mathbb{R}_+$ continuous.

First we see that $L_N^{(M)}(f)(x)$ is well defined and continuous for any $x \in [-1, 1]$. Following [6], p. 289, because $\sum_{k=1}^{N} l_{N,k}(x) = 1, \forall x \in \mathbb{R}$, for any x there exists $k \in \{1, ..., N\} : l_{N,k}(x) > 0$, hence $\bigvee_{k=1}^{N} l_{N,k}(x) > 0$. We have that $l_{N,k}(x_{N,k}) = 1$, and $l_{N,k}(x_{N,j}) = 0$, if $k \ne j$. Furthermore it holds $L_N^{(M)}(f)(x_{N,j}) = f(x_{N,j})$, all $j \in \{1, ..., N\}$, and $L_N^{(M)}(1) = 1$.

Call $I_N^+(x) = \{k \in \{1, ..., N\}; l_{N,k}(x) > 0\}$, then $I_N^+(x) \ne \emptyset$.

So for $f \in C_+([-1, 1])$ we get

$$L_N^{(M)}(f)(x) = \frac{\bigvee_{k \in I_N^+(x)} l_{N,k}(x) f(x_{N,k})}{\bigvee_{k \in I_N^+(x)} l_{N,k}(x)} \ge 0. \tag{5.76}$$

Notice here that $\left|x_{N,k} - x\right| \le 2, \forall x \in [-1, 1]$.

By [6], p. 297, we get that

$$L_N^{(M)}(|\cdot - x|)(x) = \frac{\bigvee_{k=1}^{N} l_{N,k}(x) |x_{N,k} - x|}{\bigvee_{k=1}^{N} l_{N,k}(x)} =$$

$$\frac{\bigvee_{k \in I_N^+(x)} l_{N,k}(x) |x_{N,k} - x|}{\bigvee_{k \in I_N^+(x)} l_{N,k}(x)} \le \frac{\pi^2}{6(N-1)}, \tag{5.77}$$

$N \ge 3, \forall x \in (-1, 1), N$ is odd.

We get that $(m > 1)$

$$L_N^{(M)}(|\cdot - x|^m)(x) = \frac{\bigvee_{k \in I_N^+(x)} l_{N,k}(x) |x_{N,k} - x|^m}{\bigvee_{k \in I_N^+(x)} l_{N,k}(x)} \le \frac{2^{m-1}\pi^2}{6(N-1)}, \tag{5.78}$$

$N \ge 3$ odd, $\forall x \in (-1, 1)$.

$L_N^{(M)}$ are positive sublinear operators, [6], p. 290.

We give

Theorem 5.21 *Same assumptions as in Theorem 5.19. Then*

$$\left|L_N^{(M)}(f)(x) - f(x)\right| \le \frac{\omega_1\left(D_x^\nu f, \left(\frac{2^\nu \pi^2}{6(N-1)}\right)^{\frac{1}{\nu+1}}\right)}{\Gamma(\nu+1)}. \tag{5.79}$$

$$\left[\frac{2^{\nu-1}\pi^2}{6(N-1)} + \frac{1}{(\nu+1)}\left(\frac{2^\nu \pi^2}{6(N-1)}\right)^{\frac{\nu}{\nu+1}}\right], \forall N \in \mathbb{N} : N \ge 3, \text{ odd}.$$

It holds $\lim_{N \to +\infty} L_N^{(M)}(f)(x) = f(x)$.

Proof By Theorem 5.6, choose $\delta := \left(\frac{2^\nu \pi^2}{6(N-1)}\right)^{\frac{1}{\nu+1}}$, use of (5.36) and (5.78). At ± 1 the left hand side of (5.79) is zero, thus (5.79) is trivially true. ∎

We make

Remark 5.22 Let $f \in C_+([-1, 1])$, $N \geq 4$, $N \in \mathbb{N}$, N even.
By [6], p. 298, we get

$$L_N^{(M)}(|\cdot - x|)(x) \leq \frac{4\pi^2}{3(N-1)} = \frac{2^2 \pi^2}{3(N-1)}, \forall x \in (-1, 1). \tag{5.80}$$

Hence $(m > 1)$

$$L_N^{(M)}(|\cdot - x|^m)(x) \leq \frac{2^{m+1} \pi^2}{3(N-1)}, \forall x \in (-1, 1). \tag{5.81}$$

We present

Theorem 5.23 *Same assumptions as in Theorem 5.19. Then*

$$\left| L_N^{(M)}(f)(x) - f(x) \right| \leq \frac{\omega_1 \left(D_x^\nu f, \left(\frac{2^{\nu+2} \pi^2}{3(N-1)}\right)^{\frac{1}{\nu+1}} \right)}{\Gamma(\nu+1)}. \tag{5.82}$$

$$\left[\frac{2^{\nu+1} \pi^2}{3(N-1)} + \frac{1}{(\nu+1)} \left(\frac{2^{\nu+2} \pi^2}{3(N-1)}\right)^{\frac{\nu}{\nu+1}} \right], \forall N \in \mathbb{N}, N \geq 4, N \text{ is even.}$$

It holds $\lim_{N \to +\infty} L_N^{(M)}(f)(x) = f(x)$.

Proof By Theorem 5.6, use of (5.36) and (5.81). Choose $\delta = \left(\frac{2^{\nu+2} \pi^2}{3(N-1)}\right)^{\frac{1}{\nu+1}}$, etc. ∎

We need

Definition 5.24 Let $x, x_0 \in \mathbb{R}$, $x \geq x_0$, $\nu > 0$, $\nu \notin \mathbb{N}$, $p = [\nu]$, $[\cdot]$ is the integral part, $\alpha = \nu - p$.
Let $f \in C_b^p(\mathbb{R})$, i.e. $f \in C^p(\mathbb{R})$ with $\|f^{(p)}\|_\infty < +\infty$, where $\|\cdot\|_\infty$ is the supremum norm.
Clearly $\left(J_\nu^{x_0} f\right)(x)$ can be defined via (5.8) over $[x_0, +\infty)$.
We define the subspace $C_{x_0+}^\nu(\mathbb{R})$ of $C_b^p(\mathbb{R})$:

$$C_{x_0+}^\nu(\mathbb{R}) := \left\{ f \in C_b^p(\mathbb{R}) : J_{1-\alpha}^{x_0} f^{(p)} \in C^1([x_0, +\infty)) \right\}.$$

For $f \in C^{\nu}_{x_0+}(\mathbb{R})$, we define the left generalized ν-fractional derivative of f over $[x_0, +\infty)$ as

$$D^{\nu}_{x_0+} f = \left(J^{x_0}_{1-\alpha} f^{(p)}\right)'. \tag{5.83}$$

When $\nu > 1$, clearly then the left generalized fractional Taylor formula ($f \in C^{\nu}_{x_0+}(\mathbb{R})$) (5.12) is valid.

We need

Definition 5.25 Let $x, x_0 \in \mathbb{R}$, $x \leq x_0$, $\nu > 0$, $\nu \notin \mathbb{N}$, $p = [\nu]$, $\alpha = \nu - p$. Let $f \in C^p_b(\mathbb{R})$. Clearly $\left(J^{\nu}_{x_0-} f\right)(x)$ can be defined via (5.13) over $(-\infty, x_0]$.
  We define the subspace of $C^{\nu}_{x_0-}(\mathbb{R})$ of $C^p_b(\mathbb{R})$:

$$C^{\nu}_{x_0-}(\mathbb{R}) := \left\{ f \in C^p_b(\mathbb{R}) : \left(J^{1-\alpha}_{x_0-} f^{(p)}\right) \in C^1\left((-\infty, x_0]\right) \right\}.$$

For $f \in C^{\nu}_{x_0-}(\mathbb{R})$, we define the right generalized ν-fractional derivative of f over $(-\infty, x_0]$ as

$$D^{\nu}_{x_0-} f = (-1)^{p-1} \left(J^{1-\alpha}_{x_0-} f^{(p)}\right)'. \tag{5.84}$$

When $\nu > 1$, clearly then the right generalized fractional Taylor formula ($f \in C^{\nu}_{x_0-}(\mathbb{R})$) (5.19) is valid.

We need

Definition 5.26 ([8], p. 41) Let $I \subset \mathbb{R}$ be an interval of finite or infinite length, and $f : I \to \mathbb{R}$ a bounded or uniformly continuous function. We define the first modulus of continuity

$$\omega_1 (f, \delta)_I = \sup_{\substack{x, y \in I \\ |x-y| \leq \delta}} |f(x) - f(y)|, \delta > 0. \tag{5.85}$$

Clearly, it holds $\omega_1 (f, \delta)_I < +\infty$.
  We also have

$$\omega_1 (f, r\delta)_I \leq (r+1)\, \omega_1 (f, \delta)_I, \text{ any } r \geq 0. \tag{5.86}$$

Convention 5.27 *Let a real number* $m > 1$, *from now on we assume that* $D^m_{x_0-} f$ *is either bounded or uniformly continuous function on* $(-\infty, x_0]$, *similarly from now on we assume that* $D^m_{x_0+} f$ *is either bounded or uniformly continuous function on* $[x_0, +\infty)$.

We need

Definition 5.28 Let $D^m_{x_0} f$ (real number $m > 1$) denote any of $D^m_{x_0-} f$, $D^m_{x_0+} f$ and $\delta > 0$. We set

$$\omega_1 \left(D^m_{x_0} f, \delta\right)_{\mathbb{R}} := \max \left\{ \omega_1 \left(D^m_{x_0-} f, \delta\right)_{(-\infty, x_0]}, \omega_1 \left(D^m_{x_0+} f, \delta\right)_{[x_0, +\infty)} \right\}, \tag{5.87}$$

where $x_0 \in \mathbb{R}$. Notice that $\omega_1 \left(D_{x_0}^m f, \delta \right)_{\mathbb{R}} < +\infty$.

We give

Theorem 5.29 *Let $m > 1$, $m \notin \mathbb{N}$, $p = [m]$, $x_0 \in \mathbb{R}$, and $f \in C_{x_0+}^m (\mathbb{R}) \cap C_{x_0-}^m (\mathbb{R})$. Assume that $f^{(k)} (x_0) = 0$, $k = 1, ..., p-1$, and $\left(D_{x_0+}^m f \right) (x_0) = \left(D_{x_0-}^m f \right) (x_0) = 0$. The Convention 5.27 is imposed. Then*

$$|f(x) - f(x_0)| \leq \frac{\omega_1 \left(D_{x_0}^m f, \delta \right)_{\mathbb{R}}}{\Gamma(m+1)} \left[|x - x_0|^m + \frac{|x - x_0|^{m+1}}{(m+1) \delta} \right], \delta > 0, \quad (5.88)$$

for all $x \in \mathbb{R}$.

Proof Similar to Theorem 5.3. ∎

Remark 5.30 Let $b : \mathbb{R} \to \mathbb{R}_+$ be a centered (it takes a global maximum at 0) bell-shaped function, with compact support $[-T, T]$, $T > 0$ (that is $b(x) > 0$ for all $x \in (-T, T)$) and $I = \int_{-T}^{T} b(x) dx > 0$.

The Cardaliaguet-Euvrard neural network operators are defined by (see [5])

$$C_{N,\alpha}(f)(x) = \sum_{k=-N^2}^{N^2} \frac{f \left(\frac{k}{n} \right)}{I N^{1-\alpha}} b \left(N^{1-\alpha} \left(x - \frac{k}{N} \right) \right), \quad (5.89)$$

$0 < \alpha < 1$, $N \in \mathbb{N}$ and typically here $f : \mathbb{R} \to \mathbb{R}$ is continuous and bounded or uniformly continuous on $\mathbb{R}$.

$CB(\mathbb{R})$ denotes the continuous and bounded function on $\mathbb{R}$, and

$$CB_+ (\mathbb{R}) = \{ f : \mathbb{R} \to [0, \infty); f \in CB(\mathbb{R}) \}.$$

The corresponding max-product Cardaliaguet-Euvrard neural network operators will be given by

$$C_{N,\alpha}^{(M)}(f)(x) = \frac{\bigvee_{k=-N^2}^{N^2} b \left(N^{1-\alpha} \left(x - \frac{k}{N} \right) \right) f \left(\frac{k}{N} \right)}{\bigvee_{k=-N^2}^{N^2} b \left(N^{1-\alpha} \left(x - \frac{k}{N} \right) \right)}, \quad (5.90)$$

$x \in \mathbb{R}$, typically here $f \in CB_+ (\mathbb{R})$, see also [5].

Next we follow [5].

For any $x \in \mathbb{R}$, denoting

$$J_{T,N}(x) = \left\{ k \in \mathbb{Z}; -N^2 \leq k \leq N^2, N^{1-\alpha} \left(x - \frac{k}{N} \right) \in (-T, T) \right\},$$

we can write

$$C_{N,\alpha}^{(M)}(f)(x) = \frac{\bigvee_{k \in J_{T,N}(x)} b \left(N^{1-\alpha} \left(x - \frac{k}{N} \right) \right) f \left(\frac{k}{N} \right)}{\bigvee_{k \in J_{T,N}(x)} b \left(N^{1-\alpha} \left(x - \frac{k}{N} \right) \right)}, \quad (5.91)$$

$x \in \mathbb{R}$, $N > \max \left\{ T + |x|, T^{-\frac{1}{\alpha}} \right\}$, where $J_{T,N}(x) \neq \emptyset$. Indeed, we have

$\bigvee_{k \in J_{T,N}(x)} b \left(N^{1-\alpha} \left(x - \frac{k}{N} \right) \right) > 0$, $\forall x \in \mathbb{R}$ and $N > \max \left\{ T + |x|, T^{-\frac{1}{\alpha}} \right\}$.

We have that $C_{N,\alpha}^{(M)} (1)(x) = 1$, $\forall x \in \mathbb{R}$ and $N > \max \left\{ T + |x|, T^{-\frac{1}{\alpha}} \right\}$.

See in [5] there: Lemma 2.1, Corollary 2.2 and Remarks.
We need

Theorem 5.31 ([5]) *Let $b(x)$ be a centered bell-shaped function, continuous and with compact support $[-T, T]$, $T > 0$, $0 < \alpha < 1$ and $C_{N,\alpha}^{(M)}$ be defined as in (5.90).*

(i) If $|f(x)| \leq c$ for all $x \in \mathbb{R}$ then $\left| C_{N,\alpha}^{(M)} (f)(x) \right| \leq c$, for all $x \in \mathbb{R}$ and $N > \max \left\{ T + |x|, T^{-\frac{1}{\alpha}} \right\}$ and $C_{N,\alpha}^{(M)} (f)(x)$ is continuous at any point $x \in \mathbb{R}$, for all $N > \max \left\{ T + |x|, T^{-\frac{1}{\alpha}} \right\}$;

(ii) If $f, g \in CB_+ (\mathbb{R})$ satisfy $f(x) \leq g(x)$ for all $x \in \mathbb{R}$, then $C_{N,\alpha}^{(M)} (f)(x) \leq C_{N,\alpha}^{(M)} (g)(x)$ for all $x \in \mathbb{R}$ and $N > \max \left\{ T + |x|, T^{-\frac{1}{\alpha}} \right\}$;

(iii) $C_{N,\alpha}^{(M)} (f + g)(x) \leq C_{N,\alpha}^{(M)} (f)(x) + C_{N,\alpha}^{(M)} (g)(x)$ for all $f, g \in CB_+ (\mathbb{R})$, $x \in \mathbb{R}$ and $N > \max \left\{ T + |x|, T^{-\frac{1}{\alpha}} \right\}$;

(iv) For all $f, g \in CB_+ (\mathbb{R})$, $x \in \mathbb{R}$ and $N > \max \left\{ T + |x|, T^{-\frac{1}{\alpha}} \right\}$, we have

$$\left| C_{N,\alpha}^{(M)} (f)(x) - C_{N,\alpha}^{(M)} (g)(x) \right| \leq C_{N,\alpha}^{(M)} (|f - g|)(x);$$

(v) $C_{N,\alpha}^{(M)}$ is positive homogeneous, that is $C_{N,\alpha}^{(M)} (\lambda f)(x) = \lambda C_{N,\alpha}^{(M)} (f)(x)$ for all $\lambda \geq 0$, $x \in \mathbb{R}$, $N > \max \left\{ T + |x|, T^{-\frac{1}{\alpha}} \right\}$ and $f \in CB_+ (\mathbb{R})$.

We make

Remark 5.32 We have that

$$E_{N,\alpha}(x) := C_{N,\alpha}^{(M)} (|\cdot - x|)(x) = \frac{\bigvee_{k \in J_{T,N}(x)} b \left(N^{1-\alpha} \left(x - \frac{k}{N} \right) \right) \left| x - \frac{k}{N} \right|}{\bigvee_{k \in J_{T,N}(x)} b \left(N^{1-\alpha} \left(x - \frac{k}{N} \right) \right)}, \quad (5.92)$$

$\forall x \in \mathbb{R}$, and $N > \max \left\{ T + |x|, T^{-\frac{1}{\alpha}} \right\}$.

We mention from [5] the following:

Theorem 5.33 ([5]) *Let $b(x)$ be a centered bell-shaped function, continuous and with compact support $[-T, T]$, $T > 0$ and $0 < \alpha < 1$. In addition, suppose that the following requirements are fulfilled:*

(i) There exist $0 < m_1 \leq M_1 < \infty$ such that $m_1 (T - x) \leq b(x) \leq M_1 (T - x)$, $\forall x \in [0, T]$;

(ii) There exist $0 < m_2 \leq M_2 < \infty$ *such that* $m_2 (x + T) \leq b (x) \leq M_2 (x + T)$, $\forall\, x \in [-T, 0]$.

Then for all $f \in CB_+ (\mathbb{R})$, $x \in \mathbb{R}$ *and for all* $N \in \mathbb{N}$ *satisying* $N > \max\left\{T + |x|, \left(\frac{2}{T}\right)^{\frac{1}{\alpha}}\right\}$, *we have the estimate*

$$\left|C_{N,\alpha}^{(M)} (f) (x) - f (x)\right| \leq c\omega_1 \left(f, N^{\alpha-1}\right)_{\mathbb{R}}, \tag{5.93}$$

where

$$c := 2 \left(\max\left\{\frac{T M_2}{2m_2}, \frac{T M_1}{2m_1}\right\} + 1\right),$$

and

$$\omega_1 (f, \delta)_{\mathbb{R}} := \sup_{\substack{x, y \in \mathbb{R}: \\ |x-y| \leq \delta}} |f (x) - f (y)|. \tag{5.94}$$

We make

Remark 5.34 In [5], was proved that

$$E_{N,\alpha} (x) \leq \max\left\{\frac{T M_2}{2m_2}, \frac{T M_1}{2m_1}\right\} N^{\alpha-1}, \forall N > \max\left\{T + |x|, \left(\frac{2}{T}\right)^{\frac{1}{\alpha}}\right\}. \tag{5.95}$$

That is

$$C_{N,\alpha}^{(M)} (|\cdot - x|) (x) \leq \max\left\{\frac{T M_2}{2m_2}, \frac{T M_1}{2m_1}\right\} N^{\alpha-1}, \forall N > \max\left\{T + |x|, \left(\frac{2}{T}\right)^{\frac{1}{\alpha}}\right\}. \tag{5.96}$$

From (5.92) we have that $\left|x - \frac{k}{N}\right| \leq \frac{T}{N^{1-\alpha}}$.

Hence $(m > 1)$ $(\forall\, x \in \mathbb{R}$ and $N > \max\left\{T + |x|, \left(\frac{2}{T}\right)^{\frac{1}{\alpha}}\right\})$

$$C_{N,\alpha}^{(M)} (|\cdot - x|^m) (x) = \frac{\bigvee_{k \in J_{T,N}(x)} b \left(N^{1-\alpha} \left(x - \frac{k}{N}\right)\right) \left|x - \frac{k}{N}\right|^m}{\bigvee_{k \in J_{T,N}(x)} b \left(N^{1-\alpha} \left(x - \frac{k}{N}\right)\right)} \leq \tag{5.97}$$

$$\left(\frac{T}{N^{1-\alpha}}\right)^{m-1} \max\left\{\frac{T M_2}{2m_2}, \frac{T M_1}{2m_1}\right\} N^{\alpha-1}, \forall N > \max\left\{T + |x|, \left(\frac{2}{T}\right)^{\frac{1}{\alpha}}\right\}.$$

Then $(m > 1)$ it holds

$$C_{N,\alpha}^{(M)} (|\cdot - x|^m) (x) \leq$$

$$T^{m-1} \max\left\{\frac{T M_2}{2m_2}, \frac{T M_1}{2m_1}\right\} \frac{1}{N^{m(1-\alpha)}}, \forall N > \max\left\{T + |x|, \left(\frac{2}{T}\right)^{\frac{1}{\alpha}}\right\}. \tag{5.98}$$

Call

$$\theta := \max\left\{\frac{TM_2}{2m_2}, \frac{TM_1}{2m_1}\right\} > 0. \tag{5.99}$$

Consequently $(m > 1)$ we derive

$$C_{N,\alpha}^{(M)}\left(|\cdot - x|^m\right)(x) \le \frac{\theta T^{m-1}}{N^{m(1-\alpha)}}, \forall N > \max\left\{T + |x|, \left(\frac{2}{T}\right)^{\frac{1}{\alpha}}\right\}. \tag{5.100}$$

We need

Theorem 5.35 *All here as in Theorem 5.29, where $x = x_0 \in \mathbb{R}$ is fixed. Let b be a centered bell-shaped function, continuous and with compact support $[-T, T]$, $T > 0, 0 < \alpha < 1$ and $C_{N,\alpha}^{(M)}$ be defined as in (5.90). Then*

$$\left|C_{N,\alpha}^{(M)}(f)(x) - f(x)\right| \le$$

$$\frac{\omega_1\left(D_x^m f, \delta\right)_{\mathbb{R}}}{\Gamma(m+1)}\left[C_{N,\alpha}^{(M)}\left(|\cdot - x|^m\right)(x) + \frac{C_{N,\alpha}^{(M)}\left(|\cdot - x|^{m+1}\right)(x)}{(m+1)\delta}\right], \tag{5.101}$$

$$\forall N \in \mathbb{N} : N > \max\left\{T + |x|, T^{-\frac{1}{\alpha}}\right\}.$$

Proof By Theorem 5.29 and (5.88) we get

$$|f(\cdot) - f(x)| \le \frac{\omega_1\left(D_x^m f, \delta\right)_{\mathbb{R}}}{\Gamma(m+1)}\left[|\cdot - x|^m + \frac{|\cdot - x|^{m+1}}{(m+1)\delta}\right], \delta > 0, \tag{5.102}$$

true over $\mathbb{R}$.

As in Theorem 5.31 and using similar reasoning and $C_{N,\alpha}^{(M)}(1) = 1$, we get

$$\left|C_{N,\alpha}^{(M)}(f)(x) - f(x)\right| \le C_{N,\alpha}^{(M)}\left(|f(\cdot) - f(x)|\right)(x) \overset{(5.102)}{\le}$$

$$\frac{\omega_1\left(D_x^m f, \delta\right)_{\mathbb{R}}}{\Gamma(m+1)}\left[C_{N,\alpha}^{(M)}\left(|\cdot - x|^m\right)(x) + \frac{C_{N,\alpha}^{(M)}\left(|\cdot - x|^{m+1}\right)(x)}{(m+1)\delta}\right], \tag{5.103}$$

$$\forall N \in \mathbb{N} : N > \max\left\{T + |x|, T^{-\frac{1}{\alpha}}\right\}. \quad \blacksquare$$

We continue with

Theorem 5.36 *Here all as in Theorem 5.29, where $x = x_0 \in \mathbb{R}$ is fixed and $m > 1$. Also the same assumptions as in Theorem 5.33. Then*

$$\left| C_{N,\alpha}^{(M)} (f) (x) - f (x) \right| \leq \frac{1}{\Gamma (m + 1)} \omega_1 \left(D_x^m f, \left(\frac{\theta T^m}{N^{(m+1)(1-\alpha)}} \right)^{\frac{1}{m+1}} \right)_{\mathbb{R}} \cdot$$

$$\left[\frac{\theta T^{m-1}}{N^{m(1-\alpha)}} + \frac{1}{(m + 1)} \left(\frac{\theta T^m}{N^{(m+1)(1-\alpha)}} \right)^{\frac{m}{m+1}} \right], \tag{5.104}$$

$\forall N \in \mathbb{N} : N > \max \left\{ T + |x|, \left(\frac{2}{T} \right)^{\frac{1}{\alpha}} \right\}$.

We have that $\lim_{N \to +\infty} C_{N,\alpha}^{(M)} (f) (x) = f (x)$.

Proof We apply Theorem 5.35. In (5.101) we choose

$$\delta := \left(\frac{\theta T^m}{N^{(m+1)(1-\alpha)}} \right)^{\frac{1}{m+1}},$$

thus $\delta^{m+1} = \frac{\theta T^m}{N^{(m+1)(1-\alpha)}}$, and

$$\delta^m = \left(\frac{\theta T^m}{N^{(m+1)(1-\alpha)}} \right)^{\frac{m}{m+1}}. \tag{5.105}$$

Therefore we have

$$\left| C_{N,\alpha}^{(M)} (f) (x) - f (x) \right| \overset{(5.100)}{\leq} \frac{1}{\Gamma (m + 1)} \omega_1 \left(D_x^m f, \left(\frac{\theta T^m}{N^{(m+1)(1-\alpha)}} \right)^{\frac{1}{m+1}} \right)_{\mathbb{R}} \cdot \tag{5.106}$$

$$\left[\frac{\theta T^{m-1}}{N^{m(1-\alpha)}} + \frac{1}{(m + 1) \delta} \frac{\theta T^m}{N^{(m+1)(1-\alpha)}} \right] =$$

$$\frac{1}{\Gamma (m + 1)} \omega_1 \left(D_x^m f, \left(\frac{\theta T^m}{N^{(m+1)(1-\alpha)}} \right)^{\frac{1}{m+1}} \right)_{\mathbb{R}} \left[\frac{\theta T^{m-1}}{N^{m(1-\alpha)}} + \frac{1}{(m + 1) \delta} \delta^{m+1} \right] \overset{(5.105)}{=}$$

$$\frac{1}{\Gamma (m + 1)} \omega_1 \left(D_x^m f, \left(\frac{\theta T^m}{N^{(m+1)(1-\alpha)}} \right)^{\frac{1}{m+1}} \right)_{\mathbb{R}} \cdot$$

$$\left[\frac{\theta T^{m-1}}{N^{m(1-\alpha)}} + \frac{1}{(m + 1)} \left(\frac{\theta T^m}{N^{(m+1)(1-\alpha)}} \right)^{\frac{m}{m+1}} \right], \tag{5.107}$$

$\forall N \in \mathbb{N} : N > \max \left\{ T + |x|, \left(\frac{2}{T} \right)^{\frac{1}{\alpha}} \right\}$, proving the inequality (5.104). ∎

We finish with (case of $\alpha = 1.5$)

Corollary 5.37 *Let $x \in [0, 1]$, $f : [0, 1] \to \mathbb{R}_+$ and $f \in C_{x+}^{1.5}([0, 1]) \cap C_{x-}^{1.5}([0, 1])$. Assume that $f'(x) = 0$, and $\left(D_{x+}^{1.5}f\right)(x) = \left(D_{x-}^{1.5}f\right)(x) = 0$. Then*

$$\left|B_N^{(M)}(f)(x) - f(x)\right| \leq \frac{4\omega_1\left(D_x^{1.5}f, \left(\frac{6}{\sqrt{N+1}}\right)^{\frac{2}{5}}\right)}{3\sqrt{\pi}}$$

$$\left[\frac{6}{\sqrt{N+1}} + \frac{2}{5}\left(\frac{6}{\sqrt{N+1}}\right)^{\frac{3}{5}}\right], \forall N \in \mathbb{N}. \tag{5.108}$$

Proof By Theorem 5.7, apply (5.37). ∎

Due to lack of space we do not give other example applications.

References

1. G. Anastassiou, *Fractional Differentiation Inequalities* (Springer, New York, 2009)
2. G. Anastassiou, *Intelligent Mathematics: Computational Analysis* (Springer, New York, 2011)
3. G. Anastassiou, *Canavati Fractional Approximation by Max-product Operators*. Progress in Fractional Differentiation and Applications (2017)
4. G. Anastassiou, *Caputo Fractional Approximation by Sublinear Operators* (2017, submitted)
5. G. Anastassiou, L. Coroianu, S. Gal, Approximation by a nonlinear Cardaliaguet-Euvrard neural network operator of max-product kind. J. Comput. Anal. Appl. **12**(2), 396–406 (2010)
6. B. Bede, L. Coroianu, S. Gal, *Approximation by Max-Product Type Operators* (Springer, New York, 2016)
7. J.A. Canavati, The Riemann-Liouville integral. Nieuw Archif Voor Wiskunde 5(1), 53–75 (1987)
8. R.A. DeVore, G.G. Lorentz, *Constructive Approximation* (Springer, Berlin, 1993)
9. L. Fejér, *Über Interpolation*. Göttingen Nachrichten (1916), pp. 66–91
10. G.G. Lorentz, *Bernstein Polynomials*, 2nd edn. (Chelsea Publishing Company, New York, 1986)
11. T. Popoviciu, Sur l'approximation de fonctions convexes d'order superieur. Mathematica (Cluj) **10**, 49–54 (1935)

Chapter 6
Iterated Fractional Approximations Using Max-Product Operators

Here we consider the approximation of functions by sublinear positive operators with applications to a large variety of Max-Product operators under iterated fractional differentiability. Our approach is based on our general fractional results about positive sublinear operators. We produce Jackson type inequalities under iterated fractional initial conditions. So our way is quantitative by producing inequalities with their right hand sides involving the modulus of continuity of iterated fractional derivative of the function under approximation. It follows [4].

6.1 Introduction

The inspiring motivation here is the monograph by B. Bede, L. Coroianu and S. Gal [7], 2016.

Let $N \in \mathbb{N}$, the well-known Bernstein polynomials [10] are positive linear operators, defined by the formula

$$B_N(f)(x) = \sum_{k=0}^{N} \binom{N}{k} x^k (1-x)^{N-k} f\left(\frac{k}{N}\right), \quad x \in [0, 1], \ f \in C([0, 1]).$$

(6.1)

T. Popoviciu in [12], 1935, proved for $f \in C([0, 1])$ that

$$|B_N(f)(x) - f(x)| \leq \frac{5}{4} \omega_1\left(f, \frac{1}{\sqrt{N}}\right), \quad \forall \, x \in [0, 1],$$

(6.2)

© Springer International Publishing AG, part of Springer Nature 2018
G. A. Anastassiou, *Nonlinearity: Ordinary and Fractional Approximations by Sublinear and Max-Product Operators*, Studies in Systems, Decision and Control 147, https://doi.org/10.1007/978-3-319-89509-3_6

where

$$\omega_1 (f, \delta) = \sup_{\substack{x,y \in [a,b]: \\ |x-y| \le \delta}} |f (x) - f (y)|, \ \delta > 0, \tag{6.3}$$

is the first modulus of continuity, here $[a, b] = [0, 1]$.

G.G. Lorentz in [10], 1986, p. 21, proved for $f \in C^1 ([0, 1])$ that

$$|B_N (f) (x) - f (x)| \le \frac{3}{4\sqrt{N}} \omega_1 \left(f', \frac{1}{\sqrt{N}} \right), \ \forall \, x \in [0, 1], \tag{6.4}$$

In [7], p. 10, the authors introduced the basic Max-product Bernstein operators,

$$B_N^{(M)} (f) (x) = \frac{\bigvee_{k=0}^{N} p_{N,k} (x) f \left(\frac{k}{N} \right)}{\bigvee_{k=0}^{N} p_{N,k} (x)}, \ N \in \mathbb{N}, \tag{6.5}$$

where $\bigvee$ stands for maximum, and $p_{N,k} (x) = \binom{N}{k} x^k (1 - x)^{N-k}$ and $f : [0, 1] \to \mathbb{R}_+ = [0, \infty)$.

These are nonlinear and piecewise rational operators.

The authors in [7] studied similar such nonlinear operators such as: the Max-product Favard–Szász–Mirakjan operators and their truncated version, the Max-product Baskakov operators and their truncated version, also many other similar specific operators. The study in [7] is based on presented there general theory of sublinear operators. These Max-product operators tend to converge faster to the on hand function.

So we mention from [7], p. 30, that for $f : [0, 1] \to \mathbb{R}_+$ continuous, we have the estimate

$$\left| B_N^{(M)} (f) (x) - f (x) \right| \le 12\omega_1 \left(f, \frac{1}{\sqrt{N + 1}} \right), \ \text{for all } N \in \mathbb{N}, \ x \in [0, 1]. \tag{6.6}$$

In this chapter we expand the study of [7] by considering iterated fractional smoothness of functions. So our inequalities are with respect to $\omega_1 \left(D^{(n+1)\alpha} f, \delta \right)$, $\delta > 0$, where $D^{(n+1)\alpha} f$ with $\alpha > 0$, $n \in \mathbb{N}$, is the iterated fractional derivative.

6.2   Main Results

We make

Remark 6.1 Let $f : [a, b] \to \mathbb{R}$ such that $f' \in L_\infty ([a, b])$, $x_0 \in [a, b], 0 < \alpha < 1$, the left Caputo fractional derivative of order α is defined as follows

$$\left(D^{\alpha}_{*x_0} f\right)(x) = \frac{1}{\Gamma(1-\alpha)} \int_{x_0}^{x} (x-t)^{-\alpha} f'(t) \, dt, \tag{6.7}$$

where Γ is the gamma function for all $x_0 \leq x \leq b$.

We observe that

$$\left|\left(D^{\alpha}_{*x_0} f\right)(x)\right| \leq \frac{1}{\Gamma(1-\alpha)} \int_{x_0}^{x} (x-t)^{-\alpha} \left|f'(t)\right| dt$$

$$\leq \frac{\|f'\|_{\infty}}{\Gamma(1-\alpha)} \int_{x_0}^{x} (x-t)^{-\alpha} dt = \frac{\|f'\|_{\infty}}{\Gamma(1-\alpha)} \frac{(x-x_0)^{1-\alpha}}{(1-\alpha)} = \frac{\|f'\|_{\infty} (x-x_0)^{1-\alpha}}{\Gamma(2-\alpha)}. \tag{6.8}$$

I.e.

$$\left|\left(D^{\alpha}_{*x_0} f\right)(x)\right| \leq \frac{\|f'\|_{\infty} (x-x_0)^{1-\alpha}}{\Gamma(2-\alpha)} \leq \frac{\|f'\|_{\infty} (b-x_0)^{1-\alpha}}{\Gamma(2-\alpha)} < +\infty, \tag{6.9}$$

$\forall x \in [x_0, b]$.

Clearly, then

$$\left(D^{\alpha}_{*x_0} f\right)(x_0) = 0. \tag{6.10}$$

We define $\left(D^{\alpha}_{*x_0} f\right)(x) = 0$, for $a \leq x < x_0$.

Let $n \in \mathbb{N}$, we denote the iterated fractional derivative $D^{n\alpha}_{*x_0} = D^{\alpha}_{*x_0} D^{\alpha}_{*x_0} \dots D^{\alpha}_{*x_0}$ (n-times).

Let us assume that

$$D^{k\alpha}_{*x_0} f \in C\left([x_0, b]\right), \, k = 0, 1, \dots, n+1; \, n \in \mathbb{N}, 0 < \alpha < 1.$$

By [5, 11], pp. 156–158, we have the following generalized fractional Caputo type Taylor's formula:

$$f(x) = \sum_{i=0}^{n} \frac{(x-x_0)^{i\alpha}}{\Gamma(i\alpha+1)} \left(D^{i\alpha}_{*x_0} f\right)(x_0) + \tag{6.11}$$

$$\frac{1}{\Gamma((n+1)\alpha)} \int_{x_0}^{x} (x-t)^{(n+1)\alpha-1} \left(D^{(n+1)\alpha}_{*x_0} f\right)(t) \, dt,$$

$\forall x \in [x_0, b]$.

Based on the above (6.10) and (6.11), we derive

$$f(x) - f(x_0) = \sum_{i=2}^{n} \frac{(x-x_0)^{i\alpha}}{\Gamma(i\alpha+1)} \left(D^{i\alpha}_{*x_0} f\right)(x_0) + \tag{6.12}$$

$$\frac{1}{\Gamma((n+1)\alpha)} \int_{x_0}^{x} (x-t)^{(n+1)\alpha-1} \left(D^{(n+1)\alpha}_{*x_0} f\right)(t) \, dt,$$

$\forall\, x \in [x_0, b]$, $0 < \alpha < 1$.

In case of $\left(D_{*x_0}^{i\alpha} f\right)(x_0) = 0$, $i = 2, 3, \ldots, n+1$, we get

$$f(x) - f(x_0) =$$

$$\frac{1}{\Gamma\left((n+1)\alpha\right)} \int_{x_0}^{x} (x-t)^{(n+1)\alpha-1} \left(\left(D_{*x_0}^{(n+1)\alpha} f\right)(t) - \left(D_{*x_0}^{(n+1)\alpha} f\right)(x_0)\right) dt, \quad (6.13)$$

$\forall\, x \in [x_0, b]$, $0 < \alpha < 1$.

We make

Remark 6.2 Let $f : [a, b] \to \mathbb{R}$ such that $f' \in L_\infty([a, b])$, $x_0 \in [a, b]$, $0 < \alpha < 1$, the right Caputo fractional derivative of order α is defined as follows

$$\left(D_{x_0-}^{\alpha} f\right)(x) = \frac{-1}{\Gamma(1-\alpha)} \int_{x}^{x_0} (z-x)^{-\alpha} f'(z)\, dz, \quad (6.14)$$

$\forall\, x \in [a, x_0]$.

We observe that

$$\left|\left(D_{x_0-}^{\alpha} f\right)(x)\right| \leq \frac{1}{\Gamma(1-\alpha)} \int_{x}^{x_0} (z-x)^{-\alpha} \left|f'(z)\right| dz \leq$$

$$\frac{\|f'\|_\infty}{\Gamma(1-\alpha)} \left(\int_{x}^{x_0} (z-x)^{-\alpha}\, dz\right) = \frac{\|f'\|_\infty}{\Gamma(1-\alpha)} \frac{(x_0-x)^{1-\alpha}}{(1-\alpha)} = \frac{\|f'\|_\infty}{\Gamma(2-\alpha)} (x_0-x)^{1-\alpha}. \quad (6.15)$$

That is

$$\left|\left(D_{x_0-}^{\alpha} f\right)(x)\right| \leq \frac{\|f'\|_\infty}{\Gamma(2-\alpha)} (x_0-x)^{1-\alpha} \leq \frac{\|f'\|_\infty}{\Gamma(2-\alpha)} (x_0-a)^{1-\alpha} < \infty, \quad (6.16)$$

$\forall\, x \in [a, x_0]$.

In particular we have

$$\left(D_{x_0-}^{\alpha} f\right)(x_0) = 0. \quad (6.17)$$

We define $\left(D_{x_0-}^{\alpha} f\right)(x) = 0$, for $x_0 < x \leq b$.

For $n \in \mathbb{N}$, denote the iterated fractional derivative $D_{x_0-}^{n\alpha} = D_{x_0-}^{\alpha} D_{x_0-}^{\alpha} \ldots D_{x_0-}^{\alpha}$ (n-times).

In [1], we proved the following right generalized fractional Taylor's formula: Suppose that

$$D_{x_0-}^{k\alpha} f \in C([a, x_0]), \quad \text{for } k = 0, 1, \ldots, n+1, 0 < \alpha < 1.$$

Then

$$f(x) = \sum_{i=0}^{n} \frac{(x_0-x)^{i\alpha}}{\Gamma(i\alpha+1)} \left(D_{x_0-}^{i\alpha} f\right)(x_0) + \quad (6.18)$$

$$\frac{1}{\Gamma((n+1)\alpha)} \int_x^{x_0} (z-x)^{(n+1)\alpha-1} \left(D_{x_0-}^{(n+1)\alpha} f\right)(z)\, dz,$$

$\forall\, x \in [a, x_0]$.

Based on (6.17) and (6.18), we derive

$$f(x) - f(x_0) = \sum_{i=2}^{n} \frac{(x_0 - x)^{i\alpha}}{\Gamma(i\alpha+1)} \left(D_{x_0-}^{i\alpha} f\right)(x_0) + \tag{6.19}$$

$$\frac{1}{\Gamma((n+1)\alpha)} \int_x^{x_0} (z-x)^{(n+1)\alpha-1} \left(D_{x_0-}^{(n+1)\alpha} f\right)(z)\, dz,$$

$\forall\, x \in [a, x_0],\, 0 < \alpha < 1$.

In case of $\left(D_{x_0-}^{i\alpha} f\right)(x_0) = 0$, for $i = 2, 3, ..., n+1$, we get

$$f(x) - f(x_0) =$$

$$\frac{1}{\Gamma((n+1)\alpha)} \int_x^{x_0} (z-x)^{(n+1)\alpha-1} \left(\left(D_{x_0-}^{(n+1)\alpha} f\right)(z) - \left(D_{x_0-}^{(n+1)\alpha} f\right)(x_0)\right) dz,$$
$$\tag{6.20}$$

$\forall\, x \in [a, x_0],\, 0 < \alpha < 1$.

We need

Definition 6.3 Let $D_{x_0}^{(n+1)\alpha} f$ denote any of $D_{*x_0}^{(n+1)\alpha} f$, $D_{x_0-}^{(n+1)\alpha} f$, and $\delta > 0$. We set

$$\omega_1\left(D_{x_0}^{(n+1)\alpha} f, \delta\right) = \max\left\{\omega_1\left(D_{*x_0}^{(n+1)\alpha} f, \delta\right)_{[x_0, b]}, \omega_1\left(D_{x_0-}^{(n+1)\alpha} f, \delta\right)_{[a, x_0]}\right\}, \tag{6.21}$$

where $x_0 \in [a, b]$. Here the moduli of continuity are considered over $[x_0, b]$ and $[a, x_0]$, respectively.

We present

Theorem 6.4 Let $0 < \alpha < 1$, $f : [a, b] \to \mathbb{R}$, $f' \in L_\infty([a, b])$, $x_0 \in [a, b]$. Assume that $D_{*x_0}^{k\alpha} f \in C([x_0, b])$, $k = 0, 1, ..., n+1$; $n \in \mathbb{N}$, and $\left(D_{*x_0}^{i\alpha} f\right)(x_0) = 0$, $i = 2, 3, ..., n+1$. Also, suppose that $D_{x_0-}^{k\alpha} f \in C([a, x_0])$, for $k = 0, 1, ..., n+1$, and $\left(D_{x_0-}^{i\alpha} f\right)(x_0) = 0$, for $i = 2, 3, ..., n+1$. Then

$$|f(x) - f(x_0)| \leq \frac{\omega_1\left(D_{x_0}^{(n+1)\alpha} f, \delta\right)}{\Gamma((n+1)\alpha+1)} \left[|x - x_0|^{(n+1)\alpha} + \frac{|x - x_0|^{(n+1)\alpha+1}}{\delta((n+1)\alpha+1)}\right], \tag{6.22}$$

$\forall\, x \in [a, b],\, \delta > 0$.

Proof By (6.13) we have

$$|f(x) - f(x_0)| \leq$$

$$\frac{1}{\Gamma((n+1)\alpha)} \int_{x_0}^{x} (x-t)^{(n+1)\alpha-1} \left| \left(D_{*x_0}^{(n+1)\alpha} f\right)(t) - \left(D_{*x_0}^{(n+1)\alpha} f\right)(x_0) \right| dt$$

$(\delta > 0)$

$$\leq \frac{1}{\Gamma((n+1)\alpha)} \int_{x_0}^{x} (x-t)^{(n+1)\alpha-1} \omega_1 \left(D_{*x_0}^{(n+1)\alpha} f, \frac{\delta(t-x_0)}{\delta}\right)_{[x_0,b]} dt$$

$$\leq \frac{\omega_1\left(D_{*x_0}^{(n+1)\alpha} f, \delta\right)_{[x_0,b]}}{\Gamma((n+1)\alpha)} \int_{x_0}^{x} (x-t)^{(n+1)\alpha-1} \left(1 + \frac{(t-x_0)}{\delta}\right) dt = \qquad (6.23)$$

$$\frac{\omega_1\left(D_{*x_0}^{(n+1)\alpha} f, \delta\right)_{[x_0,b]}}{\Gamma((n+1)\alpha)} \left[\frac{(x-x_0)^{(n+1)\alpha}}{(n+1)\alpha} + \frac{1}{\delta} \int_{x_0}^{x} (x-t)^{(n+1)\alpha-1} (t-x_0)^{2-1} dt\right] =$$

$$\frac{\omega_1\left(D_{*x_0}^{(n+1)\alpha} f, \delta\right)_{[x_0,b]}}{\Gamma((n+1)\alpha)} \left[\frac{(x-x_0)^{(n+1)\alpha}}{(n+1)\alpha} + \frac{1}{\delta} \frac{\Gamma((n+1)\alpha)\Gamma(2)}{\Gamma((n+1)\alpha+2)} (x-x_0)^{(n+1)\alpha+1}\right] =$$

$$\frac{\omega_1\left(D_{*x_0}^{(n+1)\alpha} f, \delta\right)_{[x_0,b]}}{\Gamma((n+1)\alpha)} \left[\frac{(x-x_0)^{(n+1)\alpha}}{(n+1)\alpha} + \frac{(x-x_0)^{(n+1)\alpha+1}}{\delta(n+1)\alpha((n+1)\alpha+1)}\right]. \qquad (6.24)$$

We have proved

$$|f(x) - f(x_0)| \leq \frac{\omega_1\left(D_{*x_0}^{(n+1)\alpha} f, \delta\right)_{[x_0,b]}}{\Gamma((n+1)\alpha+1)} \left[(x-x_0)^{(n+1)\alpha} + \frac{(x-x_0)^{(n+1)\alpha+1}}{\delta((n+1)\alpha+1)}\right], \qquad (6.25)$$

$\forall \, x \in [x_0, b], \delta > 0$.
By (6.20) we get
$$|f(x) - f(x_0)| \leq$$

$$\frac{1}{\Gamma((n+1)\alpha)} \int_{x}^{x_0} (z-x)^{(n+1)\alpha-1} \left| \left(D_{x_0-}^{(n+1)\alpha} f\right)(z) - \left(D_{x_0-}^{(n+1)\alpha} f\right)(x_0) \right| dz$$

$$\leq \frac{1}{\Gamma((n+1)\alpha)} \int_{x}^{x_0} (z-x)^{(n+1)\alpha-1} \omega_1 \left(D_{x_0-}^{(n+1)\alpha} f, \frac{\delta(x_0-z)}{\delta}\right)_{[a,x_0]} dz$$

$$\leq \frac{\omega_1\left(D_{x_0-}^{(n+1)\alpha} f, \delta\right)_{[a,x_0]}}{\Gamma((n+1)\alpha)} \left[\int_{x}^{x_0} (z-x)^{(n+1)\alpha-1} \left(1 + \frac{x_0-z}{\delta}\right) dz\right] = \qquad (6.26)$$

$$\frac{\omega_1 \left(D_{x_0-}^{(n+1)\alpha} f, \delta\right)_{[a,x_0]}}{\Gamma\left((n+1)\alpha\right)} \left[\frac{(x_0 - x)^{(n+1)\alpha}}{(n+1)\alpha} + \frac{1}{\delta}\int_x^{x_0} (x_0 - z)^{2-1} (z - x)^{(n+1)\alpha-1}\, dz\right] =$$

$$\frac{\omega_1 \left(D_{x_0-}^{(n+1)\alpha} f, \delta\right)_{[a,x_0]}}{\Gamma\left((n+1)\alpha\right)} \left[\frac{(x_0 - x)^{(n+1)\alpha}}{(n+1)\alpha} + \frac{1}{\delta}\frac{\Gamma(2)\,\Gamma\left((n+1)\alpha\right)}{\Gamma\left((n+1)\alpha+2\right)} (x_0 - x)^{(n+1)\alpha+1}\right] =$$

$$\frac{\omega_1 \left(D_{x_0-}^{(n+1)\alpha} f, \delta\right)_{[a,x_0]}}{\Gamma\left((n+1)\alpha\right)} \left[\frac{(x_0 - x)^{(n+1)\alpha}}{(n+1)\alpha} + \frac{(x_0 - x)^{(n+1)\alpha+1}}{\delta(n+1)\alpha\left((n+1)\alpha+1\right)}\right]. \qquad (6.27)$$

We have proved

$$|f(x) - f(x_0)| \le \frac{\omega_1 \left(D_{x_0-}^{(n+1)\alpha} f, \delta\right)_{[a,x_0]}}{\Gamma\left((n+1)\alpha+1\right)} \left[(x_0 - x)^{(n+1)\alpha} + \frac{(x_0 - x)^{(n+1)\alpha+1}}{\delta\left((n+1)\alpha+1\right)}\right], \qquad (6.28)$$

$\forall\, x \in [a, x_0], \delta > 0.$

By (6.25) and (6.28) we derive (6.22). ∎

We need

Definition 6.5 Here $C_+\left([a,b]\right) := \{f : [a,b] \to \mathbb{R}_+, \text{continuous functions}\}$. Let $L_N : C_+\left([a,b]\right) \to C_+\left([a,b]\right)$, operators, $\forall\, N \in \mathbb{N}$, such that
(i)

$$L_N\left(\alpha f\right) = \alpha L_N\left(f\right), \quad \forall\, \alpha \ge 0, \forall\, f \in C_+\left([a,b]\right), \qquad (6.29)$$

(ii) if $f, g \in C_+\left([a,b]\right) : f \le g$, then

$$L_N\left(f\right) \le L_N\left(g\right), \quad \forall\, N \in \mathbb{N}, \qquad (6.30)$$

(iii)

$$L_N\left(f + g\right) \le L_N\left(f\right) + L_N\left(g\right), \quad \forall\, f, g \in C_+\left([a,b]\right). \qquad (6.31)$$

We call $\{L_N\}_{N\in\mathbb{N}}$ positive sublinear operators.

We make

Remark 6.6 By [7], p. 17, we get: let $f, g \in C_+\left([a,b]\right)$, then

$$|L_N\left(f\right)(x) - L_N\left(g\right)(x)| \le L_N\left(|f - g|\right)(x), \quad \forall\, x \in [a,b]. \qquad (6.32)$$

Furthermore, we also have that

$$|L_N\left(f\right)(x) - f(x)| \le L_N\left(|f(\cdot) - f(x)|\right)(x) + |f(x)|\,|L_N\left(e_0\right)(x) - 1|, \qquad (6.33)$$

$\forall x \in [a, b]$; $e_0(t) = 1$.

From now on we assume that $L_N(1) = 1$. Hence it holds

$$|L_N(f)(x) - f(x)| \leq L_N(|f(\cdot) - f(x)|)(x), \ \forall x \in [a, b]. \tag{6.34}$$

In the assumption of Theorem 6.4 and by (6.22) and (6.34) we obtain

$$|L_N(f)(x_0) - f(x_0)| \leq \frac{\omega_1\left(D_{x_0}^{(n+1)\alpha} f, \delta\right)}{\Gamma((n+1)\alpha + 1)} \cdot \tag{6.35}$$

$$\left[L_N\left(|\cdot - x_0|^{(n+1)\alpha}\right)(x_0) + \frac{L_N\left(|\cdot - x_0|^{(n+1)\alpha+1}\right)(x_0)}{((n+1)\alpha + 1)\delta}\right], \ \delta > 0.$$

We have proved

Theorem 6.7 *Let* $\frac{1}{n+1} < \alpha < 1$, $n \in \mathbb{N}$, $f : [a, b] \to \mathbb{R}_+$, $f' \in L_\infty([a, b])$, $x_0 \in [a, b]$. *Assume that* $D_{*x_0}^{k\alpha} f \in C([x_0, b])$, $k = 0, 1, ..., n+1$, *and* $\left(D_{*x_0}^{i\alpha} f\right)(x_0) = 0$, $i = 2, 3, ..., n+1$. *Also, suppose that* $D_{x_0-}^{k\alpha} f \in C([a, x_0])$, *for* $k = 0, 1, ..., n+1$, *and* $\left(D_{x_0-}^{i\alpha} f\right)(x_0) = 0$, *for* $i = 2, 3, ..., n+1$. *Denote* $\lambda = (n+1)\alpha > 1$. *Let* $L_N : C_+([a, b]) \to C_+([a, b])$, $\forall N \in \mathbb{N}$, *be positive sublinear operators, such that* $L_N(1) = 1$, $\forall N \in \mathbb{N}$. *Then*

$$|L_N(f)(x_0) - f(x_0)| \leq \frac{\omega_1\left(D_{x_0}^{(n+1)\alpha} f, \delta\right)}{\Gamma(\lambda + 1)} \cdot$$

$$\left[L_N\left(|\cdot - x_0|^\lambda\right)(x_0) + \frac{L_N\left(|\cdot - x_0|^{\lambda+1}\right)(x_0)}{(\lambda + 1)\delta}\right], \tag{6.36}$$

$\delta > 0$, $\forall N \in \mathbb{N}$.

Note: Theorem 6.7 is also true when $0 < \alpha \leq \frac{1}{n+1}$.

6.3 Applications, Part A

Case of $(n+1)\alpha > 1$.

We give

Theorem 6.8 *Let* $\frac{1}{n+1} < \alpha < 1$, $n \in \mathbb{N}$, $f : [0, 1] \to \mathbb{R}_+$, $f' \in L_\infty([0, 1])$, $x \in [0, 1]$. *Assume that* $D_{*x}^{k\alpha} f \in C([x, 1])$, $k = 0, 1, ..., n+1$, *and* $\left(D_{*x}^{i\alpha} f\right)(x) = 0$, $i = 2, 3, ..., n+1$. *Also, suppose that* $D_{x-}^{k\alpha} f \in C([0, x])$, *for* $k = 0, 1, ..., n+1$, *and* $\left(D_{x-}^{i\alpha} f\right)(x) = 0$, *for* $i = 2, 3, ..., n+1$. *Denote* $\lambda := (n+1)\alpha > 1$. *Then*

$$\left| B_N^{(M)}(f)(x) - f(x) \right| \le \frac{\omega_1 \left(D_x^{(n+1)\alpha} f, \left(\frac{6}{\sqrt{N+1}} \right)^{\frac{1}{\lambda+1}} \right)}{\Gamma(\lambda+1)} \cdot$$

$$\left[\frac{6}{\sqrt{N+1}} + \frac{1}{(\lambda+1)} \left(\frac{6}{\sqrt{N+1}} \right)^{\frac{\lambda}{\lambda+1}} \right], \tag{6.37}$$

$\forall N \in \mathbb{N}$.

We get $\lim\limits_{N \to +\infty} B_N^{(M)}(f)(x) = f(x)$.

Proof By [3] we get that

$$B_N^{(M)}\left(|\cdot - x|^\lambda \right)(x) \le \frac{6}{\sqrt{N+1}}, \quad \forall x \in [0,1], \tag{6.38}$$

$\forall N \in \mathbb{N}, \forall \lambda > 1$.

Also $B_N^{(M)}$ maps $C_+([0,1])$ into itself, $B_N^{(M)}(1) = 1$, and it is positive sublinear operator.

We apply Theorem 6.7 and (6.36), we get

$$\left| B_N^{(M)}(f)(x) - f(x) \right| \le \frac{\omega_1 \left(D_x^{(n+1)\alpha} f, \delta \right)}{\Gamma(\lambda+1)} \left[\frac{6}{\sqrt{N+1}} + \frac{\frac{6}{\sqrt{N+1}}}{(\lambda+1)\delta} \right]. \tag{6.39}$$

Choose $\delta = \left(\frac{6}{\sqrt{N+1}} \right)^{\frac{1}{\lambda+1}}$, then $\delta^{\lambda+1} = \frac{6}{\sqrt{N+1}}$, and apply it to (6.39). Clearly we derive (6.37). ∎

We continue with

Remark 6.9 The truncated Favard–Szász–Mirakjan operators are given by

$$T_N^{(M)}(f)(x) = \frac{\bigvee_{k=0}^{N} s_{N,k}(x) f\left(\frac{k}{N}\right)}{\bigvee_{k=0}^{N} s_{N,k}(x)}, \quad x \in [0,1], \ N \in \mathbb{N}, \ f \in C_+([0,1]), \tag{6.40}$$

$s_{N,k}(x) = \frac{(Nx)^k}{k!}$, see also [7], p. 11.

By [7], pp. 178–179, we get that

$$T_N^{(M)}(|\cdot - x|)(x) \le \frac{3}{\sqrt{N}}, \quad \forall x \in [0,1], \ \forall N \in \mathbb{N}. \tag{6.41}$$

Clearly it holds

$$T_N^{(M)}\left(|\cdot - x|^{1+\beta} \right)(x) \le \frac{3}{\sqrt{N}}, \quad \forall x \in [0,1], \ \forall N \in \mathbb{N}, \forall \beta > 0. \tag{6.42}$$

The operators $T_N^{(M)}$ are positive sublinear operators mapping $C_+([0, 1])$ into itself, with $T_N^{(M)}(1) = 1$.

We continue with

Theorem 6.10 *Same assumptions as in Theorem 6.8. Then*

$$\left| T_N^{(M)}(f)(x) - f(x) \right| \leq \frac{\omega_1 \left(D_x^{(n+1)\alpha} f, \left(\frac{3}{\sqrt{N}} \right)^{\frac{1}{\lambda+1}} \right)}{\Gamma(\lambda+1)}.$$

$$\left[\frac{3}{\sqrt{N}} + \frac{1}{(\lambda+1)} \left(\frac{3}{\sqrt{N}} \right)^{\frac{\lambda}{\lambda+1}} \right], \ \forall \, N \in \mathbb{N}. \tag{6.43}$$

We get $\lim\limits_{N \to +\infty} T_N^{(M)}(f)(x) = f(x).$

Proof Use of Theorem 6.7, similar to the proof of Theorem 6.8. ∎

We make

Remark 6.11 Next we study the truncated Max-product Baskakov operators (see [7], p. 11)

$$U_N^{(M)}(f)(x) = \frac{\bigvee_{k=0}^{N} b_{N,k}(x) \, f\left(\frac{k}{N}\right)}{\bigvee_{k=0}^{N} b_{N,k}(x)}, \ x \in [0, 1], \ f \in C_+([0, 1]), \ N \in \mathbb{N}, \tag{6.44}$$

where

$$b_{N,k}(x) = \binom{N+k-1}{k} \frac{x^k}{(1+x)^{N+k}}. \tag{6.45}$$

From [7], pp. 217–218, we get ($x \in [0, 1]$)

$$\left(U_N^{(M)}(|\cdot - x|) \right)(x) \leq \frac{2\sqrt{3}\left(\sqrt{2}+2\right)}{\sqrt{N+1}}, \ N \geq 2, N \in \mathbb{N}. \tag{6.46}$$

Let $\lambda \geq 1$, clearly then it holds

$$\left(U_N^{(M)}(|\cdot - x|^\lambda) \right)(x) \leq \frac{2\sqrt{3}\left(\sqrt{2}+2\right)}{\sqrt{N+1}}, \ \forall \, N \geq 2, N \in \mathbb{N}. \tag{6.47}$$

Also it holds $U_N^{(M)}(1) = 1$, and $U_N^{(M)}$ are positive sublinear operators from $C_+([0, 1])$ into itself.

We give

Theorem 6.12 *Same assumptions as in Theorem 6.8. Then*

$$\left| U_N^{(M)}(f)(x) - f(x) \right| \leq \frac{\omega_1 \left(D_x^{(n+1)\alpha} f, \left(\frac{2\sqrt{3}(\sqrt{2}+2)}{\sqrt{N+1}} \right)^{\frac{1}{\lambda+1}} \right)}{\Gamma(\lambda+1)} \cdot \tag{6.48}$$

$$\left[\frac{2\sqrt{3}\left(\sqrt{2}+2\right)}{\sqrt{N+1}} + \frac{1}{(\lambda+1)} \left(\frac{2\sqrt{3}\left(\sqrt{2}+2\right)}{\sqrt{N+1}} \right)^{\frac{\lambda}{\lambda+1}} \right], \forall N \geq 2, N \in \mathbb{N}.$$

We get $\lim_{N \to +\infty} U_N^{(M)}(f)(x) = f(x).$

Proof Use of Theorem 6.7, similar to the proof of Theorem 6.8. ∎

We continue with

Remark 6.13 Here we study the Max-product Meyer–Köning and Zeller operators (see [7], p. 11) defined by

$$Z_N^{(M)}(f)(x) = \frac{\bigvee_{k=0}^{\infty} s_{N,k}(x) f\left(\frac{k}{N+k}\right)}{\bigvee_{k=0}^{\infty} s_{N,k}(x)}, \forall N \in \mathbb{N}, f \in C_+([0,1]), \tag{6.49}$$

$$s_{N,k}(x) = \binom{N+k}{k} x^k, x \in [0,1].$$

By [7], p. 253, we get that

$$Z_N^{(M)}(|\cdot - x|)(x) \leq \frac{8\left(1+\sqrt{5}\right)}{3} \frac{\sqrt{x}(1-x)}{\sqrt{N}}, \forall x \in [0,1], \forall N \geq 4, N \in \mathbb{N}. \tag{6.50}$$

We have that (for $\lambda \geq 1$)

$$Z_N^{(M)}\left(|\cdot - x|^{\lambda}\right)(x) \leq \frac{8\left(1+\sqrt{5}\right)}{3} \frac{\sqrt{x}(1-x)}{\sqrt{N}} := \rho(x), \tag{6.51}$$

$\forall x \in [0,1], N \geq 4, N \in \mathbb{N}.$

Also it holds $Z_N^{(M)}(1) = 1$, and $Z_N^{(M)}$ are positive sublinear operators from $C_+([0,1])$ into itself.

We give

Theorem 6.14 *Same assumptions as in Theorem 6.8. Then*

$$\left| Z_N^{(M)}(f)(x) - f(x) \right| \leq \frac{\omega_1 \left(D_x^{(n+1)\alpha} f, (\rho(x))^{\frac{1}{\lambda+1}} \right)}{\Gamma(\lambda+1)}. \tag{6.52}$$

$$\left[\rho(x) + \frac{1}{(\lambda+1)} (\rho(x))^{\frac{\lambda}{\lambda+1}} \right], \forall N \in \mathbb{N}, \ N \geq 4.$$

We get $\lim_{N \to +\infty} Z_N^{(M)}(f)(x) = f(x)$, *where* $\rho(x)$ *is as in (6.51)*.

Proof Use of Theorem 6.7, similar to the proof of Theorem 6.8. ∎

We continue with

Remark 6.15 Here we deal with the Max-product truncated sampling operators (see [7], p. 13) defined by

$$W_N^{(M)}(f)(x) = \frac{\bigvee_{k=0}^{N} \frac{\sin(Nx-k\pi)}{Nx-k\pi} f\left(\frac{k\pi}{N}\right)}{\bigvee_{k=0}^{N} \frac{\sin(Nx-k\pi)}{Nx-k\pi}}, \tag{6.53}$$

and

$$K_N^{(M)}(f)(x) = \frac{\bigvee_{k=0}^{N} \frac{\sin^2(Nx-k\pi)}{(Nx-k\pi)^2} f\left(\frac{k\pi}{N}\right)}{\bigvee_{k=0}^{N} \frac{\sin^2(Nx-k\pi)}{(Nx-k\pi)^2}}, \tag{6.54}$$

$\forall x \in [0, \pi]$, $f : [0, \pi] \to \mathbb{R}_+$ a continuous function.

Following [7], p. 343, and making the convention $\frac{\sin(0)}{0} = 1$ and denoting $s_{N,k}(x) = \frac{\sin(Nx-k\pi)}{Nx-k\pi}$, we get that $s_{N,k}\left(\frac{k\pi}{N}\right) = 1$, and $s_{N,k}\left(\frac{j\pi}{N}\right) = 0$, if $k \neq j$, furthermore $W_N^{(M)}(f)\left(\frac{j\pi}{N}\right) = f\left(\frac{j\pi}{N}\right)$, for all $j \in \{0, ..., N\}$.

Clearly $W_N^{(M)}(f)$ is a well-defined function for all $x \in [0, \pi]$, and it is continuous on $[0, \pi]$, also $W_N^{(M)}(1) = 1$.

By [7], p. 344, $W_N^{(M)}$ are positive sublinear operators.

Call $I_N^+(x) = \{k \in \{0, 1, ..., N\} ; s_{N,k}(x) > 0\}$, and set $x_{N,k} := \frac{k\pi}{N}$, $k \in \{0, 1, ..., N\}$.

We see that

$$W_N^{(M)}(f)(x) = \frac{\bigvee_{k \in I_N^+(x)} s_{N,k}(x) f(x_{N,k})}{\bigvee_{k \in I_N^+(x)} s_{N,k}(x)}. \tag{6.55}$$

By [7], p. 346, we have

$$W_N^{(M)}(|\cdot - x|)(x) \leq \frac{\pi}{2N}, \ \forall N \in \mathbb{N}, \ \forall x \in [0, \pi]. \tag{6.56}$$

Notice also $|x_{N,k} - x| \leq \pi, \forall x \in [0, \pi]$.

Therefore $(\lambda \geq 1)$ it holds

$$W_N^{(M)}(|\cdot - x|^\lambda)(x) \leq \frac{\pi^{\lambda-1}\pi}{2N} = \frac{\pi^\lambda}{2N}, \ \forall x \in [0, \pi], \forall N \in \mathbb{N}. \tag{6.57}$$

We continue with

Theorem 6.16 *Let* $\frac{1}{n+1} < \alpha < 1$, $n \in \mathbb{N}$, $f : [0, \pi] \to \mathbb{R}_+$, $f' \in L_\infty([0, \pi])$, $x \in [0, \pi]$. *Assume that* $D_{*x}^{k\alpha} f \in C([x, \pi])$, $k = 0, 1, ..., n+1$, *and* $\left(D_{*x}^{i\alpha} f\right)(x) = 0$, $i = 2, 3, ..., n+1$. *Also, suppose that* $D_{x-}^{k\alpha} f \in C([0, x])$, *for* $k = 0, 1, ..., n+1$, *and* $\left(D_{x-}^{i\alpha} f\right)(x) = 0$, *for* $i = 2, 3, ..., n+1$. *Denote* $\lambda = (n+1)\alpha > 1$. *Then*

$$\left| W_N^{(M)}(f)(x) - f(x) \right| \leq \frac{\omega_1 \left(D_x^{(n+1)\alpha} f, \left(\frac{\pi^{\lambda+1}}{2N}\right)^{\frac{1}{\lambda+1}} \right)}{\Gamma(\lambda+1)} \cdot$$

$$\left[\frac{\pi^\lambda}{2N} + \frac{1}{(\lambda+1)} \left(\frac{\pi^{\lambda+1}}{2N}\right)^{\frac{\lambda}{\lambda+1}} \right], \quad \forall N \in \mathbb{N}. \tag{6.58}$$

It holds $\lim_{N \to +\infty} W_N^{(M)}(f)(x) = f(x)$.

Proof Applying (6.36) for $W_N^{(M)}$ and using (6.57), we get

$$\left| W_N^{(M)}(f)(x) - f(x) \right| \leq \frac{\omega_1 \left(D_x^{(n+1)\alpha} f, \delta \right)}{\Gamma(\lambda+1)} \left[\frac{\pi^\lambda}{2N} + \frac{\frac{\pi^{\lambda+1}}{2N}}{(\lambda+1)\delta} \right]. \tag{6.59}$$

Choose $\delta = \left(\frac{\pi^{\lambda+1}}{2N}\right)^{\frac{1}{\lambda+1}}$, then $\delta^{\lambda+1} = \frac{\pi^{\lambda+1}}{2N}$, and $\delta^\lambda = \left(\frac{\pi^{\lambda+1}}{2N}\right)^{\frac{\lambda}{\lambda+1}}$. We use the last into (6.59) and we obtain (6.58). ∎

We make

Remark 6.17 Here we continue with the Max-product truncated sampling operators (see [7], p. 13) defined by

$$K_N^{(M)}(f)(x) = \frac{\bigvee_{k=0}^N \frac{\sin^2(Nx-k\pi)}{(Nx-k\pi)^2} f\left(\frac{k\pi}{N}\right)}{\bigvee_{k=0}^N \frac{\sin^2(Nx-k\pi)}{(Nx-k\pi)^2}}, \tag{6.60}$$

$\forall x \in [0, \pi]$, $f : [0, \pi] \to \mathbb{R}_+$ a continuous function.

Following [7], p. 350, and making the convention $\frac{\sin(0)}{0} = 1$ and denoting $s_{N,k}(x) = \frac{\sin^2(Nx-k\pi)}{(Nx-k\pi)^2}$, we get that $s_{N,k}\left(\frac{k\pi}{N}\right) = 1$, and $s_{N,k}\left(\frac{j\pi}{N}\right) = 0$, if $k \neq j$, furthermore $K_N^{(M)}(f)\left(\frac{j\pi}{N}\right) = f\left(\frac{j\pi}{N}\right)$, for all $j \in \{0, ..., N\}$.

Since $s_{N,j}\left(\frac{j\pi}{N}\right) = 1$ it follows that $\bigvee_{k=0}^N s_{N,k}\left(\frac{j\pi}{N}\right) \geq 1 > 0$, for all $j \in \{0, 1, ..., N\}$. Hence $K_N^{(M)}(f)$ is well-defined function for all $x \in [0, \pi]$, and it is continuous on $[0, \pi]$, also $K_N^{(M)}(1) = 1$. By [7], p. 350, $K_N^{(M)}$ are positive sublinear operators.

Denote $x_{N,k} := \frac{k\pi}{N}, k \in \{0, 1, ..., N\}$.
By [7], p. 352, we have

$$K_N^{(M)} \left(| \cdot - x | \right) (x) \leq \frac{\pi}{2N}, \ \forall N \in \mathbb{N}, \ \forall x \in [0, \pi] . \tag{6.61}$$

Notice also $\left| x_{N,k} - x \right| \leq \pi, \forall x \in [0, \pi]$.
    Therefore ($\lambda \geq 1$) it holds

$$K_N^{(M)} \left(| \cdot - x |^\lambda \right) (x) \leq \frac{\pi^{\lambda-1}\pi}{2N} = \frac{\pi^\lambda}{2N}, \ \forall x \in [0, \pi], \forall N \in \mathbb{N}. \tag{6.62}$$

We give

Theorem 6.18 *All as in Theorem 6.16. Then*

$$\left| K_N^{(M)} (f) (x) - f (x) \right| \leq \frac{\omega_1 \left(D_x^{(n+1)\alpha} f, \left(\frac{\pi^{\lambda+1}}{2N} \right)^{\frac{1}{\lambda+1}} \right)}{\Gamma (\lambda + 1)} \cdot$$

$$\left[\frac{\pi^\lambda}{2N} + \frac{1}{(\lambda + 1)} \left(\frac{\pi^{\lambda+1}}{2N} \right)^{\frac{\lambda}{\lambda+1}} \right], \ \forall N \in \mathbb{N}. \tag{6.63}$$

We have that $\lim\limits_{N \to +\infty} K_N^{(M)} (f) (x) = f (x)$.

Proof As in Theorem 6.16. ∎

We make

Remark 6.19 We mention the interpolation Hermite-Fejér polynomials on Chebyshev knots of the first kind (see [7], p. 4): Let $f : [-1, 1] \to \mathbb{R}$ and based on the knots $x_{N,k} = \cos \left(\frac{(2(N-k)+1)}{2(N+1)} \pi \right) \in (-1, 1), k \in \{0, ..., N\}, -1 < x_{N,0} < x_{N,1} < ... < x_{N,N} < 1$, which are the roots of the first kind Chebyshev polynomial $T_{N+1} (x) = \cos ((N + 1) \arccos x)$, we define (see Fejér [9])

$$H_{2N+1} (f) (x) = \sum_{k=0}^{N} h_{N,k} (x) f \left(x_{N,k} \right), \tag{6.64}$$

where

$$h_{N,k} (x) = \left(1 - x \cdot x_{N,k} \right) \left(\frac{T_{N+1} (x)}{(N + 1) (x - x_{N,k})} \right)^2, \tag{6.65}$$

the fundamental interpolation polynomials.

The Max-product interpolation Hermite–Fejér operators on Chebyshev knots of the first kind (see p. 12 of [7]) are defined by

$$H_{2N+1}^{(M)}(f)(x) = \frac{\bigvee_{k=0}^{N} h_{N,k}(x) f(x_{N,k})}{\bigvee_{k=0}^{N} h_{N,k}(x)}, \ \forall N \in \mathbb{N}, \tag{6.66}$$

where $f : [-1, 1] \to \mathbb{R}_+$ is continuous.
Call

$$E_N(x) := H_{2N+1}^{(M)}(|\cdot - x|)(x) = \frac{\bigvee_{k=0}^{N} h_{N,k}(x) |x_{N,k} - x|}{\bigvee_{k=0}^{N} h_{N,k}(x)}, \ x \in [-1, 1]. \tag{6.67}$$

Then by [7], p. 287 we obtain that

$$E_N(x) \leq \frac{2\pi}{N+1}, \ \forall x \in [-1, 1], \ N \in \mathbb{N}. \tag{6.68}$$

For $m > 1$, we get

$$H_{2N+1}^{(M)}(|\cdot - x|^m)(x) = \frac{\bigvee_{k=0}^{N} h_{N,k}(x) |x_{N,k} - x|^m}{\bigvee_{k=0}^{N} h_{N,k}(x)} =$$

$$\frac{\bigvee_{k=0}^{N} h_{N,k}(x) |x_{N,k} - x| |x_{N,k} - x|^{m-1}}{\bigvee_{k=0}^{N} h_{N,k}(x)} \leq 2^{m-1} \frac{\bigvee_{k=0}^{N} h_{N,k}(x) |x_{N,k} - x|}{\bigvee_{k=0}^{N} h_{N,k}(x)} \tag{6.69}$$

$$\leq \frac{2^m \pi}{N+1}, \ \forall x \in [-1, 1], \ N \in \mathbb{N}.$$

Hence it holds

$$H_{2N+1}^{(M)}(|\cdot - x|^m)(x) \leq \frac{2^m \pi}{N+1}, \ \forall x \in [-1, 1], \ m > 1, \forall N \in \mathbb{N}. \tag{6.70}$$

Furthermore we have

$$H_{2N+1}^{(M)}(1)(x) = 1, \ \forall x \in [-1, 1], \tag{6.71}$$

and $H_{2N+1}^{(M)}$ maps continuous functions to continuous functions over $[-1, 1]$ and for any $x \in \mathbb{R}$ we have $\bigvee_{k=0}^{N} h_{N,k}(x) > 0$.
We also have $h_{N,k}(x_{N,k}) = 1$, and $h_{N,k}(x_{N,j}) = 0$, if $k \neq j$, furthermore it holds $H_{2N+1}^{(M)}(f)(x_{N,j}) = f(x_{N,j})$, for all $j \in \{0, ..., N\}$, see [7], p. 282.
$H_{2N+1}^{(M)}$ are positive sublinear operators, [7], p. 282.

We give

Theorem 6.20 *Let* $\frac{1}{n+1} < \alpha < 1$, $n \in \mathbb{N}$, $f : [-1, 1] \to \mathbb{R}_+$, $f' \in L_\infty ([-1, 1])$, $x \in [-1, 1]$. *Assume that* $D_{*x}^{k\alpha} f \in C ([x, 1])$, $k = 0, 1, ..., n + 1$, *and* $\left(D_{*x}^{i\alpha} f\right) (x) = 0$, $i = 2, 3, ..., n + 1$. *Also, suppose that* $D_{x-}^{k\alpha} f \in C ([-1, x])$, *for* $k = 0, 1, ..., n + 1$, *and* $\left(D_{x-}^{i\alpha} f\right) (x) = 0$, *for* $i = 2, 3, ..., n + 1$. *Denote* $\lambda = (n + 1) \alpha > 1$. *Then*

$$\left| H_{2N+1}^{(M)} (f) (x) - f (x) \right| \le \frac{\omega_1 \left(D_x^{(n+1)\alpha} f, \left(\frac{2^{\lambda+1}\pi}{N+1}\right)^{\frac{1}{\lambda+1}} \right)}{\Gamma (\lambda + 1)}. \tag{6.72}$$

$$\left[\frac{2^\lambda \pi}{N + 1} + \frac{1}{(\lambda + 1)} \left(\frac{2^{\lambda+1}\pi}{N + 1}\right)^{\frac{\lambda}{\lambda+1}} \right], \quad \forall N \in \mathbb{N}.$$

Furthermore it holds $\lim\limits_{N\to+\infty} H_{2N+1}^{(M)} (f) (x) = f (x)$.

Proof Use of Theorem 6.7, (6.36) and (6.70). Choose $\delta := \left(\frac{2^{\lambda+1}\pi}{N+1}\right)^{\frac{1}{\lambda+1}}$, etc. ∎

We continue with

Remark 6.21 Here we deal with Lagrange interpolation polynomials on Chebyshev knots of second kind plus the endpoints ±1 (see [7], p. 5). These polynomials are linear operators attached to $f : [-1, 1] \to \mathbb{R}$ and to the knots $x_{N,k} = \cos \left(\left(\frac{N-k}{N-1}\right) \pi\right) \in [-1, 1]$, $k = 1, ..., N$, $N \in \mathbb{N}$, which are the roots of $\omega_N (x) = \sin (N - 1) t \sin t$, $x = \cos t$. Notice that $x_{N,1} = -1$ and $x_{N,N} = 1$. Their formula is given by ([7], p. 377)

$$L_N (f) (x) = \sum_{k=1}^{N} l_{N,k} (x) f \left(x_{N,k}\right), \tag{6.73}$$

where

$$l_{N,k} (x) = \frac{(-1)^{k-1} \omega_N (x)}{\left(1 + \delta_{k,1} + \delta_{k,N}\right) (N - 1) \left(x - x_{N,k}\right)}, \tag{6.74}$$

$N \ge 2$, $k = 1, ..., N$, and $\omega_N (x) = \prod_{k=1}^{N} \left(x - x_{N,k}\right)$ and $\delta_{i,j}$ denotes the Kronecher's symbol, that is $\delta_{i,j} = 1$, if $i = j$, and $\delta_{i,j} = 0$, if $i \ne j$.

The Max-product Lagrange interpolation operators on Chebyshev knots of second kind, plus the endpoints ±1, are defined by ([7], p. 12)

$$L_N^{(M)} (f) (x) = \frac{\bigvee_{k=1}^{N} l_{N,k} (x) f \left(x_{N,k}\right)}{\bigvee_{k=1}^{N} l_{N,k} (x)}, \quad x \in [-1, 1], \tag{6.75}$$

where $f : [-1, 1] \to \mathbb{R}_+$ continuous.

First we see that $L_N^{(M)}(f)(x)$ is well defined and continuous for any $x \in [-1, 1]$. Following [7], p. 289, because $\sum_{k=1}^{N} l_{N,k}(x) = 1$, $\forall x \in \mathbb{R}$, for any x there exists $k \in \{1, ..., N\} : l_{N,k}(x) > 0$, hence $\bigvee_{k=1}^{N} l_{N,k}(x) > 0$. We have that $l_{N,k}(x_{N,k}) = 1$, and $l_{N,k}(x_{N,j}) = 0$, if $k \neq j$. Furthermore it holds $L_N^{(M)}(f)(x_{N,j}) = f(x_{N,j})$, all $j \in \{1, ..., N\}$, and $L_N^{(M)}(1) = 1$.

Call $I_N^+(x) = \{k \in \{1, ..., N\}; l_{N,k}(x) > 0\}$, then $I_N^+(x) \neq \emptyset$.

So for $f \in C_+([-1, 1])$ we get

$$L_N^{(M)}(f)(x) = \frac{\bigvee_{k \in I_N^+(x)} l_{N,k}(x) f(x_{N,k})}{\bigvee_{k \in I_N^+(x)} l_{N,k}(x)} \geq 0. \tag{6.76}$$

Notice here that $|x_{N,k} - x| \leq 2$, $\forall x \in [-1, 1]$.

By [7], p. 297, we get that

$$L_N^{(M)}(|\cdot - x|)(x) = \frac{\bigvee_{k=1}^{N} l_{N,k}(x) |x_{N,k} - x|}{\bigvee_{k=1}^{N} l_{N,k}(x)} =$$

$$\frac{\bigvee_{k \in I_N^+(x)} l_{N,k}(x) |x_{N,k} - x|}{\bigvee_{k \in I_N^+(x)} l_{N,k}(x)} \leq \frac{\pi^2}{6(N-1)}, \tag{6.77}$$

$N \geq 3$, $\forall x \in (-1, 1)$, N is odd.

We get that $(m > 1)$

$$L_N^{(M)}(|\cdot - x|^m)(x) = \frac{\bigvee_{k \in I_N^+(x)} l_{N,k}(x) |x_{N,k} - x|^m}{\bigvee_{k \in I_N^+(x)} l_{N,k}(x)} \leq \frac{2^{m-1}\pi^2}{6(N-1)}, \tag{6.78}$$

$N \geq 3$ odd, $\forall x \in (-1, 1)$.

$L_N^{(M)}$ are positive sublinear operators, [7], p. 290.

We give

Theorem 6.22 *Same assumptions as in Theorem 6.20. Then*

$$\left| L_N^{(M)}(f)(x) - f(x) \right| \leq \frac{\omega_1 \left(D_x^{(n+1)\alpha} f, \left(\frac{2^\lambda \pi^2}{6(N-1)} \right)^{\frac{1}{\lambda+1}} \right)}{\Gamma(\lambda+1)}. \tag{6.79}$$

$$\left[\frac{2^{\lambda-1}\pi^2}{6(N-1)} + \frac{1}{(\lambda+1)} \left(\frac{2^\lambda \pi^2}{6(N-1)} \right)^{\frac{\lambda}{\lambda+1}} \right], \forall N \in \mathbb{N} : N \geq 3, odd.$$

It holds $\lim_{N \to +\infty} L_N^{(M)}(f)(x) = f(x)$.

Proof By Theorem 6.7, choose $\delta := \left(\frac{2^{\lambda}\pi^2}{6(N-1)}\right)^{\frac{1}{\lambda+1}}$, use of (6.36) and (6.78). At ± 1 the left hand side of (6.79) is zero, thus (6.79) is trivially true. ∎

We make

Remark 6.23 Let $f \in C_+ ([-1, 1])$, $N \geq 4$, $N \in \mathbb{N}$, N even.
By [7], p. 298, we get

$$L_N^{(M)} \left(|\cdot - x|\right)(x) \leq \frac{4\pi^2}{3(N-1)} = \frac{2^2\pi^2}{3(N-1)}, \ \forall x \in (-1, 1). \tag{6.80}$$

Hence $(m > 1)$

$$L_N^{(M)} \left(|\cdot - x|^m\right)(x) \leq \frac{2^{m+1}\pi^2}{3(N-1)}, \ \forall x \in (-1, 1). \tag{6.81}$$

We present

Theorem 6.24 *Same assumptions as in Theorem 6.20. Then*

$$\left|L_N^{(M)}(f)(x) - f(x)\right| \leq \frac{\omega_1\left(D_x^{(n+1)\alpha}f, \left(\frac{2^{\lambda+2}\pi^2}{3(N-1)}\right)^{\frac{1}{\lambda+1}}\right)}{\Gamma(\lambda+1)}. \tag{6.82}$$

$$\left[\frac{2^{\lambda+1}\pi^2}{3(N-1)} + \frac{1}{(\lambda+1)}\left(\frac{2^{\lambda+2}\pi^2}{3(N-1)}\right)^{\frac{\lambda}{\lambda+1}}\right], \ \forall N \in \mathbb{N}, \ N \geq 4, \ N \text{ is even.}$$

It holds $\lim_{N \to +\infty} L_N^{(M)}(f)(x) = f(x)$.

Proof By Theorem 6.7, use of (6.36) and (6.81). Choose $\delta = \left(\frac{2^{\lambda+2}\pi^2}{3(N-1)}\right)^{\frac{1}{\lambda+1}}$, etc. ∎

We make

Remark 6.25 Let $f : \mathbb{R} \to \mathbb{R}$ such that $f' \in L_\infty(\mathbb{R})$, $x_0 \in \mathbb{R}$, $0 < \alpha < 1$. The left Caputo fractional derivative $\left(D_{*x_0}^\alpha f\right)(x)$ is given by (6.7) for $x \geq x_0$. Clearly it holds $\left(D_{*x_0}^\alpha f\right)(x_0) = 0$, and we define $\left(D_{*x_0}^\alpha f\right)(x) = 0$, for $x < x_0$.
Let us assume that $D_{*x_0}^{k\alpha} f \in C([x_0, +\infty))$, $k = 0, 1, ..., n+1$; $n \in \mathbb{N}$.
Still (6.11)–(6.13) are valid $\forall x \in [x_0, +\infty)$.
The right Caputo fractional derivative $\left(D_{x_0-}^\alpha f\right)(x)$ is given by (6.14) for $x \leq x_0$. Clearly it holds $\left(D_{x_0-}^\alpha f\right)(x_0) = 0$, and define $\left(D_{x_0-}^\alpha f\right)(x) = 0$, for $x > x_0$.
Let us assume that $D_{x_0-}^{k\alpha} f \in C((-\infty, x_0])$, $k = 0, 1, ..., n+1$.
Still (6.18)–(6.20) are valid $\forall x \in (-\infty, x_0]$.
Here we restrict again ourselves to $\frac{1}{n+1} < \alpha < 1$, that is $\lambda := (n+1)\alpha > 1$. We denote $D_{*x_0}^\lambda f := D_{*x_0}^{(n+1)\alpha} f$, and $D_{x_0-}^\lambda f := D_{x_0-}^{(n+1)\alpha} f$.

We need

Definition 6.26 ([8], *p. 41*) Let $I \subset \mathbb{R}$ be an interval of finite or infinite length, and $f : I \to \mathbb{R}$ a bounded or uniformly continuous function. We define the first modulus of continuity

$$\omega_1 (f, \delta)_I = \sup_{\substack{x, y \in I \\ |x-y| \le \delta}} |f (x) - f (y)|, \ \delta > 0. \tag{6.83}$$

Clearly, it holds $\omega_1 (f, \delta)_I < +\infty$.

We also have

$$\omega_1 (f, r\delta)_I \le (r + 1) \omega_1 (f, \delta)_I, \ \text{any } r \ge 0. \tag{6.84}$$

Convention 6.27 *We assume that $D_{x_0-}^{\lambda} f$ is either bounded or uniformly continuous function on $(-\infty, x_0]$, similarly we assume that $D_{*x_0}^{\lambda} f$ is either bounded or uniformly continuous function on $[x_0, +\infty)$.*

We need

Definition 6.28 Let $D_{x_0}^{\lambda} f$ denote any of $D_{x_0-}^{\lambda} f$, $D_{*x_0}^{\lambda} f$ and $\delta > 0$. We set

$$\omega_1 \left(D_{x_0}^{\lambda} f, \delta \right)_{\mathbb{R}} := \max \left\{ \omega_1 \left(D_{x_0-}^{\lambda} f, \delta \right)_{(-\infty, x_0]}, \omega_1 \left(D_{*x_0}^{\lambda} f, \delta \right)_{[x_0, +\infty)} \right\}, \tag{6.85}$$

where $x_0 \in \mathbb{R}$. Notice that $\omega_1 \left(D_{x_0}^{\lambda} f, \delta \right)_{\mathbb{R}} < +\infty$.

We give

Theorem 6.29 *Let $\frac{1}{n+1} < \alpha < 1$, $n \in \mathbb{N}$, $\lambda := (n+1)\alpha > 1$, $f : \mathbb{R} \to \mathbb{R}$, $f' \in L_{\infty} (\mathbb{R})$, $x_0 \in \mathbb{R}$. Assume that $D_{*x_0}^{k\alpha} f \in C ([x_0, +\infty))$, $k = 0, 1, ..., n + 1$, and $\left(D_{*x_0}^{i\alpha} f \right) (x_0) = 0, i = 2, 3, ..., n + 1$. Suppose that $D_{x_0-}^{k\alpha} f \in C ((-\infty, x_0]),$ for $k = 0, 1, ..., n + 1$, and $\left(D_{x_0-}^{i\alpha} f \right) (x_0) = 0,$ for $i = 2, 3, ..., n + 1$. Then*

$$|f (x) - f (x_0)| \le \frac{\omega_1 \left(D_{x_0}^{\lambda} f, \delta \right)_{\mathbb{R}}}{\Gamma (\lambda + 1)} \left[|x - x_0|^{\lambda} + \frac{|x - x_0|^{\lambda+1}}{(\lambda + 1) \delta} \right], \tag{6.86}$$

$\forall x \in \mathbb{R}, \delta > 0$.

Proof Similar to Theorem 6.4. ∎

Remark 6.30 Let $b : \mathbb{R} \to \mathbb{R}_+$ be a centered (it takes a global maximum at 0) bell-shaped function, with compact support $[-T, T]$, $T > 0$ (that is $b (x) > 0$ for all $x \in (-T, T)$) and $I = \int_{-T}^{T} b (x) dx > 0$.

The Cardaliaguet–Euvrard neural network operators are defined by (see [6])

$$C_{N,\alpha} (f) (x) = \sum_{k=-N^2}^{N^2} \frac{f \left(\frac{k}{n} \right)}{I N^{1-\alpha}} b \left(N^{1-\alpha} \left(x - \frac{k}{N} \right) \right), \tag{6.87}$$

$0 < \alpha < 1$, $N \in \mathbb{N}$ and typically here $f : \mathbb{R} \to \mathbb{R}$ is continuous and bounded or uniformly continuous on $\mathbb{R}$.

$CB(\mathbb{R})$ denotes the continuous and bounded function on $\mathbb{R}$, and

$$CB_+(\mathbb{R}) = \{f : \mathbb{R} \to [0, \infty); \; f \in CB(\mathbb{R})\}.$$

The corresponding max-product Cardaliaguet–Euvrard neural network operators will be given by

$$C_{N,\alpha}^{(M)}(f)(x) = \frac{\bigvee_{k=-N^2}^{N^2} b\left(N^{1-\alpha}\left(x - \frac{k}{N}\right)\right) f\left(\frac{k}{N}\right)}{\bigvee_{k=-N^2}^{N^2} b\left(N^{1-\alpha}\left(x - \frac{k}{N}\right)\right)}, \tag{6.88}$$

$x \in \mathbb{R}$, typically here $f \in CB_+(\mathbb{R})$, see also [6].

Next we follow [6].

For any $x \in \mathbb{R}$, denoting

$$J_{T,N}(x) = \left\{k \in \mathbb{Z}; \; -N^2 \le k \le N^2, N^{1-\alpha}\left(x - \frac{k}{N}\right) \in (-T, T)\right\},$$

we can write

$$C_{N,\alpha}^{(M)}(f)(x) = \frac{\bigvee_{k \in J_{T,N}(x)} b\left(N^{1-\alpha}\left(x - \frac{k}{N}\right)\right) f\left(\frac{k}{N}\right)}{\bigvee_{k \in J_{T,N}(x)} b\left(N^{1-\alpha}\left(x - \frac{k}{N}\right)\right)}, \tag{6.89}$$

$x \in \mathbb{R}$, $N > \max\left\{T + |x|, T^{-\frac{1}{\alpha}}\right\}$, where $J_{T,N}(x) \ne \emptyset$. Indeed, we have $\bigvee_{k \in J_{T,N}(x)} b\left(N^{1-\alpha}\left(x - \frac{k}{N}\right)\right) > 0$, $\forall x \in \mathbb{R}$ and $N > \max\left\{T + |x|, T^{-\frac{1}{\alpha}}\right\}$.

We have that $C_{N,\alpha}^{(M)}(1)(x) = 1$, $\forall x \in \mathbb{R}$ and $N > \max\left\{T + |x|, T^{-\frac{1}{\alpha}}\right\}$.

See in [6] there: Lemma 2.1, Corollary 2.2 and Remarks.

We need

Theorem 6.31 ([6]) *Let $b(x)$ be a centered bell-shaped function, continuous and with compact support $[-T, T]$, $T > 0$, $0 < \alpha < 1$ and $C_{N,\alpha}^{(M)}$ be defined as in (6.88).*

(i) If $|f(x)| \le c$ for all $x \in \mathbb{R}$ then $\left|C_{N,\alpha}^{(M)}(f)(x)\right| \le c$, for all $x \in \mathbb{R}$ and $N > \max\left\{T + |x|, T^{-\frac{1}{\alpha}}\right\}$ and $C_{N,\alpha}^{(M)}(f)(x)$ is continuous at any point $x \in \mathbb{R}$, for all $N > \max\left\{T + |x|, T^{-\frac{1}{\alpha}}\right\}$;

(ii) If $f, g \in CB_+(\mathbb{R})$ satisfy $f(x) \le g(x)$ for all $x \in \mathbb{R}$, then $C_{N,\alpha}^{(M)}(f)(x) \le C_{N,\alpha}^{(M)}(g)(x)$ for all $x \in \mathbb{R}$ and $N > \max\left\{T + |x|, T^{-\frac{1}{\alpha}}\right\}$;

(iii) $C_{N,\alpha}^{(M)}(f + g)(x) \le C_{N,\alpha}^{(M)}(f)(x) + C_{N,\alpha}^{(M)}(g)(x)$ for all $f, g \in CB_+(\mathbb{R})$, $x \in \mathbb{R}$ and $N > \max\left\{T + |x|, T^{-\frac{1}{\alpha}}\right\}$;

(iv) For all $f, g \in CB_+(\mathbb{R})$, $x \in \mathbb{R}$ *and* $N > \max\left\{T + |x|, T^{-\frac{1}{\alpha}}\right\}$, *we have*

$$\left|C_{N,\alpha}^{(M)}(f)(x) - C_{N,\alpha}^{(M)}(g)(x)\right| \le C_{N,\alpha}^{(M)}(|f - g|)(x);$$

(v) $C_{N,\alpha}^{(M)}$ *is positive homogeneous, that is* $C_{N,\alpha}^{(M)}(\lambda f)(x) = \lambda C_{N,\alpha}^{(M)}(f)(x)$ *for all* $\lambda \ge 0$, $x \in \mathbb{R}$, $N > \max\left\{T + |x|, T^{-\frac{1}{\alpha}}\right\}$ *and* $f \in CB_+(\mathbb{R})$.

We make

Remark 6.32 We have that

$$E_{N,\alpha}(x) := C_{N,\alpha}^{(M)}(|\cdot - x|)(x) = \frac{\bigvee_{k \in J_{T,N}(x)} b\left(N^{1-\alpha}\left(x - \frac{k}{N}\right)\right)\left|x - \frac{k}{N}\right|}{\bigvee_{k \in J_{T,N}(x)} b\left(N^{1-\alpha}\left(x - \frac{k}{N}\right)\right)}, \quad (6.90)$$

$\forall x \in \mathbb{R}$, and $N > \max\left\{T + |x|, T^{-\frac{1}{\alpha}}\right\}$.

We mention from [6] the following:

Theorem 6.33 ([6]) *Let* $b(x)$ *be a centered bell-shaped function, continuous and with compact support* $[-T, T]$, $T > 0$ *and* $0 < \alpha < 1$. *In addition, suppose that the following requirements are fulfilled:*

(i) There exist $0 < m_1 \le M_1 < \infty$ *such that* $m_1(T - x) \le b(x) \le M_1(T - x)$, $\forall x \in [0, T]$;

(ii) There exist $0 < m_2 \le M_2 < \infty$ *such that* $m_2(x + T) \le b(x) \le M_2(x + T)$, $\forall x \in [-T, 0]$.

Then for all $f \in CB_+(\mathbb{R})$, $x \in \mathbb{R}$ *and for all* $N \in \mathbb{N}$ *satisying* $N > \max\left\{T + |x|, \left(\frac{2}{T}\right)^{\frac{1}{\alpha}}\right\}$, *we have the estimate*

$$\left|C_{N,\alpha}^{(M)}(f)(x) - f(x)\right| \le c\omega_1\left(f, N^{\alpha-1}\right)_{\mathbb{R}}, \quad (6.91)$$

where

$$c := 2\left(\max\left\{\frac{TM_2}{2m_2}, \frac{TM_1}{2m_1}\right\} + 1\right),$$

and

$$\omega_1(f, \delta)_{\mathbb{R}} := \sup_{\substack{x, y \in \mathbb{R}: \\ |x-y| \le \delta}} |f(x) - f(y)|. \quad (6.92)$$

We make

Remark 6.34 In [6], was proved that

$$E_{N,\alpha}(x) \le \max\left\{\frac{TM_2}{2m_2}, \frac{TM_1}{2m_1}\right\} N^{\alpha-1}, \ \forall N > \max\left\{T + |x|, \left(\frac{2}{T}\right)^{\frac{1}{\alpha}}\right\}. \quad (6.93)$$

That is

$$C_{N,\alpha}^{(M)}(|\cdot - x|)(x) \le \max\left\{\frac{TM_2}{2m_2}, \frac{TM_1}{2m_1}\right\} N^{\alpha-1}, \ \forall N > \max\left\{T + |x|, \left(\frac{2}{T}\right)^{\frac{1}{\alpha}}\right\}. \quad (6.94)$$

From (6.90) we have that $\left|x - \frac{k}{N}\right| \le \frac{T}{N^{1-\alpha}}$.

Hence $(\lambda > 1)$ $(\forall\, x \in \mathbb{R}$ and $N > \max\left\{T + |x|, \left(\frac{2}{T}\right)^{\frac{1}{\alpha}}\right\})$

$$C_{N,\alpha}^{(M)}(|\cdot - x|^\lambda)(x) = \frac{\bigvee_{k \in J_{T,N}(x)} b\left(N^{1-\alpha}\left(x - \frac{k}{N}\right)\right)\left|x - \frac{k}{N}\right|^\lambda}{\bigvee_{k \in J_{T,N}(x)} b\left(N^{1-\alpha}\left(x - \frac{k}{N}\right)\right)} \le \quad (6.95)$$

$$\left(\frac{T}{N^{1-\alpha}}\right)^{\lambda-1} \max\left\{\frac{TM_2}{2m_2}, \frac{TM_1}{2m_1}\right\} N^{\alpha-1}, \ \forall N > \max\left\{T + |x|, \left(\frac{2}{T}\right)^{\frac{1}{\alpha}}\right\}.$$

Then $(\lambda > 1)$ it holds
$$C_{N,\alpha}^{(M)}(|\cdot - x|^\lambda)(x) \le$$

$$T^{\lambda-1} \max\left\{\frac{TM_2}{2m_2}, \frac{TM_1}{2m_1}\right\} \frac{1}{N^{\lambda(1-\alpha)}}, \ \forall N > \max\left\{T + |x|, \left(\frac{2}{T}\right)^{\frac{1}{\alpha}}\right\}. \quad (6.96)$$

Call
$$\theta := \max\left\{\frac{TM_2}{2m_2}, \frac{TM_1}{2m_1}\right\} > 0. \quad (6.97)$$

Consequently $(\lambda > 1)$ we derive

$$C_{N,\alpha}^{(M)}(|\cdot - x|^\lambda)(x) \le \frac{\theta T^{\lambda-1}}{N^{\lambda(1-\alpha)}}, \ \forall N > \max\left\{T + |x|, \left(\frac{2}{T}\right)^{\frac{1}{\alpha}}\right\}. \quad (6.98)$$

We need

Theorem 6.35 *All here as in Theorem 6.29, where $x = x_0 \in \mathbb{R}$ is fixed. Let b be a centered bell-shaped function, continuous and with compact support $[-T, T]$, $T > 0$, $0 < \alpha < 1$ and $C_{N,\alpha}^{(M)}$ be defined as in (6.88). Then*

$$\left|C_{N,\alpha}^{(M)}(f)(x) - f(x)\right| \le$$

$$\frac{\omega_1 \left(D_x^\lambda f, \delta\right)_{\mathbb{R}}}{\Gamma\left(\lambda+1\right)} \left[C_{N,\alpha}^{(M)} \left(|\cdot - x|^\lambda\right)(x) + \frac{C_{N,\alpha}^{(M)} \left(|\cdot - x|^{\lambda+1}\right)(x)}{(\lambda+1)\,\delta}\right], \qquad (6.99)$$

$$\forall\, N \in \mathbb{N} : N > \max\left\{T + |x|, T^{-\frac{1}{\alpha}}\right\}.$$

Proof By Theorem 6.29 and (6.86) we get

$$|f\left(\cdot\right) - f\left(x\right)| \le \frac{\omega_1 \left(D_x^\lambda f, \delta\right)_{\mathbb{R}}}{\Gamma\left(\lambda+1\right)} \left[|\cdot - x|^\lambda + \frac{|\cdot - x|^{\lambda+1}}{(\lambda+1)\,\delta}\right], \quad \delta > 0, \qquad (6.100)$$

true over $\mathbb{R}$.

As in Theorem 6.31 and using similar reasoning and $C_{N,\alpha}^{(M)}\left(1\right) = 1$, we get

$$\left|C_{N,\alpha}^{(M)} \left(f\right)(x) - f\left(x\right)\right| \le C_{N,\alpha}^{(M)} \left(|f\left(\cdot\right) - f\left(x\right)|\right)(x) \overset{(6.100)}{\le}$$

$$\frac{\omega_1 \left(D_x^\lambda f, \delta\right)_{\mathbb{R}}}{\Gamma\left(\lambda+1\right)} \left[C_{N,\alpha}^{(M)} \left(|\cdot - x|^\lambda\right)(x) + \frac{C_{N,\alpha}^{(M)} \left(|\cdot - x|^{\lambda+1}\right)(x)}{(\lambda+1)\,\delta}\right], \qquad (6.101)$$

$$\forall\, N \in \mathbb{N} : N > \max\left\{T + |x|, T^{-\frac{1}{\alpha}}\right\}. \qquad \blacksquare$$

We continue with

Theorem 6.36 *Here all as in Theorem 6.29, where $x = x_0 \in \mathbb{R}$ is fixed. Also the same assumptions as in Theorem 6.33. Then*

$$\left|C_{N,\alpha}^{(M)} \left(f\right)(x) - f\left(x\right)\right| \le \frac{1}{\Gamma\left(\lambda+1\right)} \omega_1 \left(D_x^\lambda f, \left(\frac{\theta T^\lambda}{N^{(\lambda+1)(1-\alpha)}}\right)^{\frac{1}{\lambda+1}}\right)_{\mathbb{R}} \cdot$$

$$\left[\frac{\theta T^{\lambda-1}}{N^{\lambda(1-\alpha)}} + \frac{1}{(\lambda+1)} \left(\frac{\theta T^\lambda}{N^{(\lambda+1)(1-\alpha)}}\right)^{\frac{\lambda}{\lambda+1}}\right], \qquad (6.102)$$

$$\forall\, N \in \mathbb{N} : N > \max\left\{T + |x|, \left(\frac{2}{T}\right)^{\frac{1}{\alpha}}\right\}.$$

We have that $\lim\limits_{N \to +\infty} C_{N,\alpha}^{(M)} \left(f\right)(x) = f\left(x\right).$

Proof We apply Theorem 6.35. In (6.99) we choose

$$\delta := \left(\frac{\theta T^\lambda}{N^{(\lambda+1)(1-\alpha)}}\right)^{\frac{1}{\lambda+1}},$$

thus $\delta^{\lambda+1} = \frac{\theta T^\lambda}{N^{(\lambda+1)(1-\alpha)}}$, and

$$\delta^\lambda = \left(\frac{\theta T^\lambda}{N^{(\lambda+1)(1-\alpha)}}\right)^{\frac{\lambda}{\lambda+1}}. \tag{6.103}$$

Therefore we have

$$\left|C_{N,\alpha}^{(M)}(f)(x) - f(x)\right| \overset{(6.98)}{\leq} \frac{1}{\Gamma(\lambda+1)}\omega_1\left(D_x^\lambda f, \left(\frac{\theta T^\lambda}{N^{(\lambda+1)(1-\alpha)}}\right)^{\frac{1}{\lambda+1}}\right)_{\mathbb{R}} \cdot \tag{6.104}$$

$$\left[\frac{\theta T^{\lambda-1}}{N^{\lambda(1-\alpha)}} + \frac{1}{(\lambda+1)\delta}\frac{\theta T^\lambda}{N^{(\lambda+1)(1-\alpha)}}\right] =$$

$$\frac{1}{\Gamma(\lambda+1)}\omega_1\left(D_x^\lambda f, \left(\frac{\theta T^\lambda}{N^{(\lambda+1)(1-\alpha)}}\right)^{\frac{1}{\lambda+1}}\right)\left[\frac{\theta T^{\lambda-1}}{N^{\lambda(1-\alpha)}} + \frac{1}{(\lambda+1)\delta}\delta^{\lambda+1}\right] \overset{(6.103)}{=}$$

$$\frac{1}{\Gamma(\lambda+1)}\omega_1\left(D_x^\lambda f, \left(\frac{\theta T^\lambda}{N^{(\lambda+1)(1-\alpha)}}\right)^{\frac{1}{\lambda+1}}\right)_{\mathbb{R}} \cdot$$

$$\left[\frac{\theta T^{\lambda-1}}{N^{\lambda(1-\alpha)}} + \frac{1}{(\lambda+1)}\left(\frac{\theta T^\lambda}{N^{(\lambda+1)(1-\alpha)}}\right)^{\frac{\lambda}{\lambda+1}}\right], \tag{6.105}$$

$\forall N \in \mathbb{N} : N > \max\left\{T + |x|, \left(\frac{2}{7}\right)^{\frac{1}{\alpha}}\right\}$, proving the inequality (6.102). ∎

It follows an interesting application to Theorem 6.8 when $\alpha = \frac{1}{2}$, $n = 2$.

Corollary 6.37 *Let* $f : [0, 1] \to \mathbb{R}_+$, $f' \in L_\infty([0, 1])$, $x \in [0, 1]$. *Assume that* $D_{*x}^{k\frac{1}{2}}f \in C([x, 1])$, $k = 0, 1, 2, 3$, *and* $\left(D_{*x}^{i\frac{1}{2}}f\right)(x) = 0$, $i = 2, 3$. *Suppose that* $D_{x-}^{k\frac{1}{2}}f \in C([0, x])$, *for* $k = 0, 1, 2, 3$, *and* $\left(D_{x-}^{i\frac{1}{2}}f\right)(x) = 0$, *for* $i = 2, 3$. *Then*

$$\left|B_N^{(M)}(f)(x) - f(x)\right| \leq \frac{4\omega_1\left(D_x^{3\cdot\frac{1}{2}}f, \left(\frac{6}{\sqrt{N+1}}\right)^{\frac{2}{5}}\right)}{3\sqrt{\pi}}$$

$$\left[\frac{6}{\sqrt{N+1}} + \frac{2}{5}\left(\frac{6}{\sqrt{N+1}}\right)^{\frac{3}{5}}\right], \quad \forall N \in \mathbb{N}. \tag{6.106}$$

We get $\lim\limits_{N\to+\infty} B_N^{(M)}(f)(x) = f(x)$.

6.4 Applications, Part B

Case of $(n + 1) \alpha \leq 1$.

We need

Theorem 6.38 ([2]) *Let* $L : C_+ ([a, b]) \to C_+ ([a, b])$, *be a positive sublinear operator and* $f, g \in C_+ ([a, b])$, *furthermore let* $p, q > 1 : \frac{1}{p} + \frac{1}{q} = 1$. *Assume that* $L \left((f (\cdot))^p \right) (s_*)$, $L \left((g (\cdot))^q \right) (s_*) > 0$ *for some* $s_* \in [a, b]$. *Then*

$$L \left(f (\cdot) g (\cdot) \right) (s_*) \leq \left(L \left((f (\cdot))^p \right) (s_*) \right)^{\frac{1}{p}} \left(L \left((g (\cdot))^q \right) (s_*) \right)^{\frac{1}{q}} . \tag{6.107}$$

We give

Theorem 6.39 *Let* $0 < \alpha \leq \frac{1}{n+1}$, $n \in \mathbb{N}$, $f : [a, b] \to \mathbb{R}_+$, $f' \in L_\infty ([a, b])$, $x_0 \in [a, b]$. *Assume that* $D_{*x_0}^{k\alpha} f \in C ([x_0, b])$, $k = 0, 1, ..., n + 1$, *and* $\left(D_{*x_0}^{i\alpha} f \right) (x_0) = 0$, $i = 2, 3, ..., n + 1$. *Also, suppose that* $D_{x_0-}^{k\alpha} f \in C ([a, x_0])$, *for* $k = 0, 1, ..., n + 1$, *and* $\left(D_{x_0-}^{i\alpha} f \right) (x_0) = 0$, *for* $i = 2, 3, ..., n + 1$. *Denote* $\lambda := (n + 1) \alpha \leq 1$. *Let* $L_N : C_+ ([a, b]) \to C_+ ([a, b])$, $\forall N \in \mathbb{N}$, *be positive sublinear operators, such that* $L_N \left(|\cdot - x_0|^{\lambda+1} \right) (x_0) > 0$ *and* $L_N (1) = 1$, $\forall N \in \mathbb{N}$. *Then*

$$|L_N (f) (x_0) - f (x_0)| \leq \frac{\omega_1 \left(D_{x_0}^{(n+1)\alpha} f, \delta \right)}{\Gamma (\lambda + 1)} . \tag{6.108}$$

$$\left[\left(L_N \left(|\cdot - x_0|^{\lambda+1} \right) (x_0) \right)^{\frac{\lambda}{\lambda+1}} + \frac{L_N \left(|\cdot - x_0|^{\lambda+1} \right) (x_0)}{(\lambda + 1) \delta} \right],$$

$\delta > 0$, $\forall N \in \mathbb{N}$.

Proof By Theorems 6.7 and 6.38. ∎

We give

Theorem 6.40 *Let* $0 < \alpha \leq \frac{1}{n+1}$, $n \in \mathbb{N}$, $f : [a, b] \to \mathbb{R}_+$, $f' \in L_\infty ([a, b])$, $x_0 \in [a, b]$. *Assume that* $D_{*x_0}^{k\alpha} f \in C ([x_0, b])$, $k = 0, 1, ..., n + 1$, *and* $\left(D_{*x_0}^{i\alpha} f \right) (x_0) = 0$, $i = 2, 3, ..., n + 1$. *Also, suppose that* $D_{x_0-}^{k\alpha} f \in C ([a, x_0])$, *for* $k = 0, 1, ..., n + 1$, *and* $\left(D_{x_0-}^{i\alpha} f \right) (x_0) = 0$, *for* $i = 2, 3, ..., n + 1$. *Denote* $\lambda := (n + 1) \alpha \leq 1$. *Let* $L_N : C_+ ([a, b]) \to C_+ ([a, b])$, $\forall N \in \mathbb{N}$, *be positive sublinear operators, such that* $L_N \left(|\cdot - x_0|^{\lambda+1} \right) (x_0) > 0$ *and* $L_N (1) = 1$, $\forall N \in \mathbb{N}$. *Then*

$$|L_N (f) (x_0) - f (x_0)| \leq \frac{(\lambda + 2) \omega_1 \left(D_{x_0}^{(n+1)\alpha} f, \left(L_N \left(|\cdot - x_0|^{\lambda+1} \right) (x_0) \right)^{\frac{1}{\lambda+1}} \right)}{\Gamma (\lambda + 2)} .$$

$$\tag{6.109}$$

$$\left(L_N \left(|\cdot - x_0|^{\lambda+1} \right) (x_0) \right)^{\frac{\lambda}{\lambda+1}} , \forall N \in \mathbb{N}.$$

Proof In (6.108) choose $\delta := \left(L_N \left(|\cdot - x_0|^{\lambda+1}\right)(x_0)\right)^{\frac{1}{\lambda+1}}$. ■

Note: From (6.109) we get that: if $L_N \left(|\cdot - x_0|^{\lambda+1}\right)(x_0) \to 0$, as $N \to +\infty$, then $L_N(f)(x_0) \to f(x_0)$, as $N \to +\infty$.

We present

Theorem 6.41 *Let* $0 < \alpha \leq \frac{1}{n+1}$, $n \in \mathbb{N}$, $f : [0,1] \to \mathbb{R}_+$, $f' \in L_\infty([0,1])$, $x \in (0,1)$. *Assume that* $D_{*x}^{k\alpha} f \in C([x,1])$, $k = 0, 1, ..., n+1$, *and* $\left(D_{*x}^{i\alpha} f\right)(x) = 0$, $i = 2, 3, ..., n+1$. *Also, suppose that* $D_{x-}^{k\alpha} f \in C([0,x])$, *for* $k = 0, 1, ..., n+1$, *and* $\left(D_{x-}^{i\alpha} f\right)(x) = 0$, *for* $i = 2, 3, ..., n+1$. *Denote* $\lambda := (n+1)\alpha \leq 1$. *Then*

$$\left| B_N^{(M)}(f)(x) - f(x) \right| \leq \frac{(\lambda+2)\,\omega_1\left(D_x^\lambda f, \left(\frac{6}{\sqrt{N+1}}\right)^{\frac{1}{\lambda+1}}\right)}{\Gamma(\lambda+2)} \left(\frac{6}{\sqrt{N+1}}\right)^{\frac{\lambda}{\lambda+1}},$$

(6.110)

$\forall N \in \mathbb{N}$.

See that $\lim\limits_{N \to +\infty} B_N^{(M)}(f)(x) = f(x)$.

Proof The Max-product Bernstein operators $B_N^{(M)}(f)(x)$ are defined by (6.5), see also [7], p. 10; here $f : [0,1] \to \mathbb{R}_+$ is a continuous function.

We have $B_N^{(M)}(1) = 1$, and

$$B_N^{(M)}(|\cdot - x|)(x) \leq \frac{6}{\sqrt{N+1}}, \ \forall x \in [0,1], \ \forall N \in \mathbb{N},$$

(6.111)

see [7], p. 31.

$B_N^{(M)}$ are positive sublinear operators and thus they possess the monotonicity property, also since $|\cdot - x| \leq 1$, then $|\cdot - x|^\beta \leq 1$, $\forall x \in [0,1]$, $\forall \beta > 0$.

Therefore it holds

$$B_N^{(M)}\left(|\cdot - x|^{1+\beta}\right)(x) \leq \frac{6}{\sqrt{N+1}}, \ \forall x \in [0,1], \ \forall N \in \mathbb{N}, \ \forall \beta > 0.$$

(6.112)

Furthermore, clearly it holds that

$$B_N^{(M)}\left(|\cdot - x|^{1+\beta}\right)(x) > 0, \ \forall N \in \mathbb{N}, \ \forall \beta \geq 0 \text{ and any } x \in (0,1).$$

(6.113)

The operator $B_N^{(M)}$ maps $C_+([0,1])$ into itself. We apply (6.109). ■

We continue with

Remark 6.42 The truncated Favard–Szász–Mirakjan operators are given by

$$T_N^{(M)}(f)(x) = \frac{\bigvee_{k=0}^{N} s_{N,k}(x) f\left(\frac{k}{N}\right)}{\bigvee_{k=0}^{N} s_{N,k}(x)}, \ x \in [0,1], \ N \in \mathbb{N}, \ f \in C_+([0,1]),$$

(6.114)

$s_{N,k}(x) = \frac{(Nx)^k}{k!}$, see also [7], p. 11.

By [7], p. 178–179, we get that

$$T_N^{(M)}(|\cdot - x|)(x) \le \frac{3}{\sqrt{N}}, \ \forall \, x \in [0, 1], \ \forall \, N \in \mathbb{N}. \tag{6.115}$$

Clearly it holds

$$T_N^{(M)}\left(|\cdot - x|^{1+\beta}\right)(x) \le \frac{3}{\sqrt{N}}, \ \forall \, x \in [0, 1], \ \forall \, N \in \mathbb{N}, \forall \, \beta > 0. \tag{6.116}$$

The operators $T_N^{(M)}$ are positive sublinear operators mapping $C_+([0, 1])$ into itself, with $T_N^{(M)}(1) = 1$.

Furthermore it holds

$$T_N^{(M)}\left(|\cdot - x|^\lambda\right)(x) = \frac{\bigvee_{k=0}^{N} \frac{(Nx)^k}{k!} \left|\frac{k}{N} - x\right|^\lambda}{\bigvee_{k=0}^{N} \frac{(Nx)^k}{k!}} > 0, \ \forall \, x \in (0, 1], \ \forall \, \lambda \ge 1, \forall \, N \in \mathbb{N}. \tag{6.117}$$

We give

Theorem 6.43 *All as in Theorem 6.41, with $x \in (0, 1]$. Then*

$$\left|T_N^{(M)}(f)(x) - f(x)\right| \le \frac{(\lambda + 2)\, \omega_1\left(D_x^\lambda f, \left(\frac{3}{\sqrt{N}}\right)^{\frac{1}{\lambda+1}}\right)}{\Gamma(\lambda + 2)} \left(\frac{3}{\sqrt{N}}\right)^{\frac{\lambda}{\lambda+1}}, \ \forall \, N \in \mathbb{N}. \tag{6.118}$$

As $N \to +\infty$, we get $T_N^{(M)}(f)(x) \to f(x)$.

Proof We apply (6.109). ∎

We make

Remark 6.44 Next we study the truncated Max-product Baskakov operators (see [7], p. 11)

$$U_N^{(M)}(f)(x) = \frac{\bigvee_{k=0}^{N} b_{N,k}(x)\, f\left(\frac{k}{N}\right)}{\bigvee_{k=0}^{N} b_{N,k}(x)}, \ x \in [0, 1], \ f \in C_+([0, 1]), \ N \in \mathbb{N}, \tag{6.119}$$

where

$$b_{N,k}(x) = \binom{N+k-1}{k} \frac{x^k}{(1+x)^{N+k}}. \tag{6.120}$$

From [7], pp. 217–218, we get ($x \in [0, 1]$)

$$\left(U_N^{(M)}\left(|\cdot - x|\right)\right)(x) \le \frac{2\sqrt{3}\left(\sqrt{2}+2\right)}{\sqrt{N+1}}, \; N \ge 2, \, N \in \mathbb{N}. \tag{6.121}$$

Let $\beta \ge 1$, clearly then it holds

$$\left(U_N^{(M)}\left(|\cdot - x|^\beta\right)\right)(x) \le \frac{2\sqrt{3}\left(\sqrt{2}+2\right)}{\sqrt{N+1}}, \; \forall \, N \ge 2, \, N \in \mathbb{N}. \tag{6.122}$$

Also it holds $U_N^{(M)}(1) = 1$, and $U_N^{(M)}$ are positive sublinear operators from $C_+([0,1])$ into itself. Furthermore it holds

$$U_N^{(M)}\left(|\cdot - x|^\beta\right)(x) > 0, \; \forall \, x \in (0,1], \; \forall \, \beta \ge 1, \, \forall \, N \in \mathbb{N}. \tag{6.123}$$

We give

Theorem 6.45 *All as in Theorem 6.41, with $x \in (0,1]$. Then*

$$\left|U_N^{(M)}(f)(x) - f(x)\right| \le \frac{(\lambda+2)\,\omega_1\left(D_x^\lambda f, \left(\frac{2\sqrt{3}(\sqrt{2}+2)}{\sqrt{N+1}}\right)^{\frac{1}{\lambda+1}}\right)}{\Gamma(\lambda+2)} \cdot$$

$$\left(\frac{2\sqrt{3}\left(\sqrt{2}+2\right)}{\sqrt{N+1}}\right)^{\frac{\lambda}{\lambda+1}}, \; \forall \, N \ge 2, \, N \in \mathbb{N}. \tag{6.124}$$

As $N \to +\infty$, we get $U_N^{(M)}(f)(x) \to f(x)$.

Proof By Theorem 6.40. ∎

We continue with

Remark 6.46 Here we study the Max-product Meyer–Köning and Zeller operators (see [7], p. 11) defined by

$$Z_N^{(M)}(f)(x) = \frac{\bigvee_{k=0}^{\infty} s_{N,k}(x)\, f\left(\frac{k}{N+k}\right)}{\bigvee_{k=0}^{\infty} s_{N,k}(x)}, \; \forall \, N \in \mathbb{N}, \, f \in C_+([0,1]), \tag{6.125}$$

$$s_{N,k}(x) = \binom{N+k}{k} x^k, \; x \in [0,1].$$

By [7], p. 253, we get that

$$Z_N^{(M)}\left(|\cdot - x|\right)(x) \le \frac{8\left(1+\sqrt{5}\right)}{3}\frac{\sqrt{x}\,(1-x)}{\sqrt{N}}, \; \forall \, x \in [0,1], \, \forall \, N \ge 4, \, N \in \mathbb{N}. \tag{6.126}$$

We have that (for $\beta \geq 1$)

$$Z_N^{(M)} \left(|\cdot - x|^{\beta} \right) (x) \leq \frac{8 \left(1 + \sqrt{5} \right)}{3} \frac{\sqrt{x} \, (1 - x)}{\sqrt{N}} := \rho(x), \qquad (6.127)$$

$\forall \, x \in [0, 1], \, N \geq 4, \, N \in \mathbb{N}$.

Also it holds $Z_N^{(M)}(1) = 1$, and $Z_N^{(M)}$ are positive sublinear operators from $C_+([0, 1])$ into itself. Also it holds

$$Z_N^{(M)} \left(|\cdot - x|^{\beta} \right) (x) > 0, \, \forall \, x \in (0, 1), \, \forall \, \beta \geq 1, \forall \, N \in \mathbb{N}. \qquad (6.128)$$

We give

Theorem 6.47 *All as in Theorem 6.41. Then*

$$\left| Z_N^{(M)}(f)(x) - f(x) \right| \leq \frac{(\lambda + 2) \, \omega_1 \left(D_x^{\lambda} f, (\rho(x))^{\frac{1}{\lambda+1}} \right)}{\Gamma(\lambda + 2)} \, (\rho(x))^{\frac{\lambda}{\lambda+1}} \qquad (6.129)$$

$\forall \, N \geq 4, \, N \in \mathbb{N}$.

As $N \to +\infty$, we get $Z_N^{(M)}(f)(x) \to f(x)$.

Proof By Theorem 6.40. ∎

We continue with

Remark 6.48 Here we deal with the Max-product truncated sampling operators (see [7], p. 13) defined by

$$W_N^{(M)}(f)(x) = \frac{\bigvee_{k=0}^{N} \frac{\sin(Nx - k\pi)}{Nx - k\pi} f \left(\frac{k\pi}{N} \right)}{\bigvee_{k=0}^{N} \frac{\sin(Nx - k\pi)}{Nx - k\pi}}, \qquad (6.130)$$

$\forall \, x \in [0, \pi], \, f : [0, \pi] \to \mathbb{R}_+$ a continuous function. See also Remark 6.15.

By [7], p. 346, we have

$$W_N^{(M)} \left(|\cdot - x| \right) (x) \leq \frac{\pi}{2N}, \, \forall \, N \in \mathbb{N}, \, \forall \, x \in [0, \pi]. \qquad (6.131)$$

Furthermore it holds ($\beta \geq 1$)

$$W_N^{(M)} \left(|\cdot - x|^{\beta} \right) (x) \leq \frac{\pi^{\beta}}{2N}, \, \forall \, N \in \mathbb{N}, \forall \, x \in [0, \pi]. \qquad (6.132)$$

Also it holds ($\beta \geq 1$)

$$W_N^{(M)} \left(|\cdot - x|^{\beta} \right) (x) > 0, \, \forall \, x \in [0, \pi], \qquad (6.133)$$

such that $x \neq \frac{k\pi}{N}$, for any $k \in \{0, 1, ..., N\}$, see [3].

We present

Theorem 6.49 *Let* $0 < \alpha \leq \frac{1}{n+1}$, $n \in \mathbb{N}$, $x \in [0, \pi]$ *be such that* $x \neq \frac{k\pi}{N}$, $k \in \{0, 1, ..., N\}$, $\forall N \in \mathbb{N}$; $f : [0, \pi] \to \mathbb{R}_+$, $f' \in L_\infty ([0, \pi])$. *Assume that* $D_{*x}^{k\alpha} f \in C ([x, \pi])$, $k = 0, 1, ..., n + 1$, *and* $\left(D_{*x}^{i\alpha} f \right) (x) = 0$, $i = 2, 3, ..., n + 1$. *Also, suppose that* $D_{x-}^{k\alpha} f \in C ([0, x])$, *for* $k = 0, 1, ..., n + 1$, *and* $\left(D_{x-}^{i\alpha} f \right) (x) = 0$, *for* $i = 2, 3, ..., n + 1$. *Denote* $\lambda := (n + 1) \alpha \leq 1$. *Then*

$$\left| W_N^{(M)} (f) (x) - f (x) \right| \leq \frac{(\lambda + 2) \omega_1 \left(D_x^\lambda f, \left(\frac{\pi^{\lambda+1}}{2N} \right)^{\frac{1}{\lambda+1}} \right)}{\Gamma (\lambda + 2)} \left(\frac{\pi^{\lambda+1}}{2N} \right)^{\frac{\lambda}{\lambda+1}} , \quad \forall N \in \mathbb{N}.$$
(6.134)

As $N \to +\infty$, *we get* $W_N^{(M)} (f) (x) \to f (x)$.

Proof By (6.132), (6.133) and Theorem 6.40. ∎

We make

Remark 6.50 Here we continue with the Max-product truncated sampling operators (see [7], p. 13) defined by

$$K_N^{(M)} (f) (x) = \frac{\bigvee_{k=0}^N \frac{\sin^2 (Nx - k\pi)}{(Nx - k\pi)^2} f \left(\frac{k\pi}{N} \right)}{\bigvee_{k=0}^N \frac{\sin^2 (Nx - k\pi)}{(Nx - k\pi)^2}},$$
(6.135)

$\forall x \in [0, \pi]$, $f : [0, \pi] \to \mathbb{R}_+$ a continuous function.

See also Remark 6.17.

It holds ($\beta \geq 1$)

$$K_N^{(M)} \left(| \cdot - x |^\beta \right) (x) \leq \frac{\pi^\beta}{2N}, \quad \forall N \in \mathbb{N}, \forall x \in [0, \pi].$$
(6.136)

By [3], we get that ($\beta \geq 1$)

$$K_N^{(M)} \left(| \cdot - x |^\beta \right) (x) > 0, \quad \forall x \in [0, \pi],$$
(6.137)

such that $x \neq \frac{k\pi}{N}$, for any $k \in \{0, 1, ..., N\}$.

We continue with

Theorem 6.51 *All as in Theorem 6.49. Then*

$$\left| K_N^{(M)} (f) (x) - f (x) \right| \leq \frac{(\lambda + 2) \omega_1 \left(D_x^\lambda f, \left(\frac{\pi^{\lambda+1}}{2N} \right)^{\frac{1}{\lambda+1}} \right)}{\Gamma (\lambda + 1)} \left(\frac{\pi^{\lambda+1}}{2N} \right)^{\frac{\lambda}{\lambda+1}} , \quad \forall N \in \mathbb{N}.$$
(6.138)

As $N \to +\infty$, we get $K_N^{(M)}(f)(x) \to f(x)$.

Proof By (6.136), (6.137) and Theorem 6.40. ∎

We finish with

Corollary 6.52 (to Theorem 6.41, $\alpha = \frac{1}{4}$, $n = 2$, $\lambda = \frac{3}{4}$) *Let $f : [0,1] \to \mathbb{R}_+$, $f' \in L_\infty([0,1])$, $x \in (0,1)$. Assume that $D_{*x}^{k\frac{1}{4}} f \in C([x,1])$, $k = 0,1,2,3$, and $\left(D_{*x}^{i\frac{1}{4}} f\right)(x) = 0$, $i = 2,3$. Suppose that $D_{x-}^{k\frac{1}{4}} f \in C([0,x])$, for $k = 0,1,2,3$, and $\left(D_{x-}^{i\frac{1}{4}} f\right)(x) = 0$, for $i = 2,3$. Then*

$$\left| B_N^{(M)}(f)(x) - f(x) \right| \leq \tag{6.139}$$

$$(1.709)\,\omega_1\left(D_x^{3\cdot\frac{1}{4}} f, \left(\frac{6}{\sqrt{N+1}}\right)^{\frac{4}{7}}\right)\left(\frac{6}{\sqrt{N+1}}\right)^{\frac{3}{7}}, \ \forall\, N \in \mathbb{N}.$$

And $\lim\limits_{N\to+\infty} B_N^{(M)}(f)(x) = f(x)$.

Proof Use of (6.110). ∎

References

1. G. Anastassiou, Advanced fractional Taylor's formulae. J. Comput. Anal. Appl. **21**(7), 1185–1204 (2016)
2. G. Anastassiou, *Approximation by Sublinear Operators* (2017, submitted)
3. G. Anastassiou, *Caputo Fractional Approximation by Sublinear operators* (2017, submitted)
4. G. Anastassiou, *Iterated Fractional Approximation by Max-Product Operators* (2017, submitted)
5. G. Anastassiou, I. Argyros, *Intelligent Numerical Methods: Applications to Fractional Calculus* (Springer, Heidelberg, 2016)
6. G. Anastassiou, L. Coroianu, S. Gal, Approximation by a nonlinear Cardaliaguet-Euvrard neural network operator of max-product kind. J. Comput. Anal. Appl. **12**(2), 396–406 (2010)
7. B. Bede, L. Coroianu, S. Gal, *Approximation by Max-Product Type Operators* (Springer, Heidelberg, 2016)
8. R.A. DeVore, G.G. Lorentz, *Constructive Approximation* (Springer, Berlin, 1993)
9. L. Fejér, *Über Interpolation*, Göttingen Nachrichten (1916), pp. 66–91
10. G.G. Lorentz, *Bernstein Polynomials*, 2nd edn. (Chelsea Publishing Company, New York, 1986)
11. Z.M. Odibat, N.J. Shawagleh, Generalized Taylor's formula. Appl. Math. Comput. **186**, 286–293 (2007)
12. T. Popoviciu, Sur l'approximation de fonctions convexes d'order superieur. Mathematica (Cluj) **10**, 49–54 (1935)

Chapter 7
Mixed Conformable Fractional Approximation Using Positive Sublinear Operators

Here we consider the approximation of functions by positive sublinear operators with applications to a large variety of Max-Product operators under mixed conformable fractional differentiability. These are examples of positive sublinear operators. Our study is based on our general results about positive sublinear operators. We produce Jackson type inequalities under mixed conformable related basic initial conditions. So our approach is quantitative by producing inequalities with their right hand sides involving the modulus of continuity of a high order mixed conformable fractional derivative of the function under approximation. It follows [3].

7.1 Introduction

The main motivation here is the monograph by B. Bede, L. Coroianu and S. Gal [4], 2016.

Let $N \in \mathbb{N}$, the well-known Bernstein polynomials ([7]) are positive linear operators, defined by the formula

$$B_N (f) (x) = \sum_{k=0}^{N} \binom{N}{k} x^k (1 - x)^{N-k} f \left(\frac{k}{N} \right), \quad x \in [0, 1], \quad f \in C ([0, 1]).$$

(7.1)

T. Popoviciu in [8], 1935, proved for $f \in C ([0, 1])$ that

$$|B_N (f) (x) - f (x)| \leq \frac{5}{4} \omega_1 \left(f, \frac{1}{\sqrt{N}} \right), \quad \forall \, x \in [0, 1],$$

(7.2)

where

© Springer International Publishing AG, part of Springer Nature 2018

G. A. Anastassiou, *Nonlinearity: Ordinary and Fractional Approximations by Sublinear and Max-Product Operators*, Studies in Systems, Decision and Control 147, https://doi.org/10.1007/978-3-319-89509-3_7

$$\omega_1 (f, \delta) = \sup_{\substack{x, y \in [a,b]: \\ |x-y| \leq \delta}} |f(x) - f(y)|, \quad \delta > 0, \tag{7.3}$$

is the first modulus of continuity, here [a,b] = [0,1].

G.G. Lorentz in [7], 1986, p. 21, proved for $f \in C^1 ([0, 1])$ that

$$|B_N (f) (x) - f(x)| \leq \frac{3}{4\sqrt{N}} \omega_1 \left(f', \frac{1}{\sqrt{N}} \right), \quad \forall \, x \in [0, 1], \tag{7.4}$$

In [4], p. 10, the authors introduced the basic Max-product Bernstein operators,

$$B_N^{(M)} (f) (x) = \frac{\bigvee_{k=0}^{N} p_{N,k} (x) f \left(\frac{k}{N} \right)}{\bigvee_{k=0}^{N} p_{N,k} (x)}, \quad N \in \mathbb{N}, \tag{7.5}$$

where $\bigvee$ stands for maximum, and $p_{N,k} (x) = \binom{N}{k} x^k (1 - x)^{N-k}$ and $f : [0, 1] \to \mathbb{R}_+ = [0, \infty)$.

These are nonlinear and piecewise rational operators.

The authors in [4] studied similar such nonlinear operators such as: the Max-product Favard–Szász–Mirakjan operators and their truncated version, the Max-product Baskakov operators and their truncated version, also many other similar specific operators. The study in [4] is based on presented there general theory of sublinear operators. These Max-product operators tend to converge faster to the on hand function.

So we mention from [4], p. 30, that for $f : [0, 1] \to \mathbb{R}_+$ continuous, we have the estimate

$$\left| B_N^{(M)} (f) (x) - f(x) \right| \leq 12\omega_1 \left(f, \frac{1}{\sqrt{N+1}} \right), \quad \text{for all } N \in \mathbb{N}, \; x \in [0, 1]. \tag{7.6}$$

In this chapter we expand the study of [4] by considering mixed conformable fractional smoothness of functions, see [1, 6]. So our inequalities are with respect to $\omega_1 (\mathbf{T}_\alpha f, \delta)$, $\delta > 0$, where $\mathbf{T}_\alpha f$, $\alpha \in (n, n + 1]$, $n \in \mathbb{N}$, is a high order α-mixed conformable fractional derivative of f. We treat also the case of $0 < \alpha \leq 1$.

7.2 Background

Here we follow [1].

We need

Definition 7.1 ([1]) Let $a, b \in \mathbb{R}$. The left conformable fractional derivative starting from a of a function $f : [a, \infty) \to \mathbb{R}$ of order $0 < \alpha \leq 1$ is defined by

$$\left(T_\alpha^a f\right)(t) = \lim_{\varepsilon \to 0} \frac{f\left(t + \varepsilon\,(t-a)^{1-\alpha}\right) - f\,(t)}{\varepsilon}. \tag{7.7}$$

If $\left(T_\alpha^a f\right)(t)$ exists on (a, b), then

$$\left(T_\alpha^a f\right)(a) = \lim_{t \to a+} \left(T_\alpha^a f\right)(t). \tag{7.8}$$

The right conformable fractional derivative of order $0 < \alpha \le 1$ terminating at b of $f : (-\infty, b] \to \mathbb{R}$ is defined by

$$\left({}_\alpha^b T f\right)(t) = -\lim_{\varepsilon \to 0} \frac{f\left(t + \varepsilon\,(b-t)^{1-\alpha}\right) - f\,(t)}{\varepsilon}. \tag{7.9}$$

If $\left({}_\alpha^b T f\right)(t)$ exists on (a, b), then

$$\left({}_\alpha^b T f\right)(b) = \lim_{t \to b-} \left({}_\alpha^b T f\right)(t). \tag{7.10}$$

Note that if f is differentiable then

$$\left(T_\alpha^a f\right)(t) = (t-a)^{1-\alpha}\, f'\,(t), \tag{7.11}$$

and

$$\left({}_\alpha^b T f\right)(t) = -(b-t)^{1-\alpha}\, f'\,(t). \tag{7.12}$$

Denote by

$$\left(I_\alpha^a f\right)(t) = \int_a^t (x-a)^{\alpha-1}\, f\,(x)\,dx, \tag{7.13}$$

and

$$\left({}^b I_\alpha f\right)(t) = \int_t^b (b-x)^{\alpha-1}\, f\,(x)\,dx, \tag{7.14}$$

these are the left and right conformable fractional integrals of order $0 < \alpha \le 1$.

In the higher order case we can generalize things as follows:

Definition 7.2 ([1]) Let $\alpha \in (n, n+1]$, and set $\beta = \alpha - n$. Then, the left conformable fractional derivative starting from a of a function $f : [a, \infty) \to \mathbb{R}$ of order α, where $f^{(n)}(t)$ exists, is defined by

$$\left(\mathbf{T}_\alpha^a f\right)(t) = \left(T_\beta^a f^{(n)}\right)(t), \tag{7.15}$$

The right conformable fractional derivative of order α terminating at b of $f : (-\infty, b] \to \mathbb{R}$, where $f^{(n)}(t)$ exists, is defined by

$$\left({}^b_\alpha \mathbf{T} f \right)(t) = (-1)^{n+1} \left({}^b_\beta T f^{(n)} \right)(t).$$ (7.16)

If $\alpha = n + 1$ then $\beta = 1$ and $\mathbf{T}^a_{n+1} f = f^{(n+1)}$.

If n is odd, then ${}^b_{n+1} \mathbf{T} f = -f^{(n+1)}$, and if n is even, then ${}^b_{n+1} \mathbf{T} f = f^{(n+1)}$.

When $n = 0$ (or $\alpha \in (0, 1]$), then $\beta = \alpha$, and (7.15), (7.16) collapse to {(7.7), (7.8)}, {(7.9), (7.10)}, respectively.

Lemma 7.3 ([1]) *Let $f : (a, b) \rightarrow \mathbb{R}$ be continuously differentiable and $0 < \alpha \leq 1$. Then, for all $t > a$ we have*

$$I^a_\alpha T^a_\alpha (f) (t) = f (t) - f (a).$$ (7.17)

We need

Definition 7.4 (see also [1]) If $\alpha \in (n, n + 1]$, then the left fractional integral of order α starting at a is defined by

$$\left(\mathbf{I}^a_\alpha f \right)(t) = \frac{1}{n!} \int_a^t (t - x)^n (x - a)^{\beta - 1} f (x) \, dx.$$ (7.18)

Similarly, (author's definition) the right fractional integral of order α terminating at b is defined by

$$\left({}^b \mathbf{I}_\alpha f \right)(t) = \frac{1}{n!} \int_t^b (x - t)^n (b - x)^{\beta - 1} f (x) \, dx.$$ (7.19)

We need

Proposition 7.5 ([1]) *Let $\alpha \in (n, n + 1]$ and $f : [a, \infty) \rightarrow \mathbb{R}$ be $(n + 1)$ times continuously differentiable for $t > a$. Then, for all $t > a$ we have*

$$\mathbf{I}^a_\alpha T^a_\alpha (f) (t) = f (t) - \sum_{k=0}^n \frac{f^{(k)} (a) (t - a)^k}{k!}.$$ (7.20)

We also have

Proposition 7.6 *Let $\alpha \in (n, n + 1]$ and $f : (-\infty, b] \rightarrow \mathbb{R}$ be $(n + 1)$ times continuously differentiable for $t < b$. Then, for all $t < b$ we have*

$$-{}^b \mathbf{I}_\alpha {}^b_a \mathbf{T} (f) (t) = f (t) - \sum_{k=0}^n \frac{f^{(k)} (b) (t - b)^k}{k!}.$$ (7.21)

If $n = 0$ or $0 < \alpha \leq 1$, then (see also [1])

$$^b I_\alpha {}^b_\alpha T (f) (t) = f (t) - f (b).$$ (7.22)

In conclusion we derive

Theorem 7.7 *Let $\alpha \in (n, n+1]$ and $f \in C^{n+1}([a, b])$, $n \in \mathbb{N}$. Then*
(1)

$$f(t) - \sum_{k=0}^{n} \frac{f^{(k)}(a)(t-a)^k}{k!} = \frac{1}{n!} \int_a^t (t-x)^n (x-a)^{\beta-1} \left(\mathbf{T}_\alpha^a(f)\right)(x)\,dx,$$

$$(7.23)$$

and
(2)

$$f(t) - \sum_{k=0}^{n} \frac{f^{(k)}(b)(t-b)^k}{k!} = -\frac{1}{n!} \int_t^b (b-x)^{\beta-1}(x-t)^n \left({}_\alpha^b\mathbf{T}(f)\right)(x)\,dx,$$

$$(7.24)$$

$\forall\, t \in [a, b]$.

Proof of Proposition 7.6, (7.21) and (7.24).
We observe that

$$-{}^b\mathbf{I}_\alpha \, {}_\alpha^b\mathbf{T}(f)(t) = -\frac{1}{n!} \int_t^b (b-x)^{\beta-1}(x-t)^n \left({}_\alpha^b\mathbf{T}(f)\right)(x)\,dx =$$

$$-\frac{(-1)^{n+1}}{n!} \int_t^b (b-x)^{\beta-1}(x-t)^n \left({}_\beta^b\mathbf{T}f^{(n)}\right)(x)\,dx =$$

$$-\frac{(-1)^{n+2}}{n!} \int_t^b (b-x)^{\beta-1}(x-t)^n (b-x)^{1-\beta} f^{(n+1)}(x)\,dx = \qquad (7.25)$$

$$-\frac{(-1)^n}{n!} \int_t^b (x-t)^n f^{(n+1)}(x)\,dx = -\frac{1}{n!} \int_t^b (t-x)^n f^{(n+1)}(x)\,dx =$$

$$\frac{1}{n!} \int_b^t (t-x)^n f^{(n+1)}(x)\,dx,$$

then we use Taylor's formula. ∎

We make

Remark 7.8 We notice the following: let $\alpha \in (n, n+1]$ and $f \in C^{n+1}([a, b])$, $n \in \mathbb{N}$. Then ($\beta := \alpha - n$, $0 < \beta \le 1$)

$$\left(\mathbf{T}_\alpha^a(f)\right)(x) = \left(T_\beta^a f^{(n)}\right)(x) = (x-a)^{1-\beta} f^{(n+1)}(x), \qquad (7.26)$$

and

$$\left({}_\alpha^b\mathbf{T}(f)\right)(x) = (-1)^{n+1}\left({}_\beta^bTf^{(n)}\right)(x) =$$

$$(-1)^{n+1} (-1) (b - x)^{1-\beta} f^{(n+1)} (x) = (-1)^n (b - x)^{1-\beta} f^{(n+1)} (x). \qquad (7.27)$$

Consequently we get that

$$\left(\mathbf{T}_\alpha^a (f)\right) (x), \quad \left(_\alpha^b \mathbf{T} (f)\right) (x) \in C ([a, b]).$$

Furthermore it is obvious that

$$\left(\mathbf{T}_\alpha^a (f)\right) (a) = \left(_\alpha^b \mathbf{T} (f)\right) (b) = 0, \qquad (7.28)$$

when $0 < \beta < 1$, i.e. when $\alpha \in (n, n + 1)$.

If $f^{(k)} (a) = 0, k = 1, ..., n$, then

$$f (t) - f (a) = \frac{1}{n!} \int_a^t (t - x)^n (x - a)^{\beta-1} \left(\mathbf{T}_\alpha^a (f)\right) (x) \, dx, \qquad (7.29)$$

$\forall \, t \in [a, b]$.

If $f^{(k)} (b) = 0, k = 1, ..., n$, then

$$f (t) - f (b) = -\frac{1}{n!} \int_t^b (b - x)^{\beta-1} (x - t)^n \left(_\alpha^b \mathbf{T} (f)\right) (x) \, dx, \qquad (7.30)$$

$\forall \, t \in [a, b]$.

7.3 Main Results

We need

Theorem 7.9 *Let* $\alpha \in (n, n + 1)$, $n \in \mathbb{N}$, *and* $f \in C^{n+1} ([a, b])$, $x_0 \in [a, b]$ *and assume that* $f^{(k)} (x_0) = 0, k = 1, ..., n$. *Denote*

$$\omega_1 \left(^{x_0}\mathbf{T}_\alpha f, \delta\right) := \max \left\{ \omega_1 \left(\mathbf{T}_\alpha^{x_0} f, \delta\right)_{[x_0, b]}, \omega_1 \left(_\alpha^{x_0}\mathbf{T} f, \delta\right)_{[a, x_0]} \right\}, \quad \delta > 0. \qquad (7.31)$$

Then

$$|f (t) - f (x_0)| \le \omega_1 \left(^{x_0}\mathbf{T}_\alpha f, \delta\right) \left[\frac{|t - x_0|^\alpha}{\prod_{j=0}^n (\alpha - n + j)} + \frac{1}{\delta} \frac{|t - x_0|^{\alpha+1}}{\prod_{j=0}^n (\alpha - n + j + 1)} \right], \qquad (7.32)$$

$\forall \, t \in [a, b], \delta > 0$. *Equation (7.31) defines indirectly the mixed conformable fractional derivative of* f *of order* α.

Proof Here $\alpha \in (n, n + 1)$, $n \in \mathbb{N}$, and $f \in C^{n+1} ([a, b])$, $x_0 \in [a, b]$. We assume that $f^{(k)} (x_0) = 0, k = 1, ..., n$. By (7.29) and (7.28) we get

$$f(t) - f(x_0) = \frac{1}{n!} \int_{x_0}^t (t - x)^n (x - x_0)^{\beta - 1} \left[(\mathbf{T}_\alpha^{x_0} f)(x) - (\mathbf{T}_\alpha^{x_0} f)(x_0) \right] dx,$$
(7.33)

$\forall\, t \in [x_0, b]$.

Similarly, by (7.30) and (7.28) we get

$$f(t) - f(x_0) = -\frac{1}{n!} \int_t^{x_0} (x_0 - x)^{\beta - 1} (x - t)^n \left[(_\alpha^{x_0}\mathbf{T} f)(x) - (_\alpha^{x_0}\mathbf{T} f)(x_0) \right] dx,$$
(7.34)

$\forall\, t \in [a, x_0]$.

We have that $(x_0 \le t \le b)$

$$|f(t) - f(x_0)| \le \frac{1}{n!} \int_{x_0}^t (t - x)^n (x - x_0)^{\beta - 1} \left| (\mathbf{T}_\alpha^{x_0} f)(x) - (\mathbf{T}_\alpha^{x_0} f)(x_0) \right| dx \le$$

$(\delta > 0)$

$$\frac{1}{n!} \int_{x_0}^t (t - x)^n (x - x_0)^{\beta - 1} \omega_1 \left((\mathbf{T}_\alpha^{x_0} f), \frac{\delta (x - x_0)}{\delta} \right)_{[x_0, b]} dx \le$$
(7.35)

$$\frac{\omega_1 \left((\mathbf{T}_\alpha^{x_0} f), \delta \right)_{[x_0, b]}}{n!} \int_{x_0}^t (t - x)^n (x - x_0)^{\beta - 1} \left(1 + \frac{(x - x_0)}{\delta} \right) dx =$$

$$\frac{\omega_1 \left((\mathbf{T}_\alpha^{x_0} f), \delta \right)_{[x_0, b]}}{n!} \left[\int_{x_0}^t (t - x)^{(n+1)-1} (x - x_0)^{\beta - 1} dx + \right.$$

$$\left. \frac{1}{\delta} \int_{x_0}^t (t - x)^{(n+1)-1} (x - x_0)^{(\beta+1)-1} dx \right] =$$

$$\frac{\omega_1 \left((\mathbf{T}_\alpha^{x_0} f), \delta \right)_{[x_0, b]}}{n!} \left[\frac{\Gamma(n+1)\,\Gamma(\beta)}{\Gamma(n+1+\beta)} (t - x_0)^{n+\beta} + \right.$$

$$\left. \frac{1}{\delta} \frac{\Gamma(n+1)\,\Gamma(\beta+1)}{\Gamma(n+\beta+2)} (t - x_0)^{n+\beta+1} \right] =$$
(7.36)

$$\omega_1 \left(\mathbf{T}_\alpha^{x_0} f, \delta \right)_{[x_0, b]} \left[\frac{\Gamma(\beta)}{\Gamma(n+1+\beta)} (t - x_0)^{n+\beta} + \frac{1}{\delta} \frac{\Gamma(\beta+1)}{\Gamma(n+2+\beta)} (t - x_0)^{n+\beta+1} \right] =$$

$$\omega_1 \left(\mathbf{T}_\alpha^{x_0} f, \delta \right)_{[x_0, b]} \left[\frac{\Gamma(\beta)}{\Gamma(\beta) \prod_{j=0}^n (\beta + j)} (t - x_0)^{n+\beta} + \right.$$

$$\frac{1}{\delta} \frac{\Gamma(\beta+1)}{\Gamma(\beta+1) \prod_{j=0}^{n} (\beta+1+j)} (t-x_0)^{n+\beta+1} \Bigg] = \qquad (7.37)$$

$$\omega_1 \left(\mathbf{T}_\alpha^{x_0} f, \delta\right)_{[x_0,b]} \left[\frac{(t-x_0)^{n+\beta}}{\prod_{j=0}^{n} (\beta+j)} + \frac{1}{\delta} \frac{(t-x_0)^{n+\beta+1}}{\prod_{j=0}^{n} (\beta+1+j)}\right].$$

We have proved that

$$|f(t) - f(x_0)| \le \omega_1 \left(\mathbf{T}_\alpha^{x_0} f, \delta\right)_{[x_0,b]} \left[\frac{(t-x_0)^{n+\beta}}{\prod_{j=0}^{n} (\beta+j)} + \frac{1}{\delta} \frac{(t-x_0)^{n+\beta+1}}{\prod_{j=0}^{n} (\beta+1+j)}\right], \quad (7.38)$$

$\forall\, t \in [x_0, b]$, $\delta > 0$.

We also have that $(a \le t \le x_0)$

$$|f(t) - f(x_0)| \le \frac{1}{n!} \int_t^{x_0} (x_0-x)^{\beta-1} (x-t)^n \left| \left(_\alpha^{x_0}\mathbf{T}f\right)(x) - \left(_\alpha^{x_0}\mathbf{T}f\right)(x_0)\right| dx \le$$

$(\delta > 0)$

$$\frac{1}{n!} \int_t^{x_0} (x_0-x)^{\beta-1} (x-t)^n\, \omega_1 \left(_\alpha^{x_0}\mathbf{T}f, \frac{\delta(x_0-x)}{\delta}\right)_{[a,x_0]} dx \le$$

$$\frac{\omega_1 \left(_\alpha^{x_0}\mathbf{T}f, \delta\right)_{[a,x_0]}}{n!} \int_t^{x_0} (x_0-x)^{\beta-1} (x-t)^n \left(1 + \frac{(x_0-x)}{\delta}\right) dx = \qquad (7.39)$$

$$\frac{\omega_1 \left(_\alpha^{x_0}\mathbf{T}f, \delta\right)_{[a,x_0]}}{n!} \left[\int_t^{x_0} (x_0-x)^{\beta-1} (x-t)^{(n+1)-1}\, dx + \right.$$

$$\left. \frac{1}{\delta} \int_t^{x_0} (x_0-x)^{(\beta+1)-1} (x-t)^{(n+1)-1}\, dx \right] =$$

$$\frac{\omega_1 \left(_\alpha^{x_0}\mathbf{T}f, \delta\right)_{[a,x_0]}}{n!} \left[\frac{\Gamma(\beta)\,\Gamma(n+1)}{\Gamma(\beta+n+1)} (x_0-t)^{\beta+n} + \right.$$

$$\left. \frac{1}{\delta} \frac{\Gamma(\beta+1)\,\Gamma(n+1)}{\Gamma(\beta+n+2)} (x_0-t)^{\beta+n+1} \right] =$$

$$\omega_1 \left(_\alpha^{x_0}\mathbf{T}f, \delta\right)_{[a,x_0]} \left[\frac{\Gamma(\beta)}{\Gamma(\beta+n+1)} (x_0-t)^{\beta+n} + \frac{1}{\delta} \frac{\Gamma(\beta+1)}{\Gamma(\beta+n+2)} (x_0-t)^{\beta+n+1}\right] = \quad (7.40)$$

$$\omega_1 \left(_\alpha^{x_0}\mathbf{T}f, \delta\right)_{[a,x_0]} \left[\frac{(x_0-t)^{\beta+n}}{\prod_{j=0}^{n} (\beta+j)} + \frac{1}{\delta} \frac{(x_0-t)^{\beta+n+1}}{\prod_{j=0}^{n} (\beta+1+j)}\right].$$

We have proved that

$$|f(t) - f(x_0)| \le \omega_1 \left(^{x_0}_{\alpha}\mathbf{T}f, \delta\right)_{[a, x_0]} \left[\frac{(x_0 - t)^{\beta+n}}{\prod_{j=0}^{n} (\beta + j)} + \frac{1}{\delta} \frac{(x_0 - t)^{\beta+n+1}}{\prod_{j=0}^{n} (\beta + 1 + j)} \right], \quad (7.41)$$

$\forall\, t \in [a, x_0]$, $\delta > 0$.

Based on (7.38) and (7.41) we derive (7.32). ∎

We can rewrite Theorem 7.9 as follows:

Theorem 7.10 *Let* $\alpha \in (n, n+1)$, $n \in \mathbb{N}$, *and* $f \in C^{n+1}([a, b])$, $x \in [a, b]$ *and* $f^{(k)}(x) = 0$, $k = 1, ..., n$. *Denote*

$$\omega_1 \left(^{x}\mathbf{T}_\alpha f, \delta\right) := \max \left\{ \omega_1 \left(\mathbf{T}_\alpha^x f, \delta\right)_{[x, b]}, \omega_1 \left(^{x}_{\alpha}\mathbf{T}f, \delta\right)_{[a, x]} \right\}. \quad (7.42)$$

Then, over $[a, b]$, *we have*

$$|f(\cdot) - f(x)| \le \frac{\omega_1 \left(^{x}\mathbf{T}_\alpha f, \delta\right)}{\prod_{j=0}^{n-1} (\alpha - j)} \left[\frac{|\cdot - x|^\alpha}{(a - n)} + \frac{|\cdot - x|^{\alpha+1}}{(\alpha + 1) \delta} \right], \quad \delta > 0. \quad (7.43)$$

Definition 7.11 Here $C_+([a, b]) := \{f : [a, b] \to \mathbb{R}_+,$ continuous functions$\}$. Let $L_N : C_+([a, b]) \to C_+([a, b])$, operators, $\forall\, N \in \mathbb{N}$, such that
(i)
$$L_N(\alpha f) = \alpha L_N(f), \quad \forall \alpha \ge 0, \forall f \in C_+([a, b]), \quad (7.44)$$

(ii) if $f, g \in C_+([a, b]) : f \le g$, then

$$L_N(f) \le L_N(g), \forall N \in \mathbb{N}, \quad (7.45)$$

(iii)
$$L_N(f + g) \le L_N(f) + L_N(g), \quad \forall\, f, g \in C_+([a, b]). \quad (7.46)$$

We call $\{L_N\}_{N \in \mathbb{N}}$ positive sublinear operators.

We mention a Hölder's type inequality:

Theorem 7.12 (see [2]) *Let* $L : C_+([a, b]) \to C_+([a, b])$, *be a positive sublinear operator and* $f, g \in C_+([a, b])$, *furthermore let* $p, q > 1 : \frac{1}{p} + \frac{1}{q} = 1$. *Assume that* $L\left((f(\cdot))^p\right)(s_*), L\left((g(\cdot))^q\right)(s_*) > 0$ *for some* $s_* \in [a, b]$. *Then*

$$L(f(\cdot)g(\cdot))(s_*) \le \left(L\left((f(\cdot))^p\right)(s_*)\right)^{\frac{1}{p}} \left(L\left((g(\cdot))^q\right)(s_*)\right)^{\frac{1}{q}}. \quad (7.47)$$

We make

Remark 7.13 By [4], p. 17, we get: let $f, g \in C_+ ([a, b])$, then

$$|L_N (f) (x) - L_N (g) (x)| \le L_N (|f - g|) (x), \ \forall \ x \in [a, b]. \qquad (7.48)$$

Furthermore, we also have that

$$|L_N (f) (x) - f (x)| \le L_N (|f (\cdot) - f (x)|) (x) + |f (x)| |L_N (e_0) (x) - 1|, \qquad (7.49)$$

$\forall \ x \in [a, b]; e_0 (t) = 1$.

From now on we assume that $L_N (1) = 1$. Hence it holds

$$|L_N (f) (x) - f (x)| \le L_N (|f (\cdot) - f (x)|) (x), \ \forall \ x \in [a, b]. \qquad (7.50)$$

Using Theorems 7.10 and (7.43) with (7.50) we get

$$|L_N (f) (x) - f (x)| \le \frac{\omega_1 (^x\mathbf{T}_\alpha f, \delta)}{\prod_{j=0}^{n-1} (\alpha - j)} \left[\frac{L_N (|\cdot - x|^\alpha) (x)}{(a - n)} + \frac{L_N (|\cdot - x|^{\alpha+1}) (x)}{(\alpha + 1) \delta} \right], \quad (7.51)$$

$\delta > 0$.

We have proved

Theorem 7.14 *Let $\alpha \in (n, n + 1)$, $n \in \mathbb{N}$, and $f \in C^{n+1} ([a, b], \mathbb{R}_+)$, $x \in [a, b]$ and $f^{(k)} (x) = 0, k = 1, ..., n$. Let $L_N : C_+ ([a, b]) \to C_+ ([a, b]), \forall N \in \mathbb{N}$, be positive sublinear operators, such that $L_N (1) = 1, \forall \ N \in \mathbb{N}$. Then*

$$|L_N (f) (x) - f (x)| \le \frac{\omega_1 (^x\mathbf{T}_\alpha f, \delta)}{\prod_{j=0}^{n-1} (\alpha - j)} \left[\frac{L_N (|\cdot - x|^\alpha) (x)}{(a - n)} + \frac{L_N (|\cdot - x|^{\alpha+1}) (x)}{(\alpha + 1) \delta} \right], \quad (7.52)$$

$\delta > 0$.

7.4   Applications, Part A

Case of $\alpha \in (n, n + 1), n \in \mathbb{N}$.

Here we apply Theorem 7.14 to well known Max-product operators.

We make

Remark 7.15 The Max-product Bernstein operators $B_N^{(M)} (f) (x)$ are defined by (7.5), see also [4], p. 10; here $f : [0, 1] \to \mathbb{R}_+$ is a continuous function.

We have $B_N^{(M)} (1) = 1$, and

$$B_N^{(M)} (|\cdot - x|) (x) \le \frac{6}{\sqrt{N + 1}}, \ \forall \ x \in [0, 1], \ \forall \ N \in \mathbb{N}, \qquad (7.53)$$

see [4], p. 31.

$B_N^{(M)}$ are positive sublinear operators and thus they possess the monotonicity property, also since $|\cdot - x| \le 1$, then $|\cdot - x|^\beta \le 1$, $\forall\, x \in [0, 1]$, $\forall\, \beta > 0$.

Therefore it holds

$$B_N^{(M)} \left(|\cdot - x|^{1+\beta}\right)(x) \le \frac{6}{\sqrt{N+1}}, \; \forall\, x \in [0, 1], \; \forall\, N \in \mathbb{N}, \; \forall\, \beta > 0. \tag{7.54}$$

Furthermore, clearly it holds that

$$B_N^{(M)} \left(|\cdot - x|^{1+\beta}\right)(x) > 0, \; \forall\, N \in \mathbb{N}, \; \forall\, \beta \ge 0 \text{ and any } x \in (0, 1). \tag{7.55}$$

The operator $B_N^{(M)}$ maps $C_+ \left([0, 1]\right)$ into itself.

We present

Theorem 7.16 *Let $\alpha \in (n, n+1)$, $n \in \mathbb{N}$, and $f \in C^{n+1}\left([0, 1], \mathbb{R}_+\right)$, $x \in [0, 1]$ and $f^{(k)}(x) = 0$, $k = 1, ..., n$. Then*

$$\left| B_N^{(M)}(f)(x) - f(x) \right| \le \frac{\omega_1 \left({}^x\mathbf{T}_\alpha f, \left(\frac{6}{\sqrt{N+1}}\right)^{\frac{1}{\alpha+1}}\right)}{\prod_{j=0}^{n-1}(\alpha - j)}. \tag{7.56}$$

$$\left[\frac{6}{(\alpha - n)\sqrt{N+1}} + \frac{1}{(\alpha+1)} \left(\frac{6}{\sqrt{N+1}}\right)^{\frac{\alpha}{\alpha+1}} \right], \; \forall\, N \in \mathbb{N}.$$

We get $\lim\limits_{N \to +\infty} B_N^{(M)}(f)(x) = f(x)$.

Proof Here $[a, b] = [0, 1]$. We apply (7.52) for $B_N^{(M)}$ and use (7.54) to get:

$$\left| B_N^{(M)}(f)(x) - f(x) \right| \le \frac{\omega_1 \left({}^x\mathbf{T}_\alpha f, \delta\right)}{\prod_{j=0}^{n-1}(\alpha - j)} \left[\frac{B_N \left(|\cdot - x|^\alpha\right)(x)}{(\alpha - n)} + \frac{B_N \left(|\cdot - x|^{\alpha+1}\right)(x)}{(\alpha+1)\delta} \right] \tag{7.57}$$

$$\le \frac{\omega_1 \left({}^x\mathbf{T}_\alpha f, \delta\right)}{\prod_{j=0}^{n-1}(\alpha - j)} \left[\frac{6}{(\alpha - n)\sqrt{N+1}} + \frac{\left(\frac{6}{\sqrt{N+1}}\right)}{(\alpha+1)\delta} \right].$$

Choose $\delta = \left(\frac{6}{\sqrt{N+1}}\right)^{\frac{1}{\alpha+1}}$, then $\delta^{\alpha+1} = \frac{6}{\sqrt{N+1}}$ and $\delta^\alpha = \left(\frac{6}{\sqrt{N+1}}\right)^{\frac{\alpha}{\alpha+1}}$, and apply it to (7.57). Clearly we derive (7.56). ∎

We continue with

Remark 7.17 The truncated Favard–Szász–Mirakjan operators are given by

$$T_N^{(M)}(f)(x) = \frac{\bigvee_{k=0}^N s_{N,k}(x) f\left(\frac{k}{N}\right)}{\bigvee_{k=0}^N s_{N,k}(x)}, \quad x \in [0,1], \ N \in \mathbb{N}, \ f \in C_+([0,1]),$$

(7.58)

$s_{N,k}(x) = \frac{(Nx)^k}{k!}$, see also [4], p. 11.

By [4], p. 178–179, we get that

$$T_N^{(M)}(|\cdot - x|)(x) \le \frac{3}{\sqrt{N}}, \quad \forall \ x \in [0,1], \ \forall \ N \in \mathbb{N}.$$

(7.59)

Clearly it holds

$$T_N^{(M)}\left(|\cdot - x|^{1+\beta}\right)(x) \le \frac{3}{\sqrt{N}}, \quad \forall \ x \in [0,1], \ \forall \ N \in \mathbb{N}, \ \forall \ \beta > 0.$$

(7.60)

The operators $T_N^{(M)}$ are positive sublinear operators mapping $C_+([0,1])$ into itself, with $T_N^{(M)}(1) = 1$.

Furthermore it holds

$$T_N^{(M)}\left(|\cdot - x|^\lambda\right)(x) = \frac{\bigvee_{k=0}^N \frac{(Nx)^k}{k!} \left|\frac{k}{N} - x\right|^\lambda}{\bigvee_{k=0}^N \frac{(Nx)^k}{k!}} > 0, \quad \forall \ x \in (0,1], \ \forall \ \lambda \ge 1, \ \forall \ N \in \mathbb{N}.$$

(7.61)

We give

Theorem 7.18 *Same assumptions as in Theorem 7.16. Then*

$$\left|T_N^{(M)}(f)(x) - f(x)\right| \le \frac{\omega_1\left({}^x\mathbf{T}_\alpha f, \left(\frac{3}{\sqrt{N}}\right)^{\frac{1}{\alpha+1}}\right)}{\prod_{j=0}^{n-1}(\alpha - j)}.$$

(7.62)

$$\left[\frac{3}{(\alpha - n)\sqrt{N}} + \frac{1}{(\alpha + 1)}\left(\frac{3}{\sqrt{N}}\right)^{\frac{\alpha}{\alpha+1}}\right], \quad \forall \ N \in \mathbb{N}.$$

We get $\lim_{N \to +\infty} T_N^{(M)}(f)(x) = f(x)$.

Proof Use of Theorems 7.14 and (7.60), similar to the proof of Theorem 7.16. ∎

We make

Remark 7.19 Next we study the truncated Max-product Baskakov operators (see [4], p. 11)

$$U_N^{(M)}(f)(x) = \frac{\bigvee_{k=0}^N b_{N,k}(x) f\left(\frac{k}{N}\right)}{\bigvee_{k=0}^N b_{N,k}(x)}, \quad x \in [0,1], \ f \in C_+([0,1]), \ N \in \mathbb{N},$$

(7.63)

where

$$b_{N,k}(x) = \binom{N+k-1}{k} \frac{x^k}{(1+x)^{N+k}}. \tag{7.64}$$

From [4], pp. 217–218, we get ($x \in [0, 1]$)

$$\left(U_N^{(M)}(|\cdot - x|)\right)(x) \le \frac{2\sqrt{3}\left(\sqrt{2}+2\right)}{\sqrt{N+1}}, \ N \ge 2, \ N \in \mathbb{N}. \tag{7.65}$$

Let $\lambda \ge 1$, clearly then it holds

$$\left(U_N^{(M)}(|\cdot - x|^\lambda)\right)(x) \le \frac{2\sqrt{3}\left(\sqrt{2}+2\right)}{\sqrt{N+1}}, \ \forall \ N \ge 2, \ N \in \mathbb{N}. \tag{7.66}$$

Also it holds $U_N^{(M)}(1) = 1$, and $U_N^{(M)}$ are positive sublinear operators from $C_+([0, 1])$ into itself. Furthermore it holds

$$U_N^{(M)}\left(|\cdot - x|^\lambda\right)(x) > 0, \ \forall \ x \in (0, 1], \ \forall \ \lambda \ge 1, \ \forall \ N \in \mathbb{N}. \tag{7.67}$$

We give

Theorem 7.20 *Same assumptions as in Theorem 7.16. Then*

$$\left|U_N^{(M)}(f)(x) - f(x)\right| \le \frac{\omega_1\left({}^x\mathbf{T}_\alpha f, \left(\frac{2\sqrt{3}(\sqrt{2}+2)}{\sqrt{N+1}}\right)^{\frac{1}{\alpha+1}}\right)}{\prod_{j=0}^{n-1}(\alpha - j)} \cdot$$

$$\left[\frac{2\sqrt{3}\left(\sqrt{2}+2\right)}{(\alpha - n)\sqrt{N+1}} + \frac{1}{(\alpha + 1)}\left(\frac{2\sqrt{3}\left(\sqrt{2}+2\right)}{\sqrt{N+1}}\right)^{\frac{\alpha}{\alpha+1}}\right], \ \forall \ N \in \mathbb{N} - \{1\}. \tag{7.68}$$

We get $\lim_{N \to +\infty} U_N^{(M)}(f)(x) = f(x)$.

Proof Use of Theorems 7.14 and (7.66), similar to the proof of Theorem 7.16. ∎

We make

Remark 7.21 Here we study the Max-product Meyer-Köning and Zeller operators (see [4], p. 11) defined by

$$Z_N^{(M)}(f)(x) = \frac{\bigvee_{k=0}^\infty s_{N,k}(x) f\left(\frac{k}{N+k}\right)}{\bigvee_{k=0}^\infty s_{N,k}(x)}, \ \forall \ N \in \mathbb{N}, \ f \in C_+([0, 1]), \tag{7.69}$$

$$s_{N,k}(x) = \binom{N+k}{k} x^k, \, x \in [0, 1].$$

By [4], p. 253, we get that

$$Z_N^{(M)}(|\cdot - x|)(x) \le \frac{8\left(1+\sqrt{5}\right)}{3} \frac{\sqrt{x}\,(1-x)}{\sqrt{N}}, \, \forall \, x \in [0, 1], \, \forall \, N \ge 4, \, N \in \mathbb{N}.$$

(7.70)

As before we get that (for $\lambda \ge 1$)

$$Z_N^{(M)}(|\cdot - x|^\lambda)(x) \le \frac{8\left(1+\sqrt{5}\right)}{3} \frac{\sqrt{x}\,(1-x)}{\sqrt{N}} := \rho(x),$$

(7.71)

$\forall \, x \in [0, 1], \, N \ge 4, \, N \in \mathbb{N}$.

Also it holds $Z_N^{(M)}(1) = 1$, and $Z_N^{(M)}$ are positive sublinear operators from $C_+([0, 1])$ into itself. Also it holds

$$Z_N^{(M)}\left(|\cdot - x|^\lambda\right)(x) > 0, \, \forall \, x \in (0, 1), \, \forall \, \lambda \ge 1, \, \forall \, N \in \mathbb{N}.$$

(7.72)

We give

Theorem 7.22 *Same assumptions as in Theorem 7.16. Then*

$$\left| Z_N^{(M)}(f)(x) - f(x) \right| \le \frac{\omega_1\left({}^x\mathbf{T}_\alpha f, (\rho(x))^{\frac{1}{\alpha+1}} \right)}{\prod_{j=0}^{n-1}(\alpha - j)}.$$

(7.73)

$$\left[\frac{\rho(x)}{(\alpha - n)} + \frac{(\rho(x))^{\frac{\alpha}{\alpha+1}}}{(\alpha + 1)} \right], \, \forall \, N \in \mathbb{N}, \, N \ge 4.$$

We get $\lim_{N \to +\infty} Z_N^{(M)}(f)(x) = f(x)$, *where* $\rho(x)$ *is as in (7.71).*

Proof Use of Theorems 7.14 and (7.71), similar to the proof of Theorem 7.16. ∎

We make

Remark 7.23 Here we deal with the Max-product truncated sampling operators (see [4], p. 13) defined by

$$W_N^{(M)}(f)(x) = \frac{\bigvee_{k=0}^N \frac{\sin(Nx - k\pi)}{Nx - k\pi} f\left(\frac{k\pi}{N}\right)}{\bigvee_{k=0}^N \frac{\sin(Nx - k\pi)}{Nx - k\pi}},$$

(7.74)

$\forall \, x \in [0, \pi], \, f : [0, \pi] \to \mathbb{R}_+$ a continuous function.

Following [4], p. 343, and making the convention $\frac{\sin(0)}{0} = 1$ and denoting $s_{N,k}$ $(x) = \frac{\sin(Nx-k\pi)}{Nx-k\pi}$, we get that $s_{N,k}\left(\frac{k\pi}{N}\right) = 1$, and $s_{N,k}\left(\frac{j\pi}{N}\right) = 0$, if $k \neq j$, furthermore $W_N^{(M)}(f)\left(\frac{j\pi}{N}\right) = f\left(\frac{j\pi}{N}\right)$, for all $j \in \{0, ..., N\}$.

Clearly $W_N^{(M)}(f)$ is a well-defined function for all $x \in [0, \pi]$, and it is continuous on $[0, \pi]$, also $W_N^{(M)}(1) = 1$.

By [4], p. 344, $W_N^{(M)}$ are positive sublinear operators.

Call $I_N^+(x) = \{k \in \{0, 1, ..., N\}; s_{N,k}(x) > 0\}$, and set $x_{N,k} := \frac{k\pi}{N}$, $k \in \{0, 1, ..., N\}$.

We see that

$$W_N^{(M)}(f)(x) = \frac{\bigvee_{k \in I_N^+(x)} s_{N,k}(x)\, f\left(x_{N,k}\right)}{\bigvee_{k \in I_N^+(x)} s_{N,k}(x)}. \tag{7.75}$$

By [4], p. 346, we have

$$W_N^{(M)}(|\cdot - x|)(x) \leq \frac{\pi}{2N}, \quad \forall\, N \in \mathbb{N}, \quad \forall\, x \in [0, \pi]. \tag{7.76}$$

Notice also $|x_{N,k} - x| \leq \pi$, $\forall\, x \in [0, \pi]$.

Therefore $(\lambda \geq 1)$ it holds

$$W_N^{(M)}\left(|\cdot - x|^\lambda\right)(x) \leq \frac{\pi^{\lambda-1}\pi}{2N} = \frac{\pi^\lambda}{2N}, \quad \forall\, x \in [0, \pi], \quad \forall\, N \in \mathbb{N}. \tag{7.77}$$

If $x \in \left(\frac{j\pi}{N}, \frac{(j+1)\pi}{N}\right)$, with $j \in \{0, 1, ..., N\}$, we obtain $nx - j\pi \in (0, \pi)$ and thus $s_{N,j}(x) = \frac{\sin(Nx-j\pi)}{Nx-j\pi} > 0$, see [4], pp. 343–344.

Consequently it holds $(\lambda \geq 1)$

$$W_N^{(M)}\left(|\cdot - x|^\lambda\right)(x) = \frac{\bigvee_{k \in I_N^+(x)} s_{N,k}(x)\, \left|x_{N,k} - x\right|^\lambda}{\bigvee_{k \in I_N^+(x)} s_{N,k}(x)} > 0, \quad \forall\, x \in [0, \pi], \tag{7.78}$$

such that $x \neq x_{N,k}$, for any $k \in \{0, 1, ..., N\}$.

We give

Theorem 7.24 *Let* $\alpha \in (n, n + 1)$, $n \in \mathbb{N}$, *and* $f \in C^{n+1}([0, \pi], \mathbb{R}_+)$, $x \in [0, \pi]$ *and* $f^{(k)}(x) = 0$, $k = 1, ..., n$. *Denote*

$$\omega_1\left({}^x\mathbf{T}_\alpha f, \delta\right) := \max\left\{\omega_1\left(\mathbf{T}_\alpha^x f, \delta\right)_{[x,\pi]}, \omega_1\left({}_\alpha^x\mathbf{T} f, \delta\right)_{[0,x]}\right\}. \tag{7.79}$$

Then

$$\left|W_N^{(M)}(f)(x) - f(x)\right| \leq \frac{\omega_1\left({}^x\mathbf{T}_\alpha f, \left(\frac{\pi^{\alpha+1}}{2N}\right)^{\frac{1}{\alpha+1}}\right)}{\prod_{j=0}^{n-1}(\alpha - j)}. \tag{7.80}$$

$$\left[\frac{\pi^\alpha}{(\alpha - n) \, 2N} + \frac{1}{(\alpha + 1)} \left(\frac{\pi^{\alpha+1}}{2N} \right)^{\frac{\alpha}{\alpha+1}} \right], \ \forall \ N \in \mathbb{N}.$$

We get $\lim\limits_{N \to +\infty} W_N^{(M)} (f) (x) = f (x)$.

Proof We apply Theorem 7.14 for $W_N^{(M)}$ and we use (7.77), we get:

$$\left| W_N^{(M)} (f) (x) - f (x) \right| \le \frac{\omega_1 \left(^x \mathbf{T}_\alpha f, \delta \right)}{\prod_{j=0}^{n-1} (\alpha - j)} \left[\frac{\pi^\alpha}{(\alpha - n) \, 2N} + \frac{\left(\frac{\pi^{\alpha+1}}{2N} \right)}{(\alpha + 1) \, \delta} \right], \quad (7.81)$$

$\delta > 0$.

Choose $\delta = \left(\frac{\pi^{\alpha+1}}{2N} \right)^{\frac{1}{\alpha+1}}$, i.e. $\delta^{\alpha+1} = \frac{\pi^{\alpha+1}}{2N}$, and $\delta^\alpha = \left(\frac{\pi^{\alpha+1}}{2N} \right)^{\frac{\alpha}{\alpha+1}}$. We use the last into (7.81) to derive (7.80). ∎

We make

Remark 7.25 Here we continue with the Max-product truncated sampling operators (see [4], p. 13) defined by

$$K_N^{(M)} (f) (x) = \frac{\bigvee_{k=0}^{N} \frac{\sin^2(Nx - k\pi)}{(Nx - k\pi)^2} f \left(\frac{k\pi}{N} \right)}{\bigvee_{k=0}^{N} \frac{\sin^2(Nx - k\pi)}{(Nx - k\pi)^2}}, \quad (7.82)$$

$\forall \ x \in [0, \pi]$, $f : [0, \pi] \to \mathbb{R}_+$ a continuous function.

Following [4], p. 350, and making the convention $\frac{\sin(0)}{0} = 1$ and denoting $s_{N,k}$ $(x) = \frac{\sin^2(Nx - k\pi)}{(Nx - k\pi)^2}$, we get that $s_{N,k} \left(\frac{k\pi}{N} \right) = 1$, and $s_{N,k} \left(\frac{j\pi}{N} \right) = 0$, if $k \ne j$, furthermore $K_N^{(M)} (f) \left(\frac{j\pi}{N} \right) = f \left(\frac{j\pi}{N} \right)$, for all $j \in \{0, ..., N\}$.

Since $s_{N,j} \left(\frac{j\pi}{N} \right) = 1$ it follows that $\bigvee_{k=0}^{N} s_{N,k} \left(\frac{j\pi}{N} \right) \ge 1 > 0$, for all $j \in \{0, 1, ..., N\}$. Hence $K_N^{(M)} (f)$ is well-defined function for all $x \in [0, \pi]$, and it is continuous on $[0, \pi]$, also $K_N^{(M)} (1) = 1$. By [4], p. 350, $K_N^{(M)}$ are positive sublinear operators.

Denote $x_{N,k} := \frac{k\pi}{N}$, $k \in \{0, 1, ..., N\}$.

By [4], p. 352, we have

$$K_N^{(M)} (|\cdot - x|) (x) \le \frac{\pi}{2N}, \ \forall \ N \in \mathbb{N}, \ \forall \ x \in [0, \pi]. \quad (7.83)$$

Notice also $\left| x_{N,k} - x \right| \le \pi, \forall \ x \in [0, \pi]$.

Therefore ($\lambda \ge 1$) it holds

$$K_N^{(M)} \left(|\cdot - x|^\lambda \right) (x) \le \frac{\pi^{\lambda-1} \pi}{2N} = \frac{\pi^\lambda}{2N}, \ \forall \ x \in [0, \pi], \ \forall \ N \in \mathbb{N}. \quad (7.84)$$

If $x \in \left(\frac{j\pi}{N}, \frac{(j+1)\pi}{N} \right)$, with $j \in \{0, 1, ..., N\}$, we obtain $nx - j\pi \in (0, \pi)$ and thus $s_{N,j}(x) = \frac{\sin^2(Nx - j\pi)}{(Nx - j\pi)^2} > 0$, see [4], pp. 350.

Consequently it holds ($\lambda \geq 1$)

$$K_N^{(M)} \left(|\cdot - x|^\lambda \right)(x) = \frac{\bigvee_{k=0}^N s_{N,k}(x) |x_{N,k} - x|^\lambda}{\bigvee_{k=0}^N s_{N,k}(x)} > 0, \ \forall \ x \in [0, \pi], \quad (7.85)$$

such that $x \neq x_{N,k}$, for any $k \in \{0, 1, ..., N\}$.

We give

Theorem 7.26 *Let $\alpha \in (n, n+1)$, $n \in \mathbb{N}$, and $f \in C^{n+1}([0, \pi], \mathbb{R}_+)$, $x \in [0, \pi]$ and $f^{(k)}(x) = 0$, $k = 1, ..., n$. Denote*

$$\omega_1 \left({}^x\mathbf{T}_\alpha f, \delta \right) := \max \left\{ \omega_1 \left(\mathbf{T}_\alpha^x f, \delta \right)_{[x,\pi]}, \omega_1 \left({}^x_\alpha \mathbf{T} f, \delta \right)_{[0,x]} \right\}. \quad (7.86)$$

Then

$$\left| K_N^{(M)}(f)(x) - f(x) \right| \leq \frac{\omega_1 \left({}^x\mathbf{T}_\alpha f, \left(\frac{\pi^{\alpha+1}}{2N} \right)^{\frac{1}{\alpha+1}} \right)}{\prod_{j=0}^{n-1} (\alpha - j)}. \quad (7.87)$$

$$\left[\frac{\pi^\alpha}{(\alpha - n) 2N} + \frac{1}{(\alpha + 1)} \left(\frac{\pi^{\alpha+1}}{2N} \right)^{\frac{\alpha}{\alpha+1}} \right], \ \forall \ N \in \mathbb{N}.$$

We get $\lim\limits_{N \to +\infty} K_N^{(M)}(f)(x) = f(x)$.

Proof As in Theorem 7.24. ∎

We make

Remark 7.27 We mention the interpolation Hermite-Fejer polynomials on Chebyshev knots of the first kind (see [4], p. 4): Let $f : [-1, 1] \to \mathbb{R}$ and based on the knots $x_{N,k} = \cos \left(\frac{(2(N-k)+1)}{2(N+1)} \pi \right) \in (-1, 1)$, $k \in \{0, ..., N\}$, $-1 < x_{N,0} < x_{N,1} < ... < x_{N,N} < 1$, which are the roots of the first kind Chebyshev polynomial $T_{N+1}(x) = \cos((N+1) \arccos x)$, we define (see Fejér [5])

$$H_{2N+1}(f)(x) = \sum_{k=0}^N h_{N,k}(x) f(x_{N,k}), \quad (7.88)$$

where

$$h_{N,k}(x) = (1 - x \cdot x_{N,k}) \left(\frac{T_{N+1}(x)}{(N+1)(x - x_{N,k})} \right)^2, \quad (7.89)$$

the fundamental interpolation polynomials.

The Max-product interpolation Hermite-Fejér operators on Chebyshev knots of the first kind (see p. 12 of [4]) are defined by

$$H_{2N+1}^{(M)}(f)(x) = \frac{\bigvee_{k=0}^{N} h_{N,k}(x) f(x_{N,k})}{\bigvee_{k=0}^{N} h_{N,k}(x)}, \quad \forall \ N \in \mathbb{N}, \tag{7.90}$$

where $f : [-1, 1] \to \mathbb{R}_+$ is continuous.

Call

$$E_N(x) := H_{2N+1}^{(M)}(|\cdot - x|)(x) = \frac{\bigvee_{k=0}^{N} h_{N,k}(x) |x_{N,k} - x|}{\bigvee_{k=0}^{N} h_{N,k}(x)}, \quad x \in [-1, 1]. \tag{7.91}$$

Then by [4], p. 287 we obtain that

$$E_N(x) \le \frac{2\pi}{N+1}, \quad \forall \ x \in [-1, 1], \quad N \in \mathbb{N}. \tag{7.92}$$

For $m > 1$, we get

$$H_{2N+1}^{(M)}(|\cdot - x|^m)(x) = \frac{\bigvee_{k=0}^{N} h_{N,k}(x) |x_{N,k} - x|^m}{\bigvee_{k=0}^{N} h_{N,k}(x)} =$$

$$\frac{\bigvee_{k=0}^{N} h_{N,k}(x) |x_{N,k} - x| |x_{N,k} - x|^{m-1}}{\bigvee_{k=0}^{N} h_{N,k}(x)} \le 2^{m-1} \frac{\bigvee_{k=0}^{N} h_{N,k}(x) |x_{N,k} - x|}{\bigvee_{k=0}^{N} h_{N,k}(x)} \tag{7.93}$$

$$\le \frac{2^m \pi}{N+1}, \quad \forall \ x \in [-1, 1], \quad N \in \mathbb{N}.$$

Hence it holds

$$H_{2N+1}^{(M)}(|\cdot - x|^m)(x) \le \frac{2^m \pi}{N+1}, \quad \forall \ x \in [-1, 1], \quad m > 1, \forall \ N \in \mathbb{N}. \tag{7.94}$$

Furthermore we have

$$H_{2N+1}^{(M)}(1)(x) = 1, \quad \forall \ x \in [-1, 1], \tag{7.95}$$

and $H_{2N+1}^{(M)}$ maps continuous functions to continuous functions over $[-1, 1]$ and for any $x \in \mathbb{R}$ we have $\bigvee_{k=0}^{N} h_{N,k}(x) > 0$.

We also have $h_{N,k}(x_{N,k}) = 1$, and $h_{N,k}(x_{N,j}) = 0$, if $k \neq j$, furthermore it holds $H_{2N+1}^{(M)}(f)(x_{N,j}) = f(x_{N,j})$, for all $j \in \{0, ..., N\}$, see [4], p. 282.

$H_{2N+1}^{(M)}$ are positive sublinear operators, [4], p. 282.

Next we use $[a, b] = [-1, 1]$.
We give

Theorem 7.28 Let $\alpha \in (n, n+1)$, $n \in \mathbb{N}$, and $f \in C^{n+1}([-1, 1], \mathbb{R}_+)$, $x \in [-1, 1]$ and $f^{(k)}(x) = 0$, $k = 1, ..., n$. Then

$$\left| H_{2N+1}^{(M)}(f)(x) - f(x) \right| \leq \frac{\omega_1\left({}^x\mathbf{T}_\alpha f, \left(\frac{2^{\alpha+1}\pi}{N+1}\right)^{\frac{1}{\alpha+1}}\right)}{\prod_{j=0}^{n-1}(\alpha - j)} \cdot \tag{7.96}$$

$$\left[\frac{2^\alpha \pi}{(\alpha - n)(N+1)} + \frac{1}{(\alpha+1)}\left(\frac{2^{\alpha+1}\pi}{N+1}\right)^{\frac{\alpha}{\alpha+1}}\right], \quad \forall N \in \mathbb{N}.$$

Furthermore it holds $\lim_{N \to +\infty} H_{2N+1}^{(M)}(f)(x) = f(x)$.

Proof By Theorem 7.14, choose $\delta := \left(\frac{2^{\alpha+1}\pi}{N+1}\right)^{\frac{1}{\alpha+1}}$, use of (7.52) and (7.94). ∎

We continue with

Remark 7.29 Here we deal with Lagrange interpolation polynomials on Chebyshev knots of second kind plus the endpoints ± 1 (see [4], p. 5). These polynomials are linear operators attached to $f : [-1, 1] \to \mathbb{R}$ and to the knots $x_{N,k} = \cos\left(\left(\frac{N-k}{N-1}\right)\pi\right) \in [-1, 1]$, $k = 1, ..., N$, $N \in \mathbb{N}$, which are the roots of $\omega_N(x) = \sin(N-1)t \sin t$, $x = \cos t$. Notice that $x_{N,1} = -1$ and $x_{N,N} = 1$. Their formula is given by ([4], p. 377)

$$L_N(f)(x) = \sum_{k=1}^{N} l_{N,k}(x) f(x_{N,k}), \tag{7.97}$$

where

$$l_{N,k}(x) = \frac{(-1)^{k-1}\omega_N(x)}{(1 + \delta_{k,1} + \delta_{k,N})(N-1)(x - x_{N,k})}, \tag{7.98}$$

$N \geq 2$, $k = 1, ..., N$, and $\omega_N(x) = \prod_{k=1}^{N}(x - x_{N,k})$ and $\delta_{i,j}$ denotes the Kronecher's symbol, that is $\delta_{i,j} = 1$, if $i = j$, and $\delta_{i,j} = 0$, if $i \neq j$.

The Max-product Lagrange interpolation operators on Chebyshev knots of second kind, plus the endpoints ± 1, are defined by ([4], p. 12)

$$L_N^{(M)}(f)(x) = \frac{\bigvee_{k=1}^{N} l_{N,k}(x) f(x_{N,k})}{\bigvee_{k=1}^{N} l_{N,k}(x)}, \quad x \in [-1, 1], \tag{7.99}$$

where $f : [-1, 1] \to \mathbb{R}_+$ continuous.

First we see that $L_N^{(M)}(f)(x)$ is well defined and continuous for any $x \in [-1, 1]$. Following [4], p. 289, because $\sum_{k=1}^{N} l_{N,k}(x) = 1$, $\forall x \in \mathbb{R}$, for any x there exists

$k \in \{1, ..., N\} : l_{N,k}(x) > 0$, hence $\bigvee_{k=1}^{N} l_{N,k}(x) > 0$. We have that $l_{N,k}(x_{N,k}) = 1$, and $l_{N,k}(x_{N,j}) = 0$, if $k \neq j$. Furthermore it holds $L_N^{(M)}(f)(x_{N,j}) = f(x_{N,j})$, all $j \in \{1, ..., N\}$, and $L_N^{(M)}(1) = 1$.

Call $I_N^+(x) = \{k \in \{1, ..., N\}; l_{N,k}(x) > 0\}$, then $I_N^+(x) \neq \emptyset$.

So for $f \in C_+([-1, 1])$ we get

$$L_N^{(M)}(f)(x) = \frac{\bigvee_{k \in I_N^+(x)} l_{N,k}(x) f(x_{N,k})}{\bigvee_{k \in I_N^+(x)} l_{N,k}(x)} \geq 0. \tag{7.100}$$

Notice here that $|x_{N,k} - x| \leq 2, \forall x \in [-1, 1]$.

By [4], p. 297, we get that

$$L_N^{(M)}(|\cdot - x|)(x) = \frac{\bigvee_{k=1}^{N} l_{N,k}(x) |x_{N,k} - x|}{\bigvee_{k=1}^{N} l_{N,k}(x)} =$$

$$\frac{\bigvee_{k \in I_N^+(x)} l_{N,k}(x) |x_{N,k} - x|}{\bigvee_{k \in I_N^+(x)} l_{N,k}(x)} \leq \frac{\pi^2}{6(N-1)}, \tag{7.101}$$

$N \geq 3, \forall x \in (-1, 1), N$ is odd.

We get that $(m > 1)$

$$L_N^{(M)}(|\cdot - x|^m)(x) = \frac{\bigvee_{k \in I_N^+(x)} l_{N,k}(x) |x_{N,k} - x|^m}{\bigvee_{k \in I_N^+(x)} l_{N,k}(x)} \leq \frac{2^{m-1} \pi^2}{6(N-1)}, \tag{7.102}$$

$N \geq 3$ odd, $\forall x \in (-1, 1)$.

$L_N^{(M)}$ are positive sublinear operators, [4], p. 290.

We give

Theorem 7.30 *Same assumptions as in Theorem 7.28. Then*

$$\left| L_N^{(M)}(f)(x) - f(x) \right| \leq \frac{\omega_1 \left({}^x T_\alpha f, \left(\frac{2^\alpha \pi^2}{6(N-1)} \right)^{\frac{1}{\alpha+1}} \right)}{\prod_{j=0}^{n-1} (\alpha - j)}. \tag{7.103}$$

$$\left[\frac{2^{\alpha-1} \pi^2}{6(\alpha-n)(N-1)} + \frac{1}{(\alpha+1)} \left(\frac{2^\alpha \pi^2}{6(N-1)} \right)^{\frac{\alpha}{\alpha+1}} \right], \forall N \in \mathbb{N} : N \geq 3, \text{ odd}.$$

It holds $\lim_{N \to +\infty} L_N^{(M)}(f)(x) = f(x)$.

Proof By Theorem 7.14, choose $\delta := \left(\frac{2^\alpha \pi^2}{6(N-1)} \right)^{\frac{1}{\alpha+1}}$, use of (7.52) and (7.102). At ± 1 the left hand side of (7.103) is zero, thus (7.103) is trivially true. ∎

We make

Remark 7.31 Let $f \in C_+ ([-1, 1])$, $N \geq 4$, $N \in \mathbb{N}$, N even.
By [4], p. 298, we get

$$L_N^{(M)} (|\cdot - x|) (x) \leq \frac{4\pi^2}{3 (N - 1)} = \frac{2^2 \pi^2}{3 (N - 1)}, \quad \forall x \in (-1, 1). \tag{7.104}$$

Hence $(m > 1)$

$$L_N^{(M)} (|\cdot - x|^m) (x) \leq \frac{2^{m+1} \pi^2}{3 (N - 1)}, \quad \forall \, x \in (-1, 1). \tag{7.105}$$

We present

Theorem 7.32 *Same assumptions as in Theorem 7.28. Then*

$$\left| L_N^{(M)} (f) (x) - f (x) \right| \leq \frac{\omega_1 \left({}^x\mathbf{T}_\alpha f, \left(\frac{2^{\alpha+2} \pi^2}{3(N-1)} \right)^{\frac{1}{\alpha+1}} \right)}{\prod_{j=0}^{n-1} (\alpha - j)}. \tag{7.106}$$

$$\left[\frac{2^{\alpha+1} \pi^2}{3 (\alpha - n) (N - 1)} + \frac{1}{(\alpha + 1)} \left(\frac{2^{\alpha+2} \pi^2}{3 (N - 1)} \right)^{\frac{\alpha}{\alpha+1}} \right], \quad \forall \, N \in \mathbb{N}, \; N \geq 4, \; N \text{ is even.}$$

It holds $\lim_{N \to +\infty} L_N^{(M)} (f) (x) = f (x)$.

Proof By Theorem 7.14, choose $\delta := \left(\frac{2^{\alpha+2} \pi^2}{3(N-1)} \right)^{\frac{1}{\alpha+1}}$, use of (7.52) and (7.105). At ± 1, (7.106) is trivially true. ■

7.5 Applications, Part B

Case of $\alpha \in (0, 1)$.
    We start with an independent proof:
    Proof of (7.22) ($\alpha \in (0, 1]$)
    We have

$$^b I_\alpha \; {}_\alpha^b T (f) (t) = \int_t^b (b - x)^{\alpha-1} \left({}_\alpha^b T (f) \right) (x) \, dx =$$

$$\int_t^b (b - x)^{\alpha-1} (-1) (b - x)^{1-\alpha} f' (x) \, dx = - \int_t^b f' (x) \, dx =$$

$$- (f (b) - f (t)) = f (t) - f (b).$$

We state

Theorem 7.33 *Let* $f \in C^1 ([a, b])$, $\alpha \in (0, 1)$. *Then*
  (1)

$$f (t) - f (a) = \int_a^t (x - a)^{\alpha-1} \left[\left(T_\alpha^a (f) \right) (x) - \left(T_\alpha^a (f) \right) (a) \right] dx, \qquad (7.107)$$

and
  (2)

$$f (t) - f (b) = \int_t^b (b - x)^{\alpha-1} \left[\left({}_\alpha^b T (f) \right) (x) - \left({}_\alpha^b T (f) \right) (b) \right] dx, \qquad (7.108)$$

$\forall\, t \in [a, b]$.

Proof By (7.11), (7.12), (7.17) and (7.22). ■

We state the following

Theorem 7.34 *Let* $f \in C^1 ([a, b])$, $\alpha \in (0, 1)$, $x_0 \in [a, b]$. *Then*
  (1)

$$f (t) - f (x_0) = \int_{x_0}^t (x - x_0)^{\alpha-1} \left[\left(T_\alpha^{x_0} (f) \right) (x) - \left(T_\alpha^{x_0} (f) \right) (x_0) \right] dx, \qquad (7.109)$$

$\forall\, t \in [x_0, b]$,
  and
  (2)

$$f (t) - f (x_0) = \int_t^{x_0} (x_0 - x)^{\alpha-1} \left[\left({}_\alpha^{x_0} T (f) \right) (x) - \left({}_\alpha^{x_0} T (f) \right) (x_0) \right] dx, \qquad (7.110)$$

$\forall\, t \in [a, x_0]$.

Proof By Theorem 7.33. ■

We need

Theorem 7.35 *Let* $f \in C^1 ([a, b])$, $\alpha \in (0, 1)$, $x_0 \in [a, b]$. *Denote*

$$\omega_1 \left({}^{x_0} T_\alpha f, \delta \right) := \max \left\{ \omega_1 \left(T_\alpha^{x_0} f, \delta \right)_{[x_0, b]}, \omega_1 \left({}_\alpha^{x_0} T f, \delta \right)_{[a, x_0]} \right\}, \quad \delta > 0. \qquad (7.111)$$

Then

$$|f(t) - f(x_0)| \leq \omega_1 \left({}^{x_0}T_\alpha f, \delta\right) \left[\frac{|t - x_0|^\alpha}{\alpha} + \frac{|t - x_0|^{\alpha+1}}{(\alpha+1)\delta}\right], \qquad (7.112)$$

$\forall\, t \in [a, b],\, \delta > 0.$

Equation (7.111) defines indirectly the mixed conformable fractional derivative of f of order α.

Proof Clearly here $T_\alpha^{x_0}(f) \in C([x_0, b])$ and $\left({}^{x_0}_\alpha T(f)\right) \in C([a, x_0])$.

We have $(x_0 \leq t \leq b)$

$$|f(t) - f(x_0)| \leq \int_{x_0}^{t} (x - x_0)^{\alpha-1} \left|\left(T_\alpha^{x_0}(f)\right)(x) - \left(T_\alpha^{x_0}(f)\right)(x_0)\right| dx \leq$$

$(\delta > 0)$

$$\int_{x_0}^{t} (x - x_0)^{\alpha-1}\, \omega_1 \left(T_\alpha^{x_0}(f), \frac{\delta(x - x_0)}{\delta}\right)_{[x_0, b]} dx \leq \qquad (7.113)$$

$$\omega_1 \left(T_\alpha^{x_0}(f), \delta\right)_{[x_0, b]} \int_{x_0}^{t} (x - x_0)^{\alpha-1} \left(1 + \frac{(x - x_0)}{\delta}\right) dx =$$

$$\omega_1 \left(T_\alpha^{x_0}(f), \delta\right)_{[x_0, b]} \left[\frac{(t - x_0)^\alpha}{\alpha} + \frac{1}{\delta} \int_{x_0}^{t} (x - x_0)^\alpha\, dx\right] =$$

$$\omega_1 \left(T_\alpha^{x_0}(f), \delta\right)_{[x_0, b]} \left[\frac{(t - x_0)^\alpha}{\alpha} + \frac{1}{\delta} \frac{(t - x_0)^{\alpha+1}}{(\alpha+1)}\right].$$

We have proved that

$$|f(t) - f(x_0)| \leq \omega_1 \left(T_\alpha^{x_0}(f), \delta\right)_{[x_0, b]} \left[\frac{(t - x_0)^\alpha}{\alpha} + \frac{(t - x_0)^{\alpha+1}}{(\alpha+1)\delta}\right], \qquad (7.114)$$

$\forall\, t \in [x_0, b],\, \delta > 0.$

We have $(a \leq t \leq x_0)$

$$|f(t) - f(x_0)| \leq \int_{t}^{x_0} (x_0 - x)^{\alpha-1} \left|\left({}^{x_0}_\alpha T(f)\right)(x) - \left({}^{x_0}_\alpha T(f)\right)(x_0)\right| dx \leq$$

$(\delta > 0)$

$$\int_{t}^{x_0} (x_0 - x)^{\alpha-1}\, \omega_1 \left({}^{x_0}_\alpha T(f), \frac{\delta(x_0 - x)}{\delta}\right)_{[a, x_0]} dx \leq$$

$$\omega_1 \left({}^{x_0}_\alpha T(f), \delta\right)_{[a, x_0]} \int_{t}^{x_0} (x_0 - x)^{\alpha-1} \left(1 + \frac{(x_0 - x)}{\delta}\right) dx =$$

$$\omega_1 \left(^{x_0}_\alpha T\left(f\right), \delta\right)_{[a,x_0]} \left[\frac{(x_0 - t)^\alpha}{\alpha} + \frac{1}{\delta}\frac{(x_0 - t)^{\alpha+1}}{(\alpha+1)}\right].$$

We have proved that

$$|f(t) - f(x_0)| \le \omega_1 \left(^{x_0}_\alpha T\left(f\right), \delta\right)_{[a,x_0]} \left[\frac{(x_0 - t)^\alpha}{\alpha} + \frac{(x_0 - t)^{\alpha+1}}{(\alpha+1)\delta}\right], \qquad (7.115)$$

$\forall\, t \in [a, x_0]\,, \delta > 0.$

Based on (7.114) and (7.115) we derive (7.112).   ∎

We rewrite Theorem 7.35 as

Theorem 7.36 *Let* $f \in C^1\left([a, b]\right), \alpha \in (0, 1)\,, x \in [a, b].$ *Denote*

$$\omega_1 \left(^x T_\alpha f, \delta\right) := \max\left\{\omega_1 \left(T^x_\alpha f, \delta\right)_{[x,b]}, \omega_1 \left(^x_\alpha Tf, \delta\right)_{[a,x]}\right\}, \quad \delta > 0. \qquad (7.116)$$

Then over $[a, b]$ *we have*

$$|f(\cdot) - f(x)| \le \omega_1 \left(^x T_\alpha f, \delta\right) \left[\frac{|\cdot - x|^\alpha}{\alpha} + \frac{|\cdot - x|^{\alpha+1}}{(\alpha+1)\delta}\right], \quad \delta > 0. \qquad (7.117)$$

We give

Theorem 7.37 *Let* $f \in C^1\left([a, b]\,, \mathbb{R}_+\right), \alpha \in (0, 1),$ *and let* $L_N : C_+\left([a, b]\right) \to C_+\left([a, b]\right), \forall\, N \in \mathbb{N},$ *be positive sublinear operators, such that* $L_N(1) = 1, \forall\, N \in \mathbb{N}.$ *Then*

$$|L_N(f)(x) - f(x)| \le \omega_1 \left(^x T_\alpha f, \delta\right) \left[\frac{L_N\left(|\cdot - x|^\alpha\right)(x)}{\alpha} + \frac{L_N\left(|\cdot - x|^{\alpha+1}\right)(x)}{(\alpha+1)\delta}\right], \qquad (7.118)$$

$\forall\, N \in \mathbb{N}, \forall\, x \in [a, b]\,, \delta > 0.$

Proof By (7.117) and (7.50).   ∎

We need

Theorem 7.38 *Let* $f \in C^1\left([a, b]\,, \mathbb{R}_+\right), \alpha \in (0, 1)\,, x \in [a, b].$ *Let* $L_N : C_+\left([a, b]\right) \to C_+\left([a, b]\right), \forall\, N \in \mathbb{N}$ *be positive sublinear operators, such that* $L_N(1) = 1,$ *and* $L_N\left(|\cdot - x|^{\alpha+1}\right)(x) > 0, \forall\, N \in \mathbb{N}.$ *Then*

$$|L_N(f)(x) - f(x)| \le$$

$$\frac{(2\alpha + 1)}{\alpha(\alpha+1)}\omega_1 \left(^x T_\alpha f, \left(L_N\left(|\cdot - x|^{\alpha+1}\right)(x)\right)^{\frac{1}{\alpha+1}}\right)\left(L_N\left(|\cdot - x|^{\alpha+1}\right)(x)\right)^{\frac{\alpha}{\alpha+1}},$$

$$(7.119)$$

$\forall\, N \in \mathbb{N}.$

Proof By Theorem 7.12 we get

$$L_N \left(|\cdot - x|^{\alpha} \right) (x_0) \le \left(L_N \left(|\cdot - x|^{\alpha+1} \right) (x) \right)^{\frac{\alpha}{\alpha+1}} . \qquad (7.120)$$

Choose

$$\delta := \left(L_N \left(|\cdot - x|^{\alpha+1} \right) (x) \right)^{\frac{1}{\alpha+1}} > 0, \qquad (7.121)$$

that is

$$\delta^{\alpha+1} = L_N \left(|\cdot - x|^{\alpha+1} \right) (x) , \qquad (7.122)$$

and

$$\delta^{\alpha} = \left(L_N \left(|\cdot - x|^{\alpha+1} \right) (x) \right)^{\frac{\alpha}{\alpha+1}} . \qquad (7.123)$$

We apply (7.118) to get

$$|L_N (f) (x) - f(x)| \le \omega_1 \left({}^x T_\alpha f, \left(L_N \left(|\cdot - x|^{\alpha+1} \right) (x) \right)^{\frac{1}{\alpha+1}} \right) \cdot$$

$$\left[\frac{\left(L_N \left(|\cdot - x|^{\alpha+1} \right) (x) \right)^{\frac{\alpha}{\alpha+1}}}{\alpha} + \frac{\delta^{\alpha+1}}{(\alpha+1) \delta} \right] = \qquad (7.124)$$

$$\omega_1 \left({}^x T_\alpha f, \left(L_N \left(|\cdot - x|^{\alpha+1} \right) (x) \right)^{\frac{1}{\alpha+1}} \right) \cdot$$

$$\left[\frac{\left(L_N \left(|\cdot - x|^{\alpha+1} \right) (x) \right)^{\frac{\alpha}{\alpha+1}}}{\alpha} + \frac{\left(L_N \left(|\cdot - x|^{\alpha+1} \right) (x) \right)^{\frac{\alpha}{\alpha+1}}}{(\alpha+1)} \right] =$$

$$\frac{(2\alpha+1)}{\alpha (\alpha+1)} \omega_1 \left({}^x T_\alpha f, \left(L_N \left(|\cdot - x|^{\alpha+1} \right) (x) \right)^{\frac{1}{\alpha+1}} \right) \left(L_N \left(|\cdot - x|^{\alpha+1} \right) (x) \right)^{\frac{\alpha}{\alpha+1}} ,$$

proving (7.119). ∎

We give

Theorem 7.39 *Let* $f \in C^1 ([0, 1], \mathbb{R}_+)$, $\alpha \in (0, 1)$. *Then*

$$\left| B_N^{(M)} (f) (x) - f (x) \right| \le \frac{(2\alpha+1)}{\alpha (\alpha+1)} \omega_1 \left({}^x T_\alpha f, \left(\frac{6}{\sqrt{N+1}} \right)^{\frac{1}{\alpha+1}} \right) \left(\frac{6}{\sqrt{N+1}} \right)^{\frac{\alpha}{\alpha+1}} , \qquad (7.125)$$

$\forall \, N \in \mathbb{N}, \forall \, x \in (0, 1)$.
  Also $\lim\limits_{N \to +\infty} B_N^{(M)} (f) (x) = f (x)$.

Proof By Theorems 7.38, (7.54) and (7.55). ∎

Corollary 7.40 *Let* $f \in C^1([0, 1], \mathbb{R}_+)$. *Then*

$$\left| B_N^{(M)}(f)(x) - f(x) \right| \leq \frac{8}{3}\omega_1\left({}^xT_{1/2}f, \left(\frac{6}{\sqrt{N+1}} \right)^{\frac{2}{3}} \right) \left(\frac{6}{\sqrt{N+1}} \right)^{\frac{1}{3}}, \quad (7.126)$$

$\forall N \in \mathbb{N}, \forall x \in (0, 1)$.

Proof By (7.125). ∎

We continue with

Theorem 7.41 *Let* $f \in C^1([0, 1], \mathbb{R}_+)$, $\alpha \in (0, 1)$. *Then*

$$\left| T_N^{(M)}(f)(x) - f(x) \right| \leq \frac{(2\alpha + 1)}{\alpha(\alpha + 1)}\omega_1\left({}^xT_\alpha f, \left(\frac{3}{\sqrt{N}} \right)^{\frac{1}{\alpha+1}} \right) \left(\frac{3}{\sqrt{N}} \right)^{\frac{\alpha}{\alpha+1}},$$

$$(7.127)$$

$\forall N \in \mathbb{N}, x \in (0, 1].$
    Also $\lim_{N \to +\infty} T_N^{(M)}(f)(x) = f(x).$

Proof By Theorems 7.38, (7.60) and (7.73). ∎

We give

Theorem 7.42 *Let* $f \in C^1([0, 1], \mathbb{R}_+)$, $\alpha \in (0, 1)$. *Then*

$$\left| U_N^{(M)}(f)(x) - f(x) \right| \leq$$

$$\frac{(2\alpha + 1)}{\alpha(\alpha + 1)}\omega_1\left({}^xT_\alpha f, \left(\frac{2\sqrt{3}\left(\sqrt{2}+2\right)}{\sqrt{N+1}} \right)^{\frac{1}{\alpha+1}} \right) \left(\frac{2\sqrt{3}\left(\sqrt{2}+2\right)}{\sqrt{N+1}} \right)^{\frac{\alpha}{\alpha+1}}, \quad (7.128)$$

$\forall N \in \mathbb{N} - \{1\}, x \in (0, 1].$
    Notice that $\lim_{N \to +\infty} U_N^{(M)}(f)(x) = f(x).$

Proof By Theorems 7.38, (7.66) and (7.67). ∎

It follows

Theorem 7.43 *Let* $f \in C^1([0, 1], \mathbb{R}_+)$, $\alpha \in (0, 1)$. *Then*

$$\left| Z_N^{(M)}(f)(x) - f(x) \right| \leq \frac{(2\alpha + 1)}{\alpha(\alpha + 1)}\omega_1\left({}^xT_\alpha f, (\rho(x))^{\frac{1}{\alpha+1}} \right) (\rho(x))^{\frac{\alpha}{\alpha+1}}, \quad (7.129)$$

$\forall N \in \mathbb{N}, N \geq 4, \forall x \in (0, 1),$ *where* $\rho(x)$ *is as in* (7.71).
    Notice that $\lim_{N \to +\infty} Z_N^{(M)}(f)(x) = f(x).$

Proof By Theorems 7.38, (7.71) and (7.72). ∎

We continue with

Theorem 7.44 *Let* $f \in C^1 ([0, \pi], \mathbb{R}_+)$, $\alpha \in (0, 1)$ *and* $x \in [0, \pi]$ *be such that* $x \neq \frac{k\pi}{N}$, $k \in \{0, 1, ..., N\}$, $\forall N \in \mathbb{N}$. *Then*

$$\left| W_N^{(M)} (f) (x) - f (x) \right| \leq \frac{(2\alpha + 1)}{\alpha (\alpha + 1)} \omega_1 \left({}^x T_\alpha f, \left(\frac{\pi^{\alpha+1}}{2N} \right)^{\frac{1}{\alpha+1}} \right) \left(\frac{\pi^{\alpha+1}}{2N} \right)^{\frac{\alpha}{\alpha+1}},$$

$$(7.130)$$

$\forall N \in \mathbb{N}$. *As* $N \to +\infty$, *we get that* $W_N^{(M)} (f) (x) \to f (x)$.

Proof By Theorems 7.38, (7.77) and (7.78). ∎

We finish with

Theorem 7.45 *Let* $f \in C^1 ([0, \pi], \mathbb{R}_+)$, $\alpha \in (0, 1)$ *and* $x \in [0, \pi]$ *be such that* $x \neq \frac{k\pi}{N}$, $k \in \{0, 1, ..., N\}$, $\forall N \in \mathbb{N}$. *Then*

$$\left| K_N^{(M)} (f) (x) - f (x) \right| \leq \frac{(2\alpha + 1)}{\alpha (\alpha + 1)} \omega_1 \left({}^x T_\alpha f, \left(\frac{\pi^{\alpha+1}}{2N} \right)^{\frac{1}{\alpha+1}} \right) \left(\frac{\pi^{\alpha+1}}{2N} \right)^{\frac{\alpha}{\alpha+1}},$$

$$(7.131)$$

$\forall N \in \mathbb{N}$. *As* $N \to +\infty$, *we get that* $K_N^{(M)} (f) (x) \to f (x)$.

Proof By Theorems 7.38, (7.84) and (7.85). ∎

References

1. T. Abdeljawad, On conformable fractional calculus. J. Comput. Appl. Math. **279**, 57–66 (2015)
2. G. Anastassiou, *Approximation by Sublinear Operators* (2017), (submitted)
3. G. Anastassiou, *Mixed Conformable Fractional Approximation by Sublinear Operators* 2017, (submitted)
4. B. Bede, L. Coroianu, S. Gal, *Approximation by Max-Product type Operators* (Springer, Heidelberg, New York, 2016)
5. L. Fejér, *Über Interpolation*, Göttingen Nachrichten (1916), pp. 66–91
6. R. Khalil, M. Al Horani, A. Yousef, M. Sababheh, A new definition of fractional derivative. J. Comput. Appl. Math. **264**, 65–70 (2014)
7. G.G. Lorentz, *Bernstein Polynomials*, 2nd edn (Chelsea Publishing Company, New York, NY, 1986)
8. T. Popoviciu, Sur l'approximation de fonctions convexes d'order superieur. Mathematica (Cluj) **10**, 49–54 (1935)

Chapter 8
Approximation of Fuzzy Numbers Using Max-Product Operators

Here we study quantitatively the approximation of fuzzy numbers by fuzzy approximators generated by the Max-product operators of Bernstein type and Meyer-Köning and Zeller type. It follows [1].

8.1 Background

We need the following

Definition 8.1 (*see* [8]) Let $\mu : \mathbb{R} \to [0, 1]$ with the following properties:

(i) is normal, i.e., $\exists\, x_0 \in \mathbb{R}$; $\mu(x_0) = 1$.

(ii) $\mu(\lambda x + (1 - \lambda) y) \geq \min\{\mu(x), \mu(y)\}$, $\forall\, x, y \in \mathbb{R}, \forall\, \lambda \in [0, 1]$ (μ is called a convex fuzzy subset).

(iii) μ is upper semicontinuous on $\mathbb{R}$, i.e., $\forall\, x_0 \in \mathbb{R}$ and $\forall\, \varepsilon > 0$, $\exists$ neighborhood $V(x_0) : \mu(x) \leq \mu(x_0) + \varepsilon, \forall\, x \in V(x_0)$.

(iv) The set sup $p(\mu)$ is compact in $\mathbb{R}$ (where sup $p(\mu) := \{x \in \mathbb{R} : \mu(x) > 0\}$).

We call μ a fuzzy real number, or fuzzy number. Denote the set of all μ with $\mathbb{R}_{\mathcal{F}}$.

E.g. $\chi_{\{x_0\}} \in \mathbb{R}_{\mathcal{F}}$, for any $x_0 \in \mathbb{R}$, where $\chi_{\{x_0\}}$ is the characteristic function at x_0.

For $0 < r \leq 1$ and $\mu \in \mathbb{R}_{\mathcal{F}}$ define $[\mu]^r := \{x \in \mathbb{R} : \mu(x) \geq r\}$ and

$$[\mu]^0 := \overline{\{x \in \mathbb{R} : \mu(x) > 0\}}.$$

Then it is well known that for each $r \in [0, 1]$, $[\mu]^r$ is a closed and bounded interval of $\mathbb{R}$ [5]. For $u, v \in \mathbb{R}_{\mathcal{F}}$ and $\lambda \in \mathbb{R}$, we define uniquely the sum $u \oplus v$ and the product $\lambda \odot u$ by

$$[u \oplus v]^r = [u]^r + [v]^r, \quad [\lambda \odot u]^r = \lambda [u]^r, \quad \forall\, r \in [0, 1],$$

© Springer International Publishing AG, part of Springer Nature 2018
G. A. Anastassiou, *Nonlinearity: Ordinary and Fractional Approximations by Sublinear and Max-Product Operators*, Studies in Systems, Decision and Control 147, https://doi.org/10.1007/978-3-319-89509-3_8

where $[u]^r + [v]^r$ means the usual addition of two intervals (as subsets of $\mathbb{R}$) and $\lambda [u]^r$ means the usual product between a scalar and a subset of $\mathbb{R}$ (see, e.g., [8]). Notice $1 \odot u = u$ and it holds $u \oplus v = v \oplus u$, $\lambda \odot u = u \odot \lambda$. If $0 \leq r_1 \leq r_2 \leq 1$, then $[u]^{r_2} \subseteq [u]^{r_1}$. Actually $[u]^r = \left[u_-^{(r)}, u_+^{(r)} \right]$, where $u_-^{(r)} \leq u_+^{(r)}$, $u_-^{(r)}, u_+^{(r)} \in \mathbb{R}$, $\forall$ $r \in [0, 1]$. For $\lambda > 0$ one has $\lambda u_{\pm}^{(r)} = (\lambda \odot u)_{\pm}^{(r)}$, respectively.

Define
$$D : \mathbb{R}_\mathcal{F} \times \mathbb{R}_\mathcal{F} \to \mathbb{R}_+$$

by
$$D (u, v) := \sup_{r \in [0,1]} \max \left\{ \left| u_-^{(r)} - v_-^{(r)} \right|, \left| u_+^{(r)} - v_+^{(r)} \right| \right\} \tag{8.1}$$

$$= \sup_{r \in [0,1]} \text{Hausdorff distance } \left([u]^r, [v]^r \right),$$

where $[v]^r = \left[v_-^{(r)}, v_+^{(r)} \right]$; $u, v \in \mathbb{R}_\mathcal{F}$. We have that D is a metric on $\mathbb{R}_\mathcal{F}$. Then $(\mathbb{R}_\mathcal{F}, D)$ is a complete metric space, see [8, 9], with the properties

$$D (u \oplus w, v \oplus w) = D (u, v), \quad \forall u, v, w \in \mathbb{R}_\mathcal{F}, \tag{8.2}$$
$$D (k \odot u, k \odot v) = |k| D (u, v), \quad \forall u, v \in \mathbb{R}_\mathcal{F}, \forall k \in \mathbb{R},$$
$$D (u \oplus v, w \oplus e) \leq D (u, w) + D (v, e), \quad \forall u, v, w, e \in \mathbb{R}_\mathcal{F}.$$

On $\mathbb{R}_\mathcal{F}$ we define a partial order by "$\leq$" (or "$\preceq$"): $u, v \in \mathbb{R}_\mathcal{F}$, $u \leq v$ (or $u \preceq v$) iff $u_-^{(r)} \leq v_-^{(r)}$ and $u_+^{(r)} \leq v_+^{(r)}$, $\forall\, r \in [0, 1]$.

The zero element $\widetilde{0} \in \mathbb{R}_\mathcal{F}$ is defined by $\widetilde{0} := \chi_{\{0\}}$, clearly it holds $\widetilde{0}_{\pm}^{(r)} = 0$, $\forall$ $r \in [0, 1]$.

We call $u \in \mathbb{R}_\mathcal{F}$ positive, iff $u \succeq \widetilde{0}$, iff $u_-^{(r)} \geq 0$ and $u_+^{(r)} \geq 0$, $\forall\, r \in [0, 1]$.

From now on we denote $u_-^{(r)} := u^- (r)$ and $u_+^{(r)} := u^+ (r)$, $\forall\, r \in [0, 1]$. Actually we have that $u^-, u^+ : [0, 1] \to \mathbb{R}$, furthermore if $u \in \mathbb{R}_\mathcal{F}$ is positive then we get that

$$u^-, u^+ : [0, 1] \to \mathbb{R}_+.$$

We mention the important characterization.

Theorem 8.2 (Goetschel and Voxman [5]) *Let $u \in \mathbb{R}_\mathcal{F}$. Then*

(1) u^- is a bounded increasing function on $[0, 1]$,
(2) u^+ is a bounded decreasing function on $[0, 1]$.
(3) $u^- (1) \leq u^+ (1)$,
(4) u^- and u^+ are left continuous on $(0, 1]$ and right continuous at 0.
(5) If u^-, u^+ satisfy the above conditions (1)-(4), then there exists a unique $v \in \mathbb{R}_\mathcal{F}$ such that $v^- (r) = u^- (r)$ and $v^+ (r) = u^+ (r)$, $\forall\, r \in [0, 1]$.

Theorem 8.2 says that a fuzzy number $u \in \mathbb{R}_\mathcal{F}$ is completely determined by the end points of the intervals $[u]^r = \left[u^- (r), u^+ (r) \right]$, $\forall\, r \in [0, 1]$. Therefore we can identify a fuzzy number $u \in \mathbb{R}_{\mathcal{F}}$ with its parametric representation $\{(u^- (r), u^+ (r)) | 0 \le r \le 1\}$, and we can write $u = (u^-, u^+)$, and we call u^-, u^+ the level functions of u.

In this chapter we deal only with positive fuzzy numbers.

Define $C_+ ([0, 1]) := \{f : [0, 1] \to \mathbb{R}_+, \text{ continuous functions}\}$.

In [2], p. 10, the authors introduced the Max-product Bernstein operators

$$B_N^{(M)} (f) (x) = \frac{\bigvee_{k=0}^N p_{N,k} (x) f \left(\frac{k}{N}\right)}{\bigvee_{k=0}^N p_{N,k} (x)}, \quad N \in \mathbb{N}, \tag{8.3}$$

where $\bigvee$ stands for maximum, and $p_{N,k} (x) = \binom{N}{k} x^k (1 - x)^{N-k}$, $f \in C_+ ([0, 1])$, $\forall x \in [0, 1]$.

These are nonlinear and piecewise rational operators.

We notice that $B_N^M (1) = 1$, furthermore $B_N^{(M)}$ maps $C_+ ([0, 1])$ into itself, $B_N^M (f) (x) \ge 0$, and satisfies

$$B_N^M (f) (0) = f (0), \quad B_N^M (f) (1) = f (1), \quad \forall N \in \mathbb{N}, \tag{8.4}$$

where $f \in C_+ ([0, 1])$, see [2], p. 39.

Additionally we have (see [2], p. 40): if $f : [0, 1] \to \mathbb{R}_+$ is nondecreasing, then $B_N^M (f)$ is nondecreasing, and if $f : [0, 1] \to \mathbb{R}_+$ is nonincreasing, then $B_N^{(M)} (f)$ is nonincreasing, $\forall N \in \mathbb{N}$.

Next let $u \in \mathbb{R}_{\mathcal{F}}$ be a positive fuzzy number: $u = (u^-, u^+)$. We consider $B_N^{(M)} (u^-)$, $B_N^{(M)} (u^+)$ and since $B_N^{(M)}$ preserves the monotonicity it follows that $B_N^{(M)} (u^-)$ is increasing and $B_N^{(M)} (u^+)$ is decreasing over $[0, 1]$. We assume that $u^\pm$ are continuous, thus $B_N^{(M)} (u^\pm)$ are continuous too.

We further have

$$B_N^{(M)} (u^\pm) (0) = u^\pm (0), \tag{8.5}$$

$$B_N^{(M)} (u^\pm) (1) = u^\pm (1), \text{ respectively}, \forall N \in \mathbb{N}.$$

Also we have that

$$B_N^{(M)} (u^-) (1) = u^- (1) \le u^+ (1) = B_N^{(M)} (u^+) (1), \quad \forall N \in \mathbb{N}. \tag{8.6}$$

In conclusion (by Theorem 8.2)

$$\overline{B}_N^{(M)} (u) := \left(B_N^{(M)} (u^-), B_N^{(M)} (u^+) \right), \tag{8.7}$$

defines a proper fuzzy number in $\mathbb{R}_{\mathcal{F}}$, $\forall N \in \mathbb{N}$.

We mention

Theorem 8.3 (Bede–Coroianu–Gal, [2], p. 111) *Let* $u = \left(u^-, u^+\right)$ *be a positive fuzzy number with the level functions* u^- *and* u^+ *continuous. Then, denoting* $u_N :=$ $\left(u_N^-, u_N^+\right) = \overline{B}_N^{(M)}(u)$, *we have*

$$D\left(\overline{B}_N^{(M)}(u), u\right) \le 12 \max\left\{\omega_1\left(u^-, \frac{1}{\sqrt{N+1}}\right), \omega_1\left(u^+, \frac{1}{\sqrt{N+1}}\right)\right\}, \quad (8.8)$$

$\forall\, N \in \mathbb{N}$, *where for* $f \in C_+\left([0, 1]\right)$:

$$\omega_1(f, \delta) := \sup_{\substack{x, y \in [0, 1] \\ |x-y| \le \delta}} |f(x) - f(y)|, \quad \delta > 0, \quad (8.9)$$

is the first modulus of continuity of f.

In this chapter we study further the approximation of positive fuzzy numbers by Max-product operators generated sequences of positive fuzzy numbers.

8.2 Main Results

In [2], p. 11, the authors mentioned the Max-product Meyer-Köning and Zeller operators defined by

$$Z_N^{(M)}(f)(x) = \frac{\bigvee_{k=0}^{\infty} s_{N,k}(x) f\left(\frac{k}{N+k}\right)}{\bigvee_{k=0}^{\infty} s_{N,k}(x)}, \quad \forall N \in \mathbb{N}, f \in C_+\left([0, 1]\right), \quad (8.10)$$

$$s_{N,k}(x) = \binom{N+k}{k} x^k, \; x \in [0, 1].$$

It holds $Z_N^{(M)}(1) = 1$, and $Z_N^{(M)}$ maps $C_+\left([0, 1]\right)$ into itself.
We mention

Theorem 8.4 ([2], p. 248) *Let* $f \in C_+\left([0, 1]\right)$. *Then*

$$\left|Z_N^{(M)}(f)(x) - f(x)\right| \le 18\omega_1\left(f, \frac{(1-x)\sqrt{x}}{\sqrt{N}}\right), \quad (8.11)$$

$\forall\, N \in \mathbb{N}, N \ge 4, \forall\, x \in [0, 1]$.

We need

Lemma 8.5 ([2], p. 257) *For any bounded function* $f : [0, 1) \to \mathbb{R}_+$, *the Max-product operator* $Z_N^{(M)}(f)(x)$ *is nonnegative, bounded, continuous on* $[0, 1)$ *and satisfies* $Z_N^{(M)}(f)(0) = f(0)$, $\forall\, N \in \mathbb{N}$. *If, in addition,* f *is supposed to be defined and continuous on* $[0, 1]$, *then* $Z_N(f)(x)$ *is continuous at* $x = 1$ *too, and* $Z_N^{(M)}(f)(1) = f(1)$, $\forall\, N \in \mathbb{N}$.

We need

Theorem 8.6 ([2], p. 259) *If* $f : [0, 1] \to \mathbb{R}_+$ *is nondecreasing and continuous on* $[0, 1]$, *then* $Z_N^{(M)}(f)$ *is nondecreasing and continuous on* $[0, 1]$.

We need

Corollary 8.7 ([2], p. 259) *If* $f : [0, 1] \to \mathbb{R}_+$ *is continuous and nonincreasing on* $[0, 1]$, *then* $Z_N^{(M)}(f)$ *is continuous and nonincreasing on* $[0, 1]$.

Next let $u \in \mathbb{R}_{\mathcal{F}}$ be a positive fuzzy number: $u = (u^-, u^+)$. We consider $Z_N^{(M)}(u^-)$, $Z_N^{(M)}(u^+)$ and since $Z_N^{(M)}$ preserves the monotonicity, it follows that $Z_N^{(M)}(u^-)$ is increasing and $Z_N^{(M)}(u^+)$ is decreasing over $[0, 1]$. We assume that $u^{\pm}$ are continuous, thus $Z_N^{(M)}(u^{\pm})$ are continuous too.

We further have

$$Z_N^{(M)}(u^{\pm})(0) = u^{\pm}(0), \tag{8.12}$$
$$Z_N^{(M)}(u^{\pm})(1) = u^{\pm}(1), \text{ respectively}, \forall N \in \mathbb{N}.$$

Also we have that

$$Z_N^{(M)}(u^-)(1) = u^-(1) \leq u^+(1) = Z_N^{(M)}(u^+)(1), \ \forall N \in \mathbb{N}. \tag{8.13}$$

In conclusion (by Theorem 8.2)

$$\overline{Z}_N^{(M)}(u) := \left(Z_N^{(M)}(u^-), Z_N^{(M)}(u^+) \right), \tag{8.14}$$

defines a proper fuzzy number in $\mathbb{R}_{\mathcal{F}}, \forall N \in \mathbb{N}$.

We present

Theorem 8.8 *Let* $u = (u^-, u^+)$ *be a positive fuzzy number with the level functions* u^- *and* u^+ *continuous. Then, denoting* $u_N := (u_N^-, u_N^+) = \overline{Z}_N^{(M)}(u)$, *we have*

$$D\left(\overline{Z}_N^{(M)}(u), u\right) \leq 18 \max\left\{ \omega_1\left(u^-, \frac{2}{3\sqrt{3N}}\right), \omega_1\left(u^+, \frac{2}{3\sqrt{3N}}\right) \right\}, \tag{8.15}$$

$\forall N \in \mathbb{N} : N \geq 4$.

Proof We use (8.1) and (8.11). We notice the following: let $g(x) = (1 - x)\sqrt{x}$, $x \in (0, 1]$, then $g'(x) = -\sqrt{x} + (1 - x)\frac{1}{2\sqrt{x}}$, setting $g'(x) = 0$ we get the only critical number $x = \frac{1}{3} \in (0, 1]$. Furthermore we have $g''(x) = -\left(\frac{1}{\sqrt{x}} + \frac{(1-x)}{4}x^{-\frac{3}{2}}\right)$, $x \in (0, 1]$ and $g''\left(\frac{1}{3}\right) < 0$. Therefore $g(x)$ has an absolute maximum over $(0, 1]$: $g\left(\frac{1}{3}\right) = \frac{2}{3\sqrt{3}}$. ∎

We need

Definition 8.9 (*see also* [2], *pp. 20–21*) The expected interval of a fuzzy number $u \in \mathbb{R}_{\mathcal{F}}$ was introduced by Dubois and Prade [4] and Heilpern [7]. It is the real interval

$$EI(u) = \left[EI_*(u), EI^*(u) \right] = \left[\int_0^1 u^-(r)\,dr, \int_0^1 u^+(r)\,dr \right]. \qquad (8.16)$$

The expected value of u is given by

$$EV(u) = \frac{1}{2} \left(\int_0^1 u^-(r)\,dr + \int_0^1 u^+(r)\,dr \right).$$

A reducing function [3] is a nondecreasing continuous function $s : [0, 1] \to [0, 1]$ with the property that $s(0) = 0$ and $s(1) = 1$. Let $u \in \mathbb{R}_{\mathcal{F}}$, the ambiguity of u with respect to s is defined by

$$Amb_s(u) = \int_0^1 s(r) \left(u^+(r) - u^-(r) \right) dr, \qquad (8.17)$$

and the value of u with respect to s is given by

$$Val_s(u) = \int_0^1 s(r) \left(u^+(r) + u^-(r) \right) dr. \qquad (8.18)$$

If for fixed $k \in \mathbb{N}$ we have $s_k(r) = r^k$, $r \in [0, 1]$, then we denote $Amb_{s_k}(u) = Amb_k(u)$ and $Val_{s_k}(u) = Val_k(u)$, i.e.

$$Amb_k(u) = \int_0^1 r^k \left(u^+(r) - u^-(r) \right) dr, \qquad (8.19)$$

and

$$Val_k(u) = \int_0^1 r^k \left(u^+(r) + u^-(r) \right) dr. \qquad (8.20)$$

The width or the non-specificity of $u \in \mathbb{R}_{\mathcal{F}}$ is given by

$$width(u) = \int_0^1 \left(u^+(r) - u^-(r) \right) dr. \qquad (8.21)$$

We give

Theorem 8.10 *Same assumptions and notations as in Theorem 8.8. Then*

$$EI(u_N) \to EI(u), \qquad (8.22)$$

$$width\,(u_N) \to width\,(u)\,, \tag{8.23}$$

and

$$Amb_s\,(u_N) \to Amb_s\,(u)\,, \tag{8.24}$$

$$Amb_k\,(u_N) \to Amb_k\,(u)\,,\,k \in \mathbb{N}, \tag{8.25}$$

where $s : [0, 1] \to [0, 1]$ is a reduction function.

Proof Similar to [2], p. 112. Indeed for $u_N := \overline{Z}_N^{(M)}\,(u) = \left(u_N^-, u_N^+\right)$, in order to obtain the required convergence of the expected interval, width, ambiguity and of the expected value of u_N, it is enough to prove that

$$\lim_{N \to \infty} \int_0^1 s\,(r)\,u_N^-\,(r)\,dr = \int_0^1 s\,(r)\,u^-\,(r)\,dr, \tag{8.26}$$

and

$$\lim_{N \to \infty} \int_0^1 s\,(r)\,u_N^+\,(r)\,dr = \int_0^1 s\,(r)\,u^+\,(r)\,dr, \tag{8.27}$$

for any reducing function s and in particular for $s\,(r) = r^k, k \in \mathbb{N} \cup \{0\}$.

Indeed, taking $s\,(r) = r^0 = 1$, we easily get the convergence of the expected interval and of the width. For any $N \in \mathbb{N}$, we get

$$\left| \int_0^1 s\,(r)\,u_N^-\,(r)\,dr - \int_0^1 s\,(r)\,u^-\,(r)\,dr \right| \le \tag{8.28}$$

$$s\,(1) \int_0^1 \left| u_N^{(r)}\,(r) - u^{(r)}\,(r) \right| dr \le D\left(\overline{Z}_N^{(M)}\,(u)\,, u\right),$$

which by (8.15) implies that

$$\lim_{N \to \infty} \int_0^1 s\,(r)\,u_N^-\,(r)\,dr = \int_0^1 s\,(r)\,u^-\,(r)\,dr.$$

The proof of (8.27) as totally similar to (8.26) is omitted. ∎

We need

Theorem 8.11 ([2], p. 30) *Let $f \in C_+\,([0, 1])$. Then*

$$\left| B_N^{(M)}\,(f)\,(x) - f\,(x) \right| \le 12\omega_1\left(f, \frac{1}{\sqrt{N+1}}\right), \forall N \in \mathbb{N}, \forall x \in [0, 1]. \tag{8.29}$$

We also need

Corollary 8.12 ([2], p. 36) *Let* $f \in C_+$ ([0, 1]) *which is concave. Then*

$$\left| B_N^{(M)} (f) (x) - f (x) \right| \le 2\omega_1 \left(f, \frac{1}{N} \right), \ \forall N \in \mathbb{N}, \forall x \in [0, 1]. \tag{8.30}$$

Let $u, v \in \mathbb{R}_{\mathcal{F}}$, $p \ge 1$, an L_p-metric (see [6]) is given by

$$d_p (u, v) = \left(\int_0^1 \left(|u^- (r) - v^- (r)|^p + |u^+ (r) - v^+ (r)|^p \right) dr \right)^{\frac{1}{p}}. \tag{8.31}$$

We have that

$$d_p \left(\overline{B}_N^{(M)} (u), u \right) =$$

$$\left(\int_0^1 \left(\left| \left(\overline{B}_N^{(M)} (u^-) \right) (r) - u^- (r) \right|^p + \left| \left(\overline{B}_N^{(M)} (u^+) \right) (r) - u^+ (r) \right|^p \right) dr \right)^{\frac{1}{p}} \overset{(8.29)}{\le}$$

$$\tag{8.32}$$

$$12 \left(\int_0^1 \left(\left(\omega_1 \left(u^-, \frac{1}{\sqrt{N+1}} \right) \right)^p + \left(\omega_1 \left(u^+, \frac{1}{\sqrt{N+1}} \right) \right)^p \right) dr \right)^{\frac{1}{p}} =$$

$$12 \left[\left(\omega_1 \left(u^-, \frac{1}{\sqrt{N+1}} \right) \right)^p + \left(\omega_1 \left(u^+, \frac{1}{\sqrt{N+1}} \right) \right)^p \right]^{\frac{1}{p}}.$$

We have proved:

Theorem 8.13 *Let* $u = (u^-, u^+)$ *be a positive fuzzy number with the level functions* u^- *and* u^+ *continuous. Then, denoting* $u_N := (u_N^-, u_N^+) = \overline{B}_N^{(M)} (u)$, *we have*

$$d_p \left(\overline{B}_N^{(M)} (u), u \right) \le \tag{8.33}$$

$$12 \left[\left(\omega_1 \left(u^-, \frac{1}{\sqrt{N+1}} \right) \right)^p + \left(\omega_1 \left(u^+, \frac{1}{\sqrt{N+1}} \right) \right)^p \right]^{\frac{1}{p}},$$

$p \ge 1, \forall N \in \mathbb{N}$.

Similarly we get

Theorem 8.14 *All as in Theorem 8.13, plus* u^-, u^+ *are concave. Then*

$$d_p \left(\overline{B}_N^{(M)} (u), u \right) \le 2 \left[\left(\omega_1 \left(u^-, \frac{1}{N} \right) \right)^p + \left(\omega_1 \left(u^+, \frac{1}{N} \right) \right)^p \right]^{\frac{1}{p}}, \tag{8.34}$$

$p \ge 1, \forall N \in \mathbb{N}$.

We also obtain

Theorem 8.15 *All as in Theorem 8.8. Then*

$$d_p\left(\overline{Z}_N^{(M)}(u), u\right) \leq \tag{8.35}$$

$$18\left[\left(\omega_1\left(u^-, \frac{2}{3\sqrt{3N}}\right)\right)^p + \left(\omega_1\left(u^+, \frac{2}{3\sqrt{3N}}\right)\right)^p\right]^{\frac{1}{p}},$$

$p \geq 1, \forall N \in \mathbb{N} : N \geq 4.$

Finally we give

Theorem 8.16 *All as in Theorem 8.3. Additionally assume that $u^\pm$ are concave. Then*

$$D\left(\overline{B}_N^{(M)}(u), u\right) \leq 2\max\left\{\omega_1\left(u^-, \frac{1}{N}\right), \omega_1\left(u^+, \frac{1}{N}\right)\right\}, \tag{8.36}$$

$\forall N \in \mathbb{N}.$

Proof Use of (8.1) and (8.30). ∎

References

1. G. Anastassiou, *Approximation of Fuzzy Numbers by Max-Product Operators* (2017, submitted)
2. B. Bede, L. Coroianu, S. Gal, *Approximation by Max-Product Type Operators* (Springer, Heidelberg, 2016)
3. M. Delgado, M.A. Vila, W. Voxman, On a canonical representation of a fuzzy number. Fuzzy Sets Syst. **93**, 125–135 (1998)
4. D. Dubois, H. Prade, The mean value of a fuzzy number. Fuzzy Sets Syst. **24**, 279–300 (1987)
5. R. Goetschel Jr., W. Voxman, Elementary fuzzy calculus. Fuzzy Sets Syst. **18**, 31–43 (1986)
6. P. Grzegorzewski, Metrics and orders in space of fuzzy numbers. Fuzzy Sets Syst. **97**, 83–94 (1998)
7. S. Heilpern, The expected value of a fuzzy number. Fuzzy Sets Syst. **47**, 81–86 (1992)
8. C. Wu, Z. Gong, On Henstock integral of fuzzy number valued functions (I). Fuzzy Sets Syst. **120**(3), 523–532 (2001)
9. C. Wu, M. Ma, On embedding problem of fuzzy number space: part 1. Fuzzy Sets Syst. **44**, 33–38 (1991)

Chapter 9
High Order Approximation by Multivariate Sublinear and Max-Product Operators

Here we study quantitatively the approximation of multivariate function by general multivariate positive sublinear operators with applications to multivariate Max-product operators. These are of Bernstein type, of Favard–Sz ász–Mirakjan type, of Baskakov type, of sampling type, of Lagrange interpolation type and of Hermite–Fejér interpolation type. Our results are both: under the presence of smoothness and without any smoothness assumption on the function to be approximated. It follows [4].

9.1 Background

Let Q be a compact and convex subset of $\mathbb{R}^k$, $k \in \mathbb{N} - \{1\}$ and let $x_0 := (x_{01}, ..., x_{0k}) \in Q$ be fixed. Let $f \in C^n(Q)$ and suppose that each nth partial derivative $f_\alpha = \frac{\partial^\alpha f}{\partial x^\alpha}$, where $\alpha := (\alpha_1, ..., \alpha_k)$, $\alpha_i \in \mathbb{Z}^+$, $i = 1, ..., k$, and $|\alpha| := \sum_{i=1}^{k} \alpha_i = n$, has relative to Q and the l_1-norm $\|\cdot\|$, a modulus of continuity $\omega_1(f_\alpha, h) \le w$, where h and w are fixed positive numbers. Here

$$\omega_1(f_\alpha, h) := \sup_{\substack{x,y \in Q \\ \|x-y\|_{l_1} \le h}} |f_\alpha(x) - f_\alpha(y)|. \tag{9.1}$$

The jth derivative of $g_z(t) = f(x_0 + t(z - x_0))$, $(z = (z_1, ..., z_k) \in Q)$ is given by

$$g_z^{(j)}(t) = \left[\left(\sum_{i=1}^{k}(z_i - x_{0i})\frac{\partial}{\partial x_i}\right)^j f\right](x_{01} + t(z_1 - x_{01}), ..., x_{0k} + t(z_k - x_{0k})). \tag{9.2}$$

© Springer International Publishing AG, part of Springer Nature 2018
G. A. Anastassiou, *Nonlinearity: Ordinary and Fractional Approximations by Sublinear and Max-Product Operators*, Studies in Systems, Decision and Control 147, https://doi.org/10.1007/978-3-319-89509-3_9

Consequently it holds

$$f(z_1, ..., z_k) = g_z(1) = \sum_{j=0}^{n} \frac{g_z^{(j)}(0)}{j!} + R_n(z, 0),\tag{9.3}$$

where

$$R_n(z, 0) := \int_0^1 \left(\int_0^{t_1} \cdots \left(\int_0^{t_{n-1}} \left(g_z^{(n)}(t_n) - g_z^{(n)}(0) \right) dt_n \right) \cdots \right) dt_1.\tag{9.4}$$

We apply Lemma 7.1.1, [1], pp. 208–209, to $(f_\alpha(x_0 + t(z - x_0)) - f_\alpha(x_0))$ as a function of z, when $\omega_1(f_\alpha, h) \le w$.

$$|f_\alpha(x_0 + t(z - x_0)) - f_\alpha(x_0)| \le w \left\lceil \frac{t\|z - x_0\|}{h} \right\rceil,\tag{9.5}$$

all $t \ge 0$, where $\lceil \cdot \rceil$ is the ceiling function.

For $\|z - x_0\| \ne 0$, it follows from (9.2)

$$|R_n(z, 0)| \le$$

$$\int_0^1 \int_0^{t_1} \cdots \int_0^{t_{n-1}} \left(\sum_{|\alpha|=n} \frac{n!}{\alpha_1! ... \alpha_k!} |z_1 - x_{01}|^{\alpha_1} ... |z_k - x_{0k}|^{\alpha_k} w \left\lceil \frac{t_n\|z - x_0\|}{h} \right\rceil \right) dt_n ... dt_1\tag{9.6}$$

$$= \sum_{|\alpha|=n} \frac{n!}{\alpha_1! ... \alpha_k!} \frac{\prod_{i=1}^{k} |z_i - x_{0i}|^{\alpha_i}}{\|z - x_0\|^n} w \Phi_n(\|z - x_0\|) = w \Phi(\|z - x_0\|),$$

since $\|z - x_0\| = \sum_{i=1}^{k} |z_i - x_{0i}|$. Above we denote (for $h > 0$ fixed):

$$\Phi_n(x) := \int_0^{|x|} \left\lceil \frac{t}{h} \right\rceil \frac{(|x| - t)^{n-1}}{(n-1)!} dt, \ (x \in \mathbb{R}),\tag{9.7}$$

equivalently

$$\Phi_n(x) = \int_0^{|x|} \int_0^{x_1} \cdots \left(\int_0^{x_{n-1}} \left\lceil \frac{x_n}{h} \right\rceil dx_n \right) ... dx_1,\tag{9.8}$$

see [1], pp. 210–211.

Therefore we have

$$|R_n(z, 0)| \le w \Phi_n(\|z - x_0\|), \text{ for all } z \in Q.\tag{9.9}$$

Also we have $g_z(0) = f(x_0)$.

One obtains ([1], p. 210)

$$\Phi_n(x) = \frac{1}{n!}\left(\sum_{j=0}^{\infty}(|x| - jh)_+^n\right), \tag{9.10}$$

which is a polynomial spline function.

Furthermore we get ([1], pp. 210–211)

$$\Phi_n(x) \le \Phi_{*n}(x) := \left(\frac{|x|^{n+1}}{(n+1)!h} + \frac{|x|^n}{2n!} + \frac{h|x|^{n-1}}{8(n-1)!}\right), \tag{9.11}$$

with equality only at $x = 0$.

Moreover, Φ_n is convex on $\mathbb{R}$ and strictly increasing on $\mathbb{R}_+$, $n \ge 1$.

In case of $Q := \{x \in \mathbb{R}^* : \|x\| \le 1\}$, where $\|\cdot\|$ is the l_1-norm in $\mathbb{R}^k$ we have

$$0 \le \|z - x_0\| \le \|z\| + \|x_0\| \le 1 + \|x_0\|, \ \forall z \in Q,$$

hence $\Phi_n(\|z - x_0\|) \le \Phi_n(1 + \|x_0\|)$, and by convexity of Φ_n we get

$$\frac{\Phi_n(\|z - x_0\|)}{\|z - x_0\|} \le \frac{\Phi_n(1 + \|x_0\|)}{(1 + \|x_0\|)}, \tag{9.12}$$

$\forall z \in Q : \|z - x_0\| \ne 0$, and hence

$$\Phi_n(\|z - x_0\|) \le \|z - x_0\|\frac{\Phi_n(1 + \|x_0\|)}{(1 + \|x_0\|)}, \ \forall z \in Q. \tag{9.13}$$

Let Q be a compact and convex subset of $\mathbb{R}^k$, $k \in \mathbb{N} - \{1\}$, $x_0 \in Q$ fixed, $f \in C^n(Q)$. Then for $j = 1, \ldots, n$, we have

$$g_z^{(j)}(0) = \sum_{\substack{\alpha:=(\alpha_1,\ldots,\alpha_k),\ \alpha_i \in \mathbb{Z}^+, \\ i=1,\ldots,k,\ |\alpha|:=\sum_{i=1}^k \alpha_i = j}} \left(\frac{j!}{\prod_{i=1}^k \alpha_i!}\right)\left(\prod_{i=1}^k (z_i - x_{0i})^{\alpha_i}\right) f_\alpha(x_0). \tag{9.14}$$

If $f_\alpha(x_0) = 0$, for all $\alpha : |\alpha| = 1, \ldots, n$, then $g_z^{(j)}(0) = 0$, $j = 1, \ldots, n$, and by (9.3):

$$f(z) - f(x_0) = R_n(z, 0), \tag{9.15}$$

that is

$$|f(z) - f(x_0)| \le w\Phi_n(\|z - x_0\|), \ \forall z \in Q, \tag{9.16}$$

where $x_0 \in Q$ is fixed.

Using (9.11) we derive

$$\|f(z) - f(x_0)\| \leq w\left(\frac{\|z - x_0\|^{n+1}}{(n+1)!h} + \frac{\|z - x_0\|^n}{2n!} + h\frac{\|z - x_0\|^{n-1}}{8(n-1)!}\right), \forall z \in Q. \tag{9.17}$$

We have proved the following fundamental result:

Theorem 9.1 *Let* $(Q, \|\cdot\|)$, *where* $\|\cdot\|$ *is the* l_1-*norm, be a compact and convex subset of* $\mathbb{R}^k$, $k \in \mathbb{N} - \{1\}$ *and let* $x_0 \in Q$ *be fixed. Let* $f \in C^n(Q)$, $n \in \mathbb{N}$, $h > 0$. *We assume that* $f_\alpha(x_0) = 0$, *for all* $\alpha : |\alpha| = 1, ..., n$. *Then*

$$\|f(z) - f(x_0)\| \leq \left(\max_{\alpha:|\alpha=n|} \omega_1(f_\alpha, h)\right) \cdot$$

$$\left(\frac{\|z - x_0\|^{n+1}}{(n+1)!h} + \frac{\|z - x_0\|^n}{2n!} + h\frac{\|z - x_0\|^{n-1}}{8(n-1)!}\right), \forall z \in Q. \tag{9.18}$$

In conclusion we have

Theorem 9.2 *Let* $(Q, \|\cdot\|)$, *where* $\|\cdot\|$ *is the* l_1-*norm, be a compact and convex subset of* $\mathbb{R}^k$, $k \in \mathbb{N} - \{1\}$ *and let* $x \in Q$ $(x = (x_1, ..., x_k))$ *be fixed. Let* $f \in C^n(Q)$, $n \in \mathbb{N}$, $h > 0$. *We assume that* $f_\alpha(x) = 0$, *for all* $\alpha : |\alpha| = 1, ..., n$. *Then*

$$\|f(t) - f(x)\| \leq \left(\max_{\alpha:|\alpha|=n} \omega_1(f_\alpha, h)\right) \cdot \tag{9.19}$$

$$\left(\frac{\|t - x\|^{n+1}}{(n+1)!h} + \frac{\|t - x\|^n}{2n!} + h\frac{\|t - x\|^{n-1}}{8(n-1)!}\right) \leq$$

$$\left(\max_{\alpha:|\alpha|=n} \omega_1(f_\alpha, h)\right)\left(\frac{k^n\left(\sum_{i=1}^k |t_i - x_i|^{n+1}\right)}{(n+1)!h} + \frac{k^{n-1}\left(\sum_{i=1}^k |t_i - x_i|^n\right)}{2n!}\right.$$

$$\left.+\frac{hk^{n-2}}{8(n-1)!}\left(\sum_{i=1}^k |t_i - x_i|^{n-1}\right)\right), \forall t \in Q, \tag{9.20}$$

where $t = (t_1, ..., t_k)$.

Proof By Theorem 9.1 and a convexity argument. ∎

We need

Definition 9.3 *Let* Q *be a compact and convex subset of* $\mathbb{R}^k$, $k \in \mathbb{N} - \{1\}$. *Here we denote*

$$C_+(Q) = \{f : Q \to \mathbb{R}_+ \text{ and continuous}\}.$$

Let $L_N : C_+ (Q) \to C_+ (Q)$, $N \in \mathbb{N}$, be a sequence of operators satisfying the following properties:

(i) (positive homogeneous)

$$L_N (\alpha f) = \alpha L_N (f), \ \forall \, \alpha \geq 0, \ f \in C_+ (Q); \tag{9.21}$$

(ii) (monotonicity)
if $f, g \in C_+ (Q)$ satisfy $f \leq g$, then

$$L_N (f) \leq L_N (g), \ \forall \, N \in \mathbb{N}, \tag{9.22}$$

and
(iii) (subadditivity)

$$L_N (f + g) \leq L_N (f) + L_N (g), \ \forall \, f, g \in C_+ (Q). \tag{9.23}$$

We call L_N positive sublinear operators.

Remark 9.4 (*to Definition* 9.3) Let $f, g \in C_+ (Q)$. We see that $f = f - g + g \leq |f - g| + g$. Then $L_N (f) \leq L_N (|f - g|) + L_N (g)$, and $L_N (f) - L_N (g) \leq L_N (|f - g|)$.

Similarly $g = g - f + f \leq |g - f| + f$, hence $L_N (g) \leq L_N (|f - g|) + L_N (f)$, and $L_N (g) - L_N (f) \leq L_N (|f - g|)$.

Consequently it holds

$$|L_N (f) (x) - L_N (g) (x)| \leq L_N (|f - g|) (x), \ \forall \, x \in Q. \tag{9.24}$$

In this chapter we treat $L_N : L_N (1) = 1$.

We observe that

$$|L_N (f) (x) - f (x)| = |L_N (f) (x) - L_N (f (x)) (x)| \overset{(9.24)}{\leq}$$

$$L_N (|f (\cdot) - f (x)|) (x), \ \forall \, x \in Q. \tag{9.25}$$

We give

Theorem 9.5 *Let Q be a compact and convex subset of $\mathbb{R}^k$, $k \in \mathbb{N} - \{1\}$ and let $x \in Q$ be fixed. Let $f \in C^n (Q, \mathbb{R}_+)$, $n \in \mathbb{N}$, $h > 0$. We assume that $f_\alpha (x) = 0$, for all $\alpha : |\alpha| = 1, ..., n$. Let $\{L_N\}_{N \in \mathbb{N}}$ positive sublinear operators mapping $C_+ (Q)$ into itself, such that $L_N (1) = 1$. Then*

$$|L_N (f) (x) - f (x)| \leq \left(\max_{\alpha : |\alpha| = n} \omega_1 (f_\alpha, h) \right) \cdot$$

$$\left(\frac{k^n}{(n+1)!h}\left(\sum_{i=1}^{k}L_N\left(|t_i-x_i|^{n+1}\right)(x)\right)+\frac{k^{n-1}}{2n!}\left(\sum_{i=1}^{k}L_N\left(|t_i-x_i|^n\right)(x)\right)\right.$$

$$\left.+\frac{hk^{n-2}}{8(n-1)!}\left(\sum_{i=1}^{k}L_N\left(|t_i-x_i|^{n-1}\right)(x)\right)\right),\quad\forall\,N\in\mathbb{N}.\tag{9.26}$$

Proof By Theorem 9.2, see Definition 9.3, and by (9.25). ∎

We need

The Maximum Multiplicative Principle 9.6 *Here* ∨ *stands for maximum. Let* $\alpha_i >$ 0, $i = 1, ..., n$; $\beta_j > 0$, $j = 1, ..., m$. *Then*

$$\vee_{i=1}^{n}\vee_{j=1}^{m}\alpha_i\beta_j=\left(\vee_{i=1}^{n}\alpha_i\right)\left(\vee_{j=1}^{m}\beta_j\right).\tag{9.27}$$

Proof Obvious. ∎

We make

Remark 9.7 In [5], p. 10, the authors introduced the basic Max-product Bernstein operators

$$B_N^{(M)}(f)(x)=\frac{\vee_{k=0}^{N}p_{N,k}(x)\,f\left(\frac{k}{N}\right)}{\vee_{k=0}^{N}p_{N,k}(x)},\quad N\in\mathbb{N},\tag{9.28}$$

where $p_{N,k}(x) = \binom{N}{k}x^k(1-x)^{N-k}$, $x \in [0, 1]$, and $f : [0, 1] \to \mathbb{R}_+$ is continuous.

In [5], p. 31, they proved that

$$B_N^{(M)}(|\cdot-x|)(x)\le\frac{6}{\sqrt{N+1}},\quad\forall\,x\in[0,1],\forall\,N\in\mathbb{N}.\tag{9.29}$$

And in [2] was proved that

$$B_N^{(M)}\left(|\cdot-x|^m\right)(x)\le\frac{6}{\sqrt{N+1}},\quad\forall\,x\in[0,1],\forall\,m,N\in\mathbb{N}.\tag{9.30}$$

We will also use

Corollary 9.8 (to Theorem 9.5, case of $n = 1$) *Let Q be a compact and convex subset of $\mathbb{R}^k$, $k \in \mathbb{N} - \{1\}$ and let $x \in Q$. Let $f \in C^1(Q, \mathbb{R}_+)$, $h > 0$. We assume that $\frac{\partial f(x)}{\partial x_i} = 0$, for $i = 1, ..., k$. Let $\{L_N\}_{N\in\mathbb{N}}$ be positive sublinear operators from $C_+(Q)$ into $C_+(Q)$: $L_N(1) = 1$, $\forall\,N \in \mathbb{N}$. Then*

$$|L_N(f)(x)-f(x)|\le\left(\max_{i=1,...,k}\omega_1\left(\frac{\partial f}{\partial x_i},h\right)\right).$$

$$\left[\frac{k}{2h} \left(\sum_{i=1}^{k} L_N \left((t_i - x_i)^2 \right) (x) \right) + \frac{1}{2} \left(\sum_{i=1}^{k} L_N \left(|t_i - x_i| \right) (x) \right) + \frac{h}{8} \right], \quad (9.31)$$

$\forall\, N \in \mathbb{N}$.

In this chapter we study quantitatively the approximation properties of multivariate Max-product operators to the unit. These are special cases of positive sublinear operators. We give also general results regarding the convergence to the unit of positive sublinear operators. Special emphasis is given in our study about approximation under differentiability. This chapter is motivated by [5].

9.2 Main Results

From now on $Q = [0, 1]^k$, $k \in \mathbb{N} - \{1\}$, except otherwise specified.
We mention

Definition 9.9 Let $f \in C_+ \left([0, 1]^k \right)$, and $\overrightarrow{N} = (N_1, ..., N_k) \in \mathbb{N}^k$. We define the multivariate Max-product Bernstein operators as follows:

$$B_{\overrightarrow{N}}^{(M)} (f) (x) :=$$

$$\frac{\bigvee_{i_1=0}^{N_1} \bigvee_{i_2=0}^{N_2} \cdots \bigvee_{i_k=0}^{N_k} p_{N_1,i_1} (x_1) p_{N_2,i_2} (x_2) ... p_{N_k,i_k} (x_k) f \left(\frac{i_1}{N_1}, ..., \frac{i_k}{N_k} \right)}{\bigvee_{i_1=0}^{N_1} \bigvee_{i_2=0}^{N_2} \cdots \bigvee_{i_k=0}^{N_k} p_{N_1,i_1} (x_1) p_{N_2,i_2} (x_2) ... p_{N_k,i_k} (x_k)}, \quad (9.32)$$

$\forall\, x = (x_1, ..., x_k) \in [0, 1]^k$. Call $N_{\min} := \min\{N_1, ..., N_k\}$.

The operators $B_{\overrightarrow{N}}^{(M)} (f) (x)$ are positive sublinear and they map $C_+ \left([0, 1]^k \right)$ into itself, and $B_{\overrightarrow{N}}^{(M)} (1) = 1$.

See also [5], p. 123 the bivariate case. We also have

$$B_{\overrightarrow{N}}^{(M)} (f) (x) :=$$

$$\frac{\bigvee_{i_1=0}^{N_1} \bigvee_{i_2=0}^{N_2} \cdots \bigvee_{i_k=0}^{N_k} p_{N_1,i_1} (x_1) p_{N_2,i_2} (x_2) ... p_{N_k,i_k} (x_k) f \left(\frac{i_1}{N_1}, ..., \frac{i_k}{N_k} \right)}{\prod_{\lambda=1}^{k} \left(\bigvee_{i_\lambda=0}^{N_\lambda} p_{N_\lambda,i_\lambda} (x_\lambda) \right)}, \quad (9.33)$$

$\forall\, x \in [0, 1]^k$, by the maximum multiplicative principle, see (9.27).

We make

Remark 9.10 The coordinate Max-product Bernstein operators are defined as follows ($\lambda = 1, ..., k$):

$$B_{N_\lambda}^{(M)}(g)(x_\lambda) := \frac{\bigvee_{i_\lambda=0}^{N_\lambda} p_{N_\lambda,i_\lambda}(x_\lambda)\, g\left(\frac{i_\lambda}{N_\lambda}\right)}{\bigvee_{i_\lambda=0}^{N_\lambda} p_{N_\lambda,i_\lambda}(x_\lambda)}, \tag{9.34}$$

$\forall\, N_\lambda \in \mathbb{N}$, and $\forall\, x_\lambda \in [0, 1]$, $\forall\, g \in C_+([0, 1]) := \{g : [0, 1] \to \mathbb{R}_+ \text{ continuous}\}$.
   Here we have

$$p_{N_\lambda,i_\lambda}(x_\lambda) = \binom{N_\lambda}{i_\lambda} x_\lambda^{i_\lambda}(1 - x_\lambda)^{N_\lambda - i_\lambda}, \text{ for all } \lambda = 1, ..., k; \; x_\lambda \in [0, 1]. \tag{9.35}$$

In case of $f \in C_+([0, 1]^k)$ is such that $f(x) := g(x_\lambda)$, $\forall\, x \in [0, 1]^k$, where $x = (x_1, ..., x_\lambda, ..., x_k)$ and $g \in C_+([0, 1])$, we get that

$$B_{\overrightarrow{N}}^{(M)}(f)(x) = B_{N_\lambda}^{(M)}(g)(x_\lambda), \tag{9.36}$$

by the maximum multiplicative principle (9.27) and simplification of (9.33).
   Clearly it holds that

$$B_{\overrightarrow{N}}^{(M)}(f)(x) = f(x), \; \forall\, x = (x_1, ..., x_k) \in [0, 1]^k : x_\lambda \in \{0, 1\}, \; \lambda = 1, ..., k. \tag{9.37}$$

We present

Theorem 9.11 *Let $x \in [0, 1]^k$, $k \in \mathbb{N} - \{1\}$, be fixed, and let $f \in C^n([0, 1]^k, \mathbb{R}_+)$, $n \in \mathbb{N} - \{1\}$. We assume that $f_\alpha(x) = 0$, for all $\alpha : |\alpha| = 1, ..., n$. Then*

$$\left|B_{\overrightarrow{N}}^{(M)}(f)(x) - f(x)\right| \le 6 \left(\max_{\alpha:|\alpha|=n}\left(\omega_1\left(f_\alpha, \left(\frac{1}{\sqrt{N_{\min}+1}}\right)^{\frac{1}{n+1}}\right)\right)\right). \tag{9.38}$$

$$\left[\frac{k^{n+1}}{(n+1)!}\left(\frac{1}{\sqrt{N_{\min}+1}}\right)^{\frac{n}{n+1}} + \frac{k^n}{2n!}\left(\frac{1}{\sqrt{N_{\min}+1}}\right) + \frac{k^{n-1}}{8(n-1)!}\left(\frac{1}{\sqrt{N_{\min}+1}}\right)^{\frac{n+2}{n+1}}\right],$$

$\forall\, \overrightarrow{N} \in \mathbb{N}^k$, *where $N_{\min} := \min\{N_1, ..., N_k\}$.*
   We have that $\lim\limits_{\overrightarrow{N}\to(\infty,...,\infty)} B_{\overrightarrow{N}}^{(M)}(f)(x) = f(x)$.

Proof By (9.26) we get:

$$\left|B_{\overrightarrow{N}}^{(M)}(f)(x) - f(x)\right| \overset{(9.36)}{\le} \left(\max_{\alpha:|\alpha|=n}\omega_1(f_\alpha, h)\right) \cdot$$

$$\left[\frac{k^n}{(n+1)!h}\left(\sum_{i=1}^{k} B_{N_i}^{(M)}\left(|t_i - x_i|^{n+1}\right)(x_i)\right) + \frac{k^{n-1}}{2n!}\left(\sum_{i=1}^{k} B_{N_i}^{(M)}\left(|t_i - x_i|^n\right)(x_i)\right)\right] \tag{9.39}$$

$$+ \frac{hk^{n-2}}{8(n-1)!} \left(\sum_{i=1}^{k} B_{N_i}^{(M)} \left(|t_i - x_i|^{n-1} \right) (x_i) \right) \Bigg] \overset{(9.30)}{\leq}$$

$$\left(\frac{6}{\sqrt{N_{\min}+1}} \right) \left(\max_{\alpha:|\alpha|=n} \omega_1 \left(f_\alpha, h \right) \right) \left[\frac{k^{n+1}}{(n+1)!h} + \frac{k^n}{2n!} + \frac{hk^{n-1}}{8(n-1)!} \right] =: (\xi) .$$

Above notice $\sum_{i=1}^{k} B_{N_i}^{(M)} \left(|t_i - x_i|^n \right) (x_i) \overset{(9.30)}{\leq} \sum_{i=1}^{k} \frac{6}{\sqrt{N_i+1}} \leq \frac{6k}{\sqrt{N_{\min}+1}}$, etc.

Next we choose $h := \left(\frac{1}{\sqrt{N_{\min}+1}} \right)^{\frac{1}{n+1}}$, then $h^n = \left(\frac{1}{\sqrt{N_{\min}+1}} \right)^{\frac{n}{n+1}}$ and $h^{n+1} = \frac{1}{\sqrt{N_{\min}+1}}$.
We have

$$(\xi) = 6 \left(\max_{\alpha:|\alpha|=n} \left(\omega_1 \left(f_\alpha, \left(\frac{1}{\sqrt{N_{\min}+1}} \right)^{\frac{1}{n+1}} \right) \right) \right). \tag{9.40}$$

$$\left[\frac{k^{n+1}}{(n+1)!} \left(\frac{1}{\sqrt{N_{\min}+1}} \right)^{\frac{n}{n+1}} + \frac{k^n}{2n!} \left(\frac{1}{\sqrt{N_{\min}+1}} \right) + \frac{k^{n-1}}{8(n-1)!} \left(\frac{1}{\sqrt{N_{\min}+1}} \right)^{\frac{n+2}{n+1}} \right],$$

proving the claim. ∎

We also give

Proposition 9.12 *Let* $x \in [0,1]^k$, $k \in \mathbb{N} - \{1\}$, *be fixed and let* $f \in C^1 \left([0,1]^k, \mathbb{R}_+ \right)$.
We assume that $\frac{\partial f(x)}{\partial x_i} = 0$, *for* $i = 1, ..., k$. *Then*

$$\left| B_{\overrightarrow{N}}^{(M)} (f) (x) - f (x) \right| \leq 3 \left(\max_{i=1,...,k} \omega_1 \left(\frac{\partial f}{\partial x_i}, \frac{1}{\sqrt[4]{N_{\min}+1}} \right) \right). \tag{9.41}$$

$$\left[\frac{k^2}{\sqrt[4]{N_{\min}+1}} + \frac{k}{\sqrt{N_{\min}+1}} + \frac{1}{4 \left(\sqrt[4]{N_{\min}+1} \right)^3} \right],$$

$\forall \; \overrightarrow{N} \in \mathbb{N}^k$, *where* $N_{\min} := \min\{N_1, ..., N_k\}$.
Also it holds $\lim_{\overrightarrow{N} \to (\infty,...,\infty)} B_{\overrightarrow{N}}^{(M)} (f) (x) = f (x)$.

Proof By (9.31) we get:

$$\left| B_{\overrightarrow{N}}^{(M)} (f) (x) - f (x) \right| \overset{(9.36)}{\leq} \left(\max_{i=1,...,k} \omega_1 \left(\frac{\partial f}{\partial x_i}, h \right) \right).$$

$$\left[\frac{k}{2h} \left(\sum_{i=1}^{k} B_{N_i}^{(M)} \left((t_i - x_i)^2 \right) (x_i) \right) + \frac{1}{2} \left(\sum_{i=1}^{k} B_{N_i}^{(M)} \left(|t_i - x_i| \right) (x_i) \right) + \frac{h}{8} \right] \overset{(9.30)}{\leq}$$

$$\tag{9.42}$$

$$\left(\frac{6}{\sqrt{N_{\min}+1}} \right) \left(\max_{i=1,...,k} \omega_1 \left(\frac{\partial f}{\partial x_i}, h \right) \right) \left[\frac{k^2}{2h} + \frac{k}{2} + \frac{h}{8} \right] =: (\psi) .$$

Next we choose $h := \left(\frac{1}{\sqrt{N_{\min}+1}}\right)^{\frac{1}{2}}$, then $h^2 = \frac{1}{\sqrt{N_{\min}+1}}$.

We have that

$$(\psi) = 6\left(\max_{i=1,\dots,k}\omega_1\left(\frac{\partial f}{\partial x_i},\left(\frac{1}{\sqrt{N_{\min}+1}}\right)^{\frac{1}{2}}\right)\right). \tag{9.43}$$

$$\left[\frac{k^2}{2}\left(\frac{1}{\sqrt{N_{\min}+1}}\right)^{\frac{1}{2}}+\frac{k}{2}\left(\frac{1}{\sqrt{N_{\min}+1}}\right)+\frac{1}{8}\left(\frac{1}{\sqrt{N_{\min}+1}}\right)^{\frac{3}{2}}\right],$$

proving the claim.   ∎

We need

Theorem 9.13 *Let Q with $\|\cdot\|$ the l_1-norm, be a compact and convex subset of $\mathbb{R}^k$, $k \in \mathbb{N} - \{1\}$, and $f \in C_+(Q)$; $h > 0$. We denote $\omega_1(f,h) := \sup\limits_{\substack{x,y\in Q:\\ \|x-y\|\le h}} |f(x) - f(y)|$, the modulus of continuity of f. Let $\{L_N\}_{N\in\mathbb{N}}$ be positive sublinear operators from $C_+(Q)$ into itself such that $L_N(1) = 1$, $\forall N \in \mathbb{N}$. Then*

$$|L_N(f)(x) - f(x)| \le \omega_1(f,h)\left(1+\frac{1}{h}L_N(\|t-x\|)(x)\right) \le$$

$$\omega_1(f,h)\left(1+\frac{1}{h}\left(\sum_{i=1}^{k}L_N(|t_i-x_i|)(x)\right)\right), \tag{9.44}$$

$\forall N \in \mathbb{N}$, $\forall x \in Q$, where $x := (x_1, \dots, x_k)$; $t = (t_1, \dots, t_k) \in Q$.

Proof We have that ([1], pp. 208–209)

$$|f(t) - f(x)| \le \omega_1(f,h)\left\lceil\frac{\|t-x\|}{h}\right\rceil \le \omega_1(f,h)\left(1+\frac{\|t-x\|}{h}\right), \tag{9.45}$$

$\forall t, x \in Q$.

By (9.25) we get:

$$|L_N(f)(x) - f(x)| \le L_N(|f(t) - f(x)|)(x) \le \tag{9.46}$$

$$\omega_1(f,h)\left(1+\frac{1}{h}L_N(\|t-x\|)(x)\right), \forall N \in \mathbb{N},$$

proving the claim.   ∎

We give

Theorem 9.14 *Let* $f \in C_+\left([0,1]^k\right)$, $k \in \mathbb{N} - \{1\}$. *Then*

$$\left| B_{\overrightarrow{N}}^{(M)}(f)(x) - f(x) \right| \leq (6k+1)\,\omega_1\left(f, \frac{1}{\sqrt{N_{\min}+1}}\right), \qquad (9.47)$$

$\forall\, x \in [0,1]^k$, $\forall\, \overrightarrow{N} \in \mathbb{N}^k$, *where* $N_{\min} := \min\{N_1, ..., N_k\}$.
   That is

$$\left\| B_{\overrightarrow{N}}^{(M)}(f) - f \right\|_{\infty} \leq (6k+1)\,\omega_1\left(f, \frac{1}{\sqrt{N_{\min}+1}}\right). \qquad (9.48)$$

It holds that $\displaystyle\lim_{\overrightarrow{N}\to(\infty,...,\infty)} B_{\overrightarrow{N}}^{(M)}(f)(x) = f(x)$, *uniformly.*

Proof We get that (use of (9.44))

$$\left| B_{\overrightarrow{N}}^{(M)}(f)(x) - f(x) \right| \overset{(9.36)}{\leq} \omega_1(f,h)\left(1 + \frac{1}{h}\left(\sum_{i=1}^{k} B_{N_i}^{(M)}\left(|t_i - x_i|\right)(x)\right)\right)$$

$$\overset{(9.29)}{\leq} \omega_1(f,h)\left(1 + \frac{1}{h}\left(\frac{6k}{\sqrt{N_{\min}+1}}\right)\right) \qquad (9.49)$$

(setting $h := \frac{1}{\sqrt{N_{\min}+1}}$)

$$= \omega_1\left(f, \frac{1}{\sqrt{N_{\min}+1}}\right)(6k+1), \quad \forall\, x \in [0,1]^k, \quad \forall\, \overrightarrow{N} \in \mathbb{N}^k,$$

proving the claim. ∎

We continue with

Definition 9.15 ([5], p. 123) We define the bivariate Max-product Bernstein type operators:

$$A_N^{(M)}(f)(x,y) := \frac{\bigvee_{i=0}^{N} \bigvee_{j=0}^{N-i} \binom{N}{i}\binom{N-i}{j} x^i y^j (1-x-y)^{N-i-j} f\left(\frac{i}{N}, \frac{j}{N}\right)}{\bigvee_{i=0}^{N} \bigvee_{j=0}^{N-i} \binom{N}{i}\binom{N-i}{j} x^i y^j (1-x-y)^{N-i-j}},$$

$$(9.50)$$

$\forall\, (x,y) \in \Delta := \{(x,y) : x \geq 0, y \geq 0, x+y \leq 1\}$, $\forall\, N \in \mathbb{N}$, and $\forall\, f \in C_+(\Delta)$.

Remark 9.16 By [5], p. 137, Theorem 2.7.5 there, $A_N^{(M)}$ is a positive sublinear operator mapping $C_+(\Delta)$ into itself and $A_N^{(M)}(1) = 1$, furthermore it holds

$$\left| A_N^{(M)}(f) - A_N^{(M)}(g) \right| \le A_N^{(M)}(|f - g|), \ \forall \ f, g \in C_+(\Delta), \forall \ N \in \mathbb{N}. \quad (9.51)$$

By [5], p. 125 we get that $A_N^{(M)}(f)(1, 0) = f(1, 0)$, $A_N^{(M)}(f)(0, 1) = f(0, 1)$, and $A_N^{(M)}(f)(0, 0) = f(0, 0)$.

By [5], p. 139, we have that $((x, y) \in \Delta)$:

$$A_N^{(M)}(|\cdot - x|)(x, y) = B_N^{(M)}(|\cdot - x|)(x), \quad (9.52)$$

and

$$A_N^{(M)}(|\cdot - y|)(x, y) = B_N^{(M)}(|\cdot - y|)(y). \quad (9.53)$$

Working exactly the same way as (9.52), (9.53) are proved we also derive ($m \in \mathbb{N}$, $(x, y) \in \Delta$):

$$A_N^{(M)}\left(|\cdot - x|^m\right)(x, y) = B_N^{(M)}\left(|\cdot - x|^m\right)(x), \quad (9.54)$$

and

$$A_N^{(M)}\left(|\cdot - y|^m\right)(x, y) = B_N^{(M)}\left(|\cdot - y|^m\right)(y). \quad (9.55)$$

We present

Theorem 9.17 *Let $x := (x_1, x_2) \in \Delta$ be fixed, and $f \in C^n(\Delta, \mathbb{R}_+)$, $n \in \mathbb{N} - \{1\}$. We assume that $f_\alpha(x) = 0$, for all $\alpha : |\alpha| = 1, ..., n$. Then*

$$\left| A_N^{(M)}(f)(x_1, x_2) - f(x_1, x_2) \right| \le 6 \left(\max_{\alpha : |\alpha| = n} \omega_1 \left(f_\alpha, \left(\frac{1}{\sqrt{N+1}} \right)^{\frac{1}{n+1}} \right) \right) \cdot \quad (9.56)$$

$$\left[\frac{2^{n+1}}{(n+1)!} \left(\frac{1}{\sqrt{N+1}} \right)^{\frac{n}{n+1}} + \frac{2^{n-1}}{n!} \left(\frac{1}{\sqrt{N+1}} \right) + \frac{2^{n-4}}{(n-1)!} \left(\frac{1}{\sqrt{N+1}} \right)^{\frac{n+2}{n+1}} \right],$$

$\forall \ N \in \mathbb{N}$.

It holds $\lim_{N \to \infty} A_N^{(M)}(f)(x_1, x_2) = f(x_1, x_2)$.

Proof By (9.26) we get (here $x := (x_1, x_2) \in \Delta$):

$$\left| A_N^{(M)}(f)(x_1, x_2) - f(x_1, x_2) \right| \le \left(\max_{\alpha : |\alpha| = n} \omega_1(f_\alpha, h) \right) \cdot$$

$$\left[\frac{2^n}{(n+1)!h} \left(\sum_{i=1}^{2} A_N^{(M)}\left(|t_i - x_i|^{n+1}\right)(x) \right) + \frac{2^{n-2}}{n!} \left(\sum_{i=1}^{2} A_N^{(M)}\left(|t_i - x_i|^n\right)(x) \right) \right.$$

$$\left. + \frac{h 2^{n-5}}{(n-1)!} \left(\sum_{i=1}^{2} A_N^{(M)}\left(|t_i - x_i|^{n-1}\right)(x) \right) \right] \overset{\text{(by (9.54), (9.55))}}{=}$$

$\quad (9.57)$

$$\left(\max_{\alpha:|\alpha|=n} \omega_1 \left(f_\alpha, h \right) \right) \left[\frac{2^n}{(n+1)!h} \left(\sum_{i=1}^{2} B_N^{(M)} \left(|t_i - x_i|^{n+1} \right) (x_i) \right) + \right.$$

$$\left. \frac{2^{n-2}}{n!} \left(\sum_{i=1}^{2} B_N^{(M)} \left(|t_i - x_i|^n \right) (x_i) \right) + \frac{h2^{n-5}}{(n-1)!} \left(\sum_{i=1}^{2} B_N^{(M)} \left(|t_i - x_i|^{n-1} \right) (x_i) \right) \right]$$

$$\tag{9.58}$$

$$\overset{(9.30)}{\leq} \frac{6 \left(\max_{\alpha:|\alpha|=n} \omega_1 \left(f_\alpha, h \right) \right)}{\sqrt{N+1}} \left[\frac{2^{n+1}}{(n+1)!h} + \frac{2^{n-1}}{n!} + \frac{h2^{n-4}}{(n-1)!} \right] =: (\xi).$$

Next we choose $h := \left(\frac{1}{\sqrt{N+1}} \right)^{\frac{1}{n+1}}$, then $h^n = \left(\frac{1}{\sqrt{N+1}} \right)^{\frac{n}{n+1}}$ and $h^{n+1} = \frac{1}{\sqrt{N+1}}$.

We have

$$(\xi) = 6 \left(\max_{\alpha:|\alpha|=n} \omega_1 \left(f_\alpha, \left(\frac{1}{\sqrt{N+1}} \right)^{\frac{1}{n+1}} \right) \right). \tag{9.59}$$

$$\left[\frac{2^{n+1}}{(n+1)!} \left(\frac{1}{\sqrt{N+1}} \right)^{\frac{n}{n+1}} + \frac{2^{n-1}}{n!} \left(\frac{1}{\sqrt{N+1}} \right) + \frac{2^{n-4}}{(n-1)!} \left(\frac{1}{\sqrt{N+1}} \right)^{\frac{n+2}{n+1}} \right],$$

proving the claim. ∎

We also give

Theorem 9.18 *Let* $x := (x_1, x_2) \in \Delta$ *be fixed, and* $f \in C^1 (\Delta, \mathbb{R}_+)$. *We assume that* $\frac{\partial f}{\partial x_i} (x) = 0$, *for* $i = 1, 2$. *Then*

$$\left| A_N^{(M)} (f) (x_1, x_2) - f (x_1, x_2) \right| \leq 6 \left(\max_{i=1,2} \omega_1 \left(\frac{\partial f}{\partial x_i}, \frac{1}{\sqrt[4]{N+1}} \right) \right). \tag{9.60}$$

$$\left[\frac{2}{\sqrt[4]{N+1}} + \frac{1}{\sqrt{N+1}} + \frac{1}{8} \left(\frac{1}{\sqrt{N+1}} \right)^{\frac{3}{2}} \right],$$

$\forall N \in \mathbb{N}$.

It holds $\lim_{N \to \infty} A_N^{(M)} (f) (x_1, x_2) = f (x_1, x_2)$.

Proof By (9.31) we get (here $x := (x_1, x_2) \in \Delta$):

$$\left| A_N^{(M)} (f) (x_1, x_2) - f (x_1, x_2) \right| \leq \left(\max_{i=1,2} \omega_1 \left(\frac{\partial f}{\partial x_i}, h \right) \right).$$

$$\left[\frac{1}{h} \left(\sum_{i=1}^{2} A_N^{(M)} \left((t_i - x_i)^2 \right) (x) \right) + \frac{1}{2} \left(\sum_{i=1}^{2} A_N^{(M)} \left(|t_i - x_i| \right) (x) \right) + \frac{h}{8} \right] \tag{9.61}$$

$$\overset{\text{(by (9.54), (9.55))}}{=} \left(\max\omega_1 \left(\frac{\partial f}{\partial x_i}, h\right)\right) \left[\frac{1}{h}\left(\sum_{i=1}^{2} B_N^{(M)}\left((t_i - x_i)^2\right)(x_i)\right) + \right.$$

$$\left. \frac{1}{2}\left(\sum_{i=1}^{2} B_N^{(M)}\left(|t_i - x_i|\right)(x_i)\right) + \frac{h}{8}\right] \overset{\text{(9.30)}}{\leq}$$

$$\frac{6\left(\max\omega_1\left(\frac{\partial f}{\partial x_i}, h\right)\right)}{\sqrt{N+1}} \left[\frac{2}{h} + 1 + \frac{h}{8}\right] =: (\psi).$$

Next we choose $h := \left(\frac{1}{\sqrt{N+1}}\right)^{\frac{1}{2}}$, then $h^2 = \frac{1}{\sqrt{N+1}}$.

Hence we have

$$(\psi) = 6\left(\max_{i=1,2}\omega_1\left(\frac{\partial f}{\partial x_i}, \left(\frac{1}{\sqrt{N+1}}\right)^{\frac{1}{2}}\right)\right). \tag{9.62}$$

$$\left[2\left(\frac{1}{\sqrt{N+1}}\right)^{\frac{1}{2}} + \left(\frac{1}{\sqrt{N+1}}\right) + \frac{1}{8}\left(\frac{1}{\sqrt{N+1}}\right)^{\frac{3}{2}}\right],$$

proving the claim.   ∎

We further obtain

Theorem 9.19 *Let* $f \in C_+(\Delta)$. *Then*

$$\left|A_N^{(M)}(f)(x_1, x_2) - f(x_1, x_2)\right| \leq 13\omega_1\left(f, \frac{1}{\sqrt{N+1}}\right), \tag{9.63}$$

$\forall (x_1, x_2) \in \Delta, \forall N \in \mathbb{N}.$

That is

$$\left\|A_N^{(M)}(f) - f\right\|_{\infty, \Delta} \leq 13\omega_1\left(f, \frac{1}{\sqrt{N+1}}\right), \tag{9.64}$$

$\forall N \in \mathbb{N}.$

It holds that $\lim_{N \to \infty} A_N^{(M)}(f) = f$, *uniformly,* $\forall f \in C_+(\Delta)$.

Proof Using (9.44) ($x := (x_1, x_2) \in \Delta$) we get:

$$\left|A_N^{(M)}(f)(x_1, x_2) - f(x_1, x_2)\right| \leq$$

$$\omega_1\left(f,h\right)\left(1+\frac{1}{h}\left(\sum_{i=1}^{2}A_N^{(M)}\left(|t_i-x_i|\right)(x)\right)\right)\overset{\text{(by (9.52), (9.53))}}{=}$$

$$\omega_1\left(f,h\right)\left(1+\frac{1}{h}\left(\sum_{i=1}^{2}B_N^{(M)}\left(|t_i-x_i|\right)(x_i)\right)\right)\overset{(9.29)}{\le}$$

$$\omega_1\left(f,h\right)\left(1+\frac{2}{h}\cdot\frac{6}{\sqrt{N+1}}\right)\tag{9.65}$$

(setting $h:=\frac{1}{\sqrt{N+1}}$)

$$=13\omega_1\left(f,\frac{1}{\sqrt{N+1}}\right),\ \forall\ (x_1,x_2)\in\Delta,\ \forall\ N\in\mathbb{N},$$

proving the claim. ∎

We make

Remark 9.20 The Max-product truncated Favard–Szász–Mirakjan operators

$$T_N^{(M)}\left(f\right)(x)=\frac{\bigvee_{k=0}^{N}s_{N,k}\left(x\right)f\left(\frac{k}{N}\right)}{\bigvee_{k=0}^{N}s_{N,k}\left(x\right)},\ x\in[0,1],\ N\in\mathbb{N},\ f\in C_+\left([0,1]\right),\tag{9.66}$$

$s_{N,k}\left(x\right)=\frac{(Nx)^k}{k!}$, see also [5], p. 11.

By [5], pp. 178–179, we get that

$$T_N^{(M)}\left(|\cdot-x|\right)(x)\le\frac{3}{\sqrt{N}},\ \forall\ x\in[0,1],\ \forall\ N\in\mathbb{N}.\tag{9.67}$$

And from [2] we have

$$T_N^{(M)}\left(|\cdot-x|^m\right)(x)\le\frac{3}{\sqrt{N}},\ \forall\ x\in[0,1],\ \forall\ N,m\in\mathbb{N}.\tag{9.68}$$

We make

Definition 9.21 Let $f\in C_+\left([0,1]^k\right)$, $k\in\mathbb{N}-\{1\}$, and $\overrightarrow{N}=(N_1,...,N_k)\in\mathbb{N}^k$. We define the multivariate Max-product truncated Favard–Sz ász–Mirakjan operators as follows:
$$T_{\overrightarrow{N}}^{(M)}\left(f\right)(x):=$$

$$\frac{\bigvee_{i_1=0}^{N_1}\bigvee_{i_2=0}^{N_2}\cdots\bigvee_{i_k=0}^{N_k}s_{N_1,i_1}\left(x_1\right)s_{N_2,i_2}\left(x_2\right)...s_{N_k,i_k}\left(x_k\right)f\left(\frac{i_1}{N_1},...,\frac{i_k}{N_k}\right)}{\bigvee_{i_1=0}^{N_1}\bigvee_{i_2=0}^{N_2}\cdots\bigvee_{i_k=0}^{N_k}s_{N_1,i_1}\left(x_1\right)s_{N_2,i_2}\left(x_2\right)...s_{N_k,i_k}\left(x_k\right)},\tag{9.69}$$

$\forall\ x=(x_1,...,x_k)\in[0,1]^k$. Call $N_{\min}:=\min\{N_1,...,N_k\}$.

The operators $T_{\overrightarrow{N}}^{(M)}(f)(x)$ are positive sublinear mapping $C_+\left([0, 1]^k\right)$ into itself, and $T_{\overrightarrow{N}}^{(M)}(1) = 1$.

We also have

$$T_{\overrightarrow{N}}^{(M)}(f)(x) :=$$

$$\frac{\bigvee_{i_1=0}^{N_1} \bigvee_{i_2=0}^{N_2} \cdots \bigvee_{i_k=0}^{N_k} s_{N_1,i_1}(x_1)\, s_{N_2,i_2}(x_2)\, \cdots s_{N_k,i_k}(x_k)\, f\left(\frac{i_1}{N_1}, \ldots, \frac{i_k}{N_k}\right)}{\prod_{\lambda=1}^{k}\left(\bigvee_{i_\lambda=0}^{N_\lambda} s_{N_\lambda,i_\lambda}(x_\lambda)\right)}, \tag{9.70}$$

$\forall\, x \in [0, 1]^k$, by the maximum multiplicative principle, see (9.27).

We make

Remark 9.22 The coordinate Max-product truncated Favard–Szász–Mirakjan operators are defined as follows ($\lambda = 1, \ldots, k$):

$$T_{N_\lambda}^{(M)}(g)(x_\lambda) := \frac{\bigvee_{i_\lambda=0}^{N_\lambda} s_{N_\lambda,i_\lambda}(x_\lambda)\, g\left(\frac{i_\lambda}{N_\lambda}\right)}{\bigvee_{i_\lambda=0}^{N_\lambda} s_{N_\lambda,i_\lambda}(x_\lambda)}, \tag{9.71}$$

$\forall\, N_\lambda \in \mathbb{N}$, and $\forall\, x_\lambda \in [0, 1]$, $\forall\, g \in C_+\left([0, 1]\right)$.

Here we have

$$s_{N_\lambda,i_\lambda}(x_\lambda) = \frac{(N_\lambda x_\lambda)^{i_\lambda}}{i_\lambda!}, \quad \lambda = 1, \ldots, k;\ x_\lambda \in [0, 1]. \tag{9.72}$$

In case of $f \in C_+\left([0, 1]^k\right)$ such that $f(x) := g(x_\lambda)$, $\forall\, x \in [0, 1]^k$, where $x = (x_1, \ldots, x_\lambda, \ldots, x_k)$ and $g \in C_+([0, 1])$, we get that

$$T_{\overrightarrow{N}}^{(M)}(f)(x) = T_{N_\lambda}^{(M)}(g)(x_\lambda), \tag{9.73}$$

by the maximum multiplicative principle (9.27) and simplification of (9.70).

We present

Theorem 9.23 *Let* $x \in [0, 1]^k$, $k \in \mathbb{N} - \{1\}$, *be fixed, and let* $f \in C^n\left([0, 1]^k, \mathbb{R}_+\right)$, $n \in \mathbb{N} - \{1\}$. *We assume that* $f_\alpha(x) = 0$, *for all* $\alpha : |\alpha| = 1, \ldots, n$. *Then*

$$\left| T_{\overrightarrow{N}}^{(M)}(f)(x) - f(x) \right| \le 3 \left(\max_{\alpha:|\alpha|=n} \left(\omega_1 \left(f_\alpha, \left(\frac{1}{\sqrt{N_{\min}}} \right)^{\frac{1}{n+1}} \right) \right) \right) \cdot$$

$$\left[\frac{k^{n+1}}{(n+1)!} \left(\frac{1}{\sqrt{N_{\min}}} \right)^{\frac{n}{n+1}} + \frac{k^n}{2n!} \left(\frac{1}{\sqrt{N_{\min}}} \right) + \frac{k^{n-1}}{8(n-1)!} \left(\frac{1}{\sqrt{N_{\min}}} \right)^{\frac{n+2}{n+1}} \right], \tag{9.74}$$

$\forall\, \overrightarrow{N} \in \mathbb{N}^k$, *where* $N_{\min} := \min\{N_1, \ldots, N_k\}$.

We have that $\lim\limits_{\vec{N} \to (\infty,...,\infty)} T_{\vec{N}}^{(M)} (f) (x) = f (x)$.

Proof By (9.26) we get:

$$\left| T_{\vec{N}}^{(M)} (f) (x) - f (x) \right| \overset{(9.73)}{\leq} \left(\max_{\alpha:|\alpha|=n} \omega_1 (f_\alpha, h) \right) \cdot$$

$$\left[\frac{k^n}{(n+1)!h} \left(\sum_{i=1}^{k} T_{N_i}^{(M)} \left(|t_i - x_i|^{n+1} \right) (x_i) \right) + \frac{k^{n-1}}{2n!} \left(\sum_{i=1}^{k} T_{N_i}^{(M)} \left(|t_i - x_i|^n \right) (x_i) \right) \right.$$

$$\left. + \frac{hk^{n-2}}{8(n-1)!} \left(\sum_{i=1}^{k} T_{N_i}^{(M)} \left(|t_i - x_i|^{n-1} \right) (x_i) \right) \right] \overset{(9.68)}{\leq}$$

$$\frac{3}{\sqrt{N_{\min}}} \left(\max_{\alpha:|\alpha|=n} \omega_1 (f_\alpha, h) \right) \left[\frac{k^{n+1}}{(n+1)!h} + \frac{k^n}{2n!} + \frac{hk^{n-1}}{8(n-1)!} \right] =: (\xi) .$$

Above notice that $\sum_{i=1}^{k} T_{N_i}^{(M)} \left(|t_i - x_i|^n \right) (x_i) \overset{(9.68)}{\leq} \sum_{i=1}^{k} \frac{3}{\sqrt{N_i}} \leq \frac{3k}{\sqrt{N_{\min}}}$, etc.

Next we choose $h := \left(\frac{1}{\sqrt{N_{\min}}} \right)^{\frac{1}{n+1}}$, then $h^n = \left(\frac{1}{\sqrt{N_{\min}}} \right)^{\frac{n}{n+1}}$ and $h^{n+1} = \frac{1}{\sqrt{N_{\min}}}$.
We have

$$(\xi) = 3 \left(\max_{\alpha:|\alpha|=n} \left(\omega_1 \left(f_\alpha, \left(\frac{1}{\sqrt{N_{\min}}} \right)^{\frac{1}{n+1}} \right) \right) \right) \cdot$$

$$\left[\frac{k^{n+1}}{(n+1)!} \left(\frac{1}{\sqrt{N_{\min}}} \right)^{\frac{n}{n+1}} + \frac{k^n}{2n!} \left(\frac{1}{\sqrt{N_{\min}}} \right) + \frac{k^{n-1}}{8(n-1)!} \left(\frac{1}{\sqrt{N_{\min}}} \right)^{\frac{n+2}{n+1}} \right], \quad (9.76)$$

proving the claim. ∎

We also give

Proposition 9.24 *Let* $x \in [0,1]^k, k \in \mathbb{N} - \{1\}$, *be fixed and let* $f \in C^1 \left([0,1]^k, \mathbb{R}_+ \right)$.
We assume that $\frac{\partial f(x)}{\partial x_i} = 0$, *for* $i = 1, ..., k$. *Then*

$$\left| T_{\vec{N}}^{(M)} (f) (x) - f (x) \right| \leq \frac{3}{2} \left(\max_{i=1,...,k} \omega_1 \left(\frac{\partial f}{\partial x_i}, \frac{1}{\sqrt[4]{N_{\min}}} \right) \right) \cdot$$

$$\left[\frac{k^2}{\sqrt[4]{N_{\min}}} + \frac{k}{\sqrt{N_{\min}}} + \frac{1}{4 \left(\sqrt[4]{N_{\min}} \right)^3} \right], \quad (9.77)$$

$\forall \ \vec{N} \in \mathbb{N}^k$, *where* $N_{\min} := \min\{N_1, ..., N_k\}$.
Also it holds $\lim\limits_{\vec{N} \to (\infty,...,\infty)} T_{\vec{N}}^{(M)} (f) (x) = f (x)$.

Proof By (9.31) we get:

$$\left| T_{\overrightarrow{N}}^{(M)}(f)(x) - f(x) \right| \overset{(9.73)}{\leq} \left(\max_{i=1,...,k} \omega_1 \left(\frac{\partial f}{\partial x_i}, h \right) \right) \cdot$$

$$\left[\frac{k}{2h} \left(\sum_{i=1}^{k} T_{N_i}^{(M)} \left((t_i - x_i)^2 \right)(x_i) \right) + \frac{1}{2} \left(\sum_{i=1}^{k} T_{N_i}^{(M)} \left(|t_i - x_i| \right)(x_i) \right) + \frac{h}{8} \right] \overset{(9.68)}{\leq}$$
$$\tag{9.78}$$

$$\frac{3}{\sqrt{N_{\min}}} \left(\max_{i=1,...,k} \omega_1 \left(\frac{\partial f}{\partial x_i}, h \right) \right) \left[\frac{k^2}{2h} + \frac{k}{2} + \frac{h}{8} \right] =: (\psi).$$

Next we choose $h := \left(\frac{1}{\sqrt{N_{\min}}} \right)^{\frac{1}{2}}$, then $h^2 = \frac{1}{\sqrt{N_{\min}}}$.

We have that

$$(\psi) = 3 \left(\max_{i=1,...,k} \omega_1 \left(\frac{\partial f}{\partial x_i}, \frac{1}{\sqrt[4]{N_{\min}}} \right) \right) \cdot$$

$$\left[\frac{k^2}{2} \left(\frac{1}{\sqrt[4]{N_{\min}}} \right) + \frac{k}{2} \left(\frac{1}{\sqrt{N_{\min}}} \right) + \frac{1}{8} \left(\frac{1}{\sqrt{N_{\min}}} \right)^{\frac{3}{2}} \right], \tag{9.79}$$

proving the claim. ∎

It follows

Theorem 9.25 *Let* $f \in C_+ \left([0,1]^k \right)$, $k \in \mathbb{N} - \{1\}$. *Then*

$$\left| T_{\overrightarrow{N}}^{(M)}(f)(x) - f(x) \right| \leq (3k+1) \omega_1 \left(f, \frac{1}{\sqrt{N_{\min}}} \right), \tag{9.80}$$

$\forall\, x \in [0,1]^k$, $\forall\, \overrightarrow{N} \in \mathbb{N}^k$, *where* $N_{\min} := \min\{N_1, ..., N_k\}$.

That is

$$\left\| T_{\overrightarrow{N}}^{(M)}(f) - f \right\|_\infty \leq (3k+1) \omega_1 \left(f, \frac{1}{\sqrt{N_{\min}}} \right). \tag{9.81}$$

It holds that $\displaystyle\lim_{\overrightarrow{N} \to (\infty,...,\infty)} T_{\overrightarrow{N}}^{(M)}(f) = f$, *uniformly.*

Proof We get that (use of (9.44))

$$\left| T_{\overrightarrow{N}}^{(M)}(f)(x) - f(x) \right| \overset{(9.73)}{\leq} \omega_1(f,h) \left(1 + \frac{1}{h} \left(\sum_{i=1}^{k} T_{N_i}^{(M)} \left(|t_i - x_i| \right)(x) \right) \right)$$

$$\overset{(9.67)}{\leq} \omega_1(f,h) \left(1 + \frac{1}{h} \left(\frac{3k}{\sqrt{N_{\min}}} \right) \right) \tag{9.82}$$

(setting $h := \frac{1}{\sqrt{N_{\min}}}$)

$$= \omega_1 \left(f, \frac{1}{\sqrt{N_{\min}}} \right) (3k + 1), \ \forall \, x \in [0, 1]^k, \ \forall \, \overrightarrow{N} \in \mathbb{N}^k,$$

proving the claim. ∎

We make

Remark 9.26 We mention the truncated Max-product Baskakov operator (see [5], p. 11)

$$U_N^{(M)} (f) (x) = \frac{\bigvee_{k=0}^N b_{N,k} (x) f \left(\frac{k}{N} \right)}{\bigvee_{k=0}^N b_{N,k} (x)}, \ x \in [0, 1], \ f \in C_+ ([0, 1]), \ \forall \, N \in \mathbb{N},$$

$$(9.83)$$

where

$$b_{N,k} (x) = \binom{N + k - 1}{k} \frac{x^k}{(1 + x)^{N+k}}.$$

$$(9.84)$$

From [5], pp. 217–218, we get ($x \in [0, 1]$)

$$\left(U_N^{(M)} (|\cdot - x|) \right) (x) \le \frac{12}{\sqrt{N + 1}}, \ N \ge 2, N \in \mathbb{N}.$$

$$(9.85)$$

And as in [2], we obtain ($m \in \mathbb{N}$)

$$\left(U_N^{(M)} (|\cdot - x|^m) \right) (x) \le \frac{12}{\sqrt{N + 1}}, \ N \ge 2, N \in \mathbb{N}, \ \forall \, x \in [0, 1].$$

$$(9.86)$$

Definition 9.27 Let $f \in C_+ \left([0, 1]^k \right)$, $k \in \mathbb{N} - \{1\}$, and $\overrightarrow{N} = (N_1, ..., N_k) \in \mathbb{N}^k$. We define the multivariate Max-product truncated Baskakov operators as follows:

$$U_{\overrightarrow{N}}^{(M)} (f) (x) :=$$

$$\frac{\bigvee_{i_1=0}^{N_1} \bigvee_{i_2=0}^{N_2} \cdots \bigvee_{i_k=0}^{N_k} b_{N_1,i_1} (x_1) b_{N_2,i_2} (x_2) ... b_{N_k,i_k} (x_k) f \left(\frac{i_1}{N_1}, ..., \frac{i_k}{N_k} \right)}{\bigvee_{i_1=0}^{N_1} \bigvee_{i_2=0}^{N_2} \cdots \bigvee_{i_k=0}^{N_k} b_{N_1,i_1} (x_1) b_{N_2,i_2} (x_2) ... b_{N_k,i_k} (x_k)},$$

$$(9.87)$$

$\forall \, x = (x_1, ..., x_k) \in [0, 1]^k$. Call $N_{\min} := \min\{N_1, ..., N_k\}$.

The operators $U_{\overrightarrow{N}}^{(M)} (f) (x)$ are positive sublinear mapping $C_+ \left([0, 1]^k \right)$ into itself, and $U_{\overrightarrow{N}}^{(M)} (1) = 1$.

We also have

$$U_{\overrightarrow{N}}^{(M)} (f) (x) :=$$

$$\frac{\vee_{i_1=0}^{N_1} \vee_{i_2=0}^{N_2} \dots \vee_{i_k=0}^{N_k} b_{N_1,i_1}(x_1)\, b_{N_2,i_2}(x_2) \dots b_{N_k,i_k}(x_k)\, f\left(\frac{i_1}{N_1}, \dots, \frac{i_k}{N_k}\right)}{\prod_{\lambda=1}^{k}\left(\vee_{i_\lambda=0}^{N_\lambda} b_{N_\lambda,i_\lambda}(x_\lambda)\right)}, \tag{9.88}$$

$\forall\, x \in [0, 1]^k$, by the maximum multiplicative principle, see (9.27).

We make

Remark 9.28 The coordinate Max-product truncated Baskakov operators are defined as follows ($\lambda = 1, \dots, k$):

$$U_{N_\lambda}^{(M)}(g)(x_\lambda) := \frac{\vee_{i_\lambda=0}^{N_\lambda} b_{N_\lambda,i_\lambda}(x_\lambda)\, g\left(\frac{i_\lambda}{N_\lambda}\right)}{\vee_{i_\lambda=0}^{N_\lambda} b_{N_\lambda,i_\lambda}(x_\lambda)}, \tag{9.89}$$

$\forall\, N_\lambda \in \mathbb{N}$, and $\forall\, x_\lambda \in [0, 1]$, $\forall\, g \in C_+([0, 1])$.
Here we have

$$b_{N_\lambda,i_\lambda}(x_\lambda) = \binom{N_\lambda + i_\lambda - 1}{i_\lambda} \frac{x_\lambda^{i_\lambda}}{(1+x_\lambda)^{N+i_\lambda}}, \quad \lambda = 1, \dots, k; \; x_\lambda \in [0, 1].$$

In case of $f \in C_+\left([0, 1]^k\right)$ such that $f(x) := g(x_\lambda)$, $\forall\, x \in [0, 1]^k$, where $x = (x_1, \dots, x_\lambda, \dots, x_k)$ and $g \in C_+([0, 1])$, we get that

$$U_{\vec{N}}^{(M)}(f)(x) = U_{N_\lambda}^{(M)}(g)(x_\lambda), \tag{9.90}$$

by the maximum multiplicative principle (9.27) and simplification of (9.89).

We present

Theorem 9.29 *Let $x \in [0, 1]^k$, $k \in \mathbb{N} - \{1\}$, be fixed, and let $f \in C^n\left([0, 1]^k, \mathbb{R}_+\right)$, $n \in \mathbb{N} - \{1\}$. We assume that $f_\alpha(x) = 0$, for all $\alpha : |\alpha| = 1, \dots, n$. Then*

$$\left|U_{\vec{N}}^{(M)}(f)(x) - f(x)\right| \le 12 \left(\max_{\alpha:|\alpha|=n}\left(\omega_1\left(f_\alpha, \left(\frac{1}{\sqrt{N_{\min}+1}}\right)^{\frac{1}{n+1}}\right)\right)\right) \cdot$$

$$\left[\frac{k^{n+1}}{(n+1)!}\left(\frac{1}{\sqrt{N_{\min}+1}}\right)^{\frac{n}{n+1}} + \frac{k^n}{2n!}\left(\frac{1}{\sqrt{N_{\min}+1}}\right) + \frac{k^{n-1}}{8(n-1)!}\left(\frac{1}{\sqrt{N_{\min}+1}}\right)^{\frac{n+2}{n+1}}\right], \tag{9.91}$$

$\forall\, \vec{N} \in (\mathbb{N} - \{1\})^k$, *where* $N_{\min} := \min\{N_1, \dots, N_k\}$.
We have that $\lim\limits_{\vec{N}\to(\infty,\dots,\infty)} U_{\vec{N}}^{(M)}(f)(x) = f(x)$.

Proof By (9.26) we get:

$$\left|U_{\vec{N}}^{(M)}(f)(x) - f(x)\right| \overset{(9.90)}{\le} \left(\max_{\alpha:|\alpha|=n} \omega_1\left(f_\alpha, h\right)\right) \cdot$$

$$\left[\frac{k^n}{(n+1)!h} \left(\sum_{i=1}^{k} U_{N_i}^{(M)} \left(|t_i - x_i|^{n+1} \right) (x_i) \right) + \frac{k^{n-1}}{2n!} \left(\sum_{i=1}^{k} U_{N_i}^{(M)} \left(|t_i - x_i|^{n} \right) (x_i) \right) \right.$$

$$\left. + \frac{hk^{n-2}}{8(n-1)!} \left(\sum_{i=1}^{k} U_{N_i}^{(M)} \left(|t_i - x_i|^{n-1} \right) (x_i) \right) \right] \overset{(9.86)}{\leq} \tag{9.92}$$

$$\frac{12}{\sqrt{N_{\min}+1}} \left(\max_{\alpha:|\alpha|=n} \omega_1 \left(f_\alpha, h \right) \right) \left[\frac{k^{n+1}}{(n+1)!h} + \frac{k^n}{2n!} + \frac{hk^{n-1}}{8(n-1)!} \right] =: (\xi).$$

Above notice that $\sum_{i=1}^{k} U_{N_i}^{(M)} \left(|t_i - x_i|^{n} \right) (x_i) \overset{(9.86)}{\leq} \sum_{i=1}^{k} \frac{12}{\sqrt{N_i+1}} \leq \frac{12k}{\sqrt{N_{\min}+1}}$, etc.

Next we choose $h := \left(\frac{1}{\sqrt{N_{\min}+1}} \right)^{\frac{1}{n+1}}$, then $h^n = \left(\frac{1}{\sqrt{N_{\min}+1}} \right)^{\frac{n}{n+1}}$ and $h^{n+1} = \frac{1}{\sqrt{N_{\min}+1}}$.
We have

$$(\xi) = 12 \left(\max_{\alpha:|\alpha|=n} \left(\omega_1 \left(f_\alpha, \left(\frac{1}{\sqrt{N_{\min}+1}} \right)^{\frac{1}{n+1}} \right) \right) \right).$$

$$\left[\frac{k^{n+1}}{(n+1)!} \left(\frac{1}{\sqrt{N_{\min}+1}} \right)^{\frac{n}{n+1}} + \frac{k^n}{2n!} \left(\frac{1}{\sqrt{N_{\min}+1}} \right) + \frac{k^{n-1}}{8(n-1)!} \left(\frac{1}{\sqrt{N_{\min}+1}} \right)^{\frac{n+2}{n+1}} \right], \tag{9.93}$$

proving the claim. ∎

We also give

Proposition 9.30 Let $x \in [0,1]^k$, $k \in \mathbb{N} - \{1\}$, be fixed and let $f \in C^1 \left([0,1]^k, \mathbb{R}_+ \right)$. We assume that $\frac{\partial f(x)}{\partial x_i} = 0$, for $i = 1, ..., k$. Then

$$\left| U_{\vec{N}}^{(M)} (f) (x) - f (x) \right| \leq 6 \left(\max_{i=1,...,k} \omega_1 \left(\frac{\partial f}{\partial x_i}, \frac{1}{\sqrt[4]{N_{\min}+1}} \right) \right). \tag{9.94}$$

$$\left[\frac{k^2}{\sqrt[4]{N_{\min}+1}} + \frac{k}{\sqrt{N_{\min}+1}} + \frac{1}{4 \left(\sqrt[4]{N_{\min}+1} \right)^3} \right],$$

$\forall \vec{N} \in (\mathbb{N} - \{1\})^k$, where $N_{\min} := \min\{N_1, ..., N_k\}$.
Also it holds $\lim_{\vec{N} \to (\infty,...,\infty)} U_{\vec{N}}^{(M)} (f) (x) = f (x)$.

Proof By (9.31) we get:

$$\left| U_{\vec{N}}^{(M)} (f) (x) - f (x) \right| \overset{(9.90)}{\leq} \left(\max_{i=1,...,k} \omega_1 \left(\frac{\partial f}{\partial x_i}, h \right) \right).$$

$$\left[\frac{k}{2h} \left(\sum_{i=1}^{k} U_{N_i}^{(M)} \left((t_i - x_i)^2 \right) (x_i) \right) + \frac{1}{2} \left(\sum_{i=1}^{k} U_{N_i}^{(M)} \left(|t_i - x_i| \right) (x_i) \right) + \frac{h}{8} \right] \leq \qquad (9.85)$$

$$(9.95)$$

$$\frac{12}{\sqrt{N_{\min} + 1}} \left(\max_{i=1,\dots,k} \omega_1 \left(\frac{\partial f}{\partial x_i}, h \right) \right) \left[\frac{k^2}{2h} + \frac{k}{2} + \frac{h}{8} \right] =: (\psi).$$

Next we choose $h := \left(\frac{1}{\sqrt{N_{\min}+1}} \right)^{\frac{1}{2}}$, then $h^2 = \frac{1}{\sqrt{N_{\min}+1}}$.

We have that

$$(\psi) = 12 \left(\max_{i=1,\dots,k} \omega_1 \left(\frac{\partial f}{\partial x_i}, \frac{1}{\sqrt[4]{N_{\min} + 1}} \right) \right) \cdot$$

$$\left[\frac{k^2}{2} \left(\frac{1}{\sqrt[4]{N_{\min} + 1}} \right) + \frac{k}{2} \left(\frac{1}{\sqrt{N_{\min} + 1}} \right) + \frac{1}{8} \left(\frac{1}{\sqrt{N_{\min} + 1}} \right)^{\frac{3}{2}} \right], \qquad (9.96)$$

proving the claim. ∎

It follows

Theorem 9.31 Let $f \in C_+ \left([0, 1]^k \right)$, $k \in \mathbb{N} - \{1\}$. Then

$$\left| U_{\vec{N}}^{(M)} (f) (x) - f (x) \right| \leq (12k + 1) \, \omega_1 \left(f, \frac{1}{\sqrt{N_{\min} + 1}} \right), \qquad (9.97)$$

$\forall \, x \in [0, 1]^k$, $\forall \, \vec{N} \in (\mathbb{N} - \{1\})^k$, where $N_{\min} := \min\{N_1, \dots, N_k\}$.

That is

$$\left\| U_{\vec{N}}^{(M)} (f) - f \right\|_{\infty} \leq (12k + 1) \, \omega_1 \left(f, \frac{1}{\sqrt{N_{\min} + 1}} \right). \qquad (9.98)$$

It holds that $\lim\limits_{\vec{N} \to (\infty,\dots,\infty)} U_{\vec{N}}^{(M)} (f) = f$, uniformly.

Proof We get that (use of (9.44))

$$\left| U_{\vec{N}}^{(M)} (f) (x) - f (x) \right| \overset{(9.90)}{\leq} \omega_1 (f, h) \left(1 + \frac{1}{h} \left(\sum_{i=1}^{k} U_{N_i}^{(M)} \left(|t_i - x_i| \right) (x_i) \right) \right)$$

$$\overset{(9.85)}{\leq} \omega_1 (f, h) \left(1 + \frac{1}{h} \left(\frac{12k}{\sqrt{N_{\min} + 1}} \right) \right) \qquad (9.99)$$

(setting $h := \frac{1}{\sqrt{N_{\min}+1}}$)

$$= \omega_1 \left(f, \frac{1}{\sqrt{N_{\min} + 1}} \right) (12k + 1), \; \forall \, x \in [0, 1]^k, \; \forall \, \vec{N} \in (\mathbb{N} - \{1\})^k,$$

proving the claim. ∎

We make

Remark 9.32 Here we mention the Max-product truncated sampling operators (see [5], p. 13) defined by

$$W_N^{(M)}(f)(x) := \frac{\bigvee_{k=0}^{N} \frac{\sin(Nx-k\pi)}{Nx-k\pi} f\left(\frac{k\pi}{N}\right)}{\bigvee_{k=0}^{N} \frac{\sin(Nx-k\pi)}{Nx-k\pi}}, \quad x \in [0,\pi], \tag{9.100}$$

$f : [0, \pi] \to \mathbb{R}_+$, continuous, and

$$K_N^{(M)}(f)(x) := \frac{\bigvee_{k=0}^{N} \frac{\sin^2(Nx-k\pi)}{(Nx-k\pi)^2} f\left(\frac{k\pi}{N}\right)}{\bigvee_{k=0}^{N} \frac{\sin^2(Nx-k\pi)}{(Nx-k\pi)^2}}, \quad x \in [0,\pi], \tag{9.101}$$

$f : [0, \pi] \to \mathbb{R}_+$, continuous.

By convention we talk $\frac{\sin(0)}{0} = 1$, which implies for every $x = \frac{k\pi}{N}, k \in \{0, 1, ..., N\}$ that we have $\frac{\sin(Nx-k\pi)}{Nx-k\pi} = 1$.

We define the Max-product truncated combined sampling operators

$$M_N^{(M)}(f)(x) := \frac{\bigvee_{k=0}^{N} \rho_{N,k}(x) f\left(\frac{k\pi}{N}\right)}{\bigvee_{k=0}^{N} \rho_{N,k}(x)}, \quad x \in [0,\pi], \tag{9.102}$$

$f \in C_+([0,\pi])$, where

$$M_N^{(M)}(f)(x) := \begin{cases} W_N^{(M)}(f)(x), \text{ if } \rho_{N,k}(x) := \frac{\sin(Nx-k\pi)}{Nx-k\pi}, \\ K_N^{(M)}(f)(x), \text{ if } \rho_{N,k}(x) := \left(\frac{\sin(Nx-k\pi)}{Nx-k\pi}\right)^2. \end{cases} \tag{9.103}$$

By [5], p. 346 and p. 352 we get

$$\left(M_N^{(M)}(|\cdot - x|)\right)(x) \le \frac{\pi}{2N}, \tag{9.104}$$

and by [3] $(m \in \mathbb{N})$ we have

$$\left(M_N^{(M)}(|\cdot - x|^m)\right)(x) \le \frac{\pi^m}{2N}, \quad \forall x \in [0,\pi], \forall N \in \mathbb{N}. \tag{9.105}$$

We give

Definition 9.33 Let $f \in C_+\left([0,\pi]^k\right)$, $k \in \mathbb{N} - \{1\}$, and $\overrightarrow{N} = (N_1, ..., N_k) \in \mathbb{N}^k$. We define the multivariate Max-product truncated combined sampling operators as follows:

$$M_{\overrightarrow{N}}^{(M)}(f)(x) :=$$

$$\frac{\bigvee_{i_1=0}^{N_1} \bigvee_{i_2=0}^{N_2} \cdots \bigvee_{i_k=0}^{N_k} \rho_{N_1,i_1}\left(x_1\right) \rho_{N_2,i_2}\left(x_2\right) \cdots \rho_{N_k,i_k}\left(x_k\right) f\left(\frac{i_1\pi}{N_1}, \frac{i_2\pi}{N_2}, \ldots, \frac{i_k\pi}{N_k}\right)}{\bigvee_{i_1=0}^{N_1} \bigvee_{i_2=0}^{N_2} \cdots \bigvee_{i_k=0}^{N_k} \rho_{N_1,i_1}\left(x_1\right) \rho_{N_2,i_2}\left(x_2\right) \cdots \rho_{N_k,i_k}\left(x_k\right)},$$

(9.106)

$\forall \, x = (x_1, \ldots, x_k) \in [0, \pi]^k$. Call $N_{\min} := \min\{N_1, \ldots, N_k\}$.

The operators $M_{\overrightarrow{N}}^{(M)}(f)(x)$ are positive sublinear mapping $C_+\left([0, \pi]^k\right)$ into itself, and $M_{\overrightarrow{N}}^{(M)}(1) = 1$.

We also have

$$M_{\overrightarrow{N}}^{(M)}(f)(x) :=$$

$$\frac{\bigvee_{i_1=0}^{N_1} \bigvee_{i_2=0}^{N_2} \cdots \bigvee_{i_k=0}^{N_k} \rho_{N_1,i_1}\left(x_1\right) \rho_{N_2,i_2}\left(x_2\right) \cdots \rho_{N_k,i_k}\left(x_k\right) f\left(\frac{i_1\pi}{N_1}, \frac{i_2\pi}{N_2}, \ldots, \frac{i_k\pi}{N_k}\right)}{\prod_{\lambda=1}^{k}\left(\bigvee_{i_\lambda=0}^{N_\lambda} \rho_{N_\lambda,i_\lambda}\left(x_\lambda\right)\right)},$$

(9.107)

$\forall \, x \in [0, \pi]^k$, by the maximum multiplicative principle, see (9.27).

We make

Remark 9.34 The coordinate Max-product truncated combined sampling operators are defined as follows ($\lambda = 1, \ldots, k$):

$$M_{N_\lambda}^{(M)}(g)(x_\lambda) := \frac{\bigvee_{i_\lambda=0}^{N_\lambda} \rho_{N_\lambda,i_\lambda}\left(x_\lambda\right) g\left(\frac{i_\lambda\pi}{N_\lambda}\right)}{\bigvee_{i_\lambda=0}^{N_\lambda} \rho_{N_\lambda,i_\lambda}\left(x_\lambda\right)},$$

(9.108)

$\forall \, N_\lambda \in \mathbb{N}$, and $\forall \, x_\lambda \in [0, \pi], \forall \, g \in C_+\left([0, \pi]\right)$.

Here we have ($\lambda = 1, \ldots, k; \, x_\lambda \in [0, \pi]$)

$$\rho_{N_\lambda,i_\lambda}\left(x_\lambda\right) = \begin{cases} \frac{\sin(N_\lambda x_\lambda - i_\lambda\pi)}{N_\lambda x_\lambda - i_\lambda\pi}, & \text{if } M_{N_\lambda}^{(M)} = W_{N_\lambda}^{(M)}, \\ \left(\frac{\sin(N_\lambda x_\lambda - i_\lambda\pi)}{N_\lambda x_\lambda - i_\lambda\pi}\right)^2, & \text{if } M_{N_\lambda}^{(M)} = K_{N_\lambda}^{(M)}. \end{cases}$$

(9.109)

In case of $f \in C_+\left([0, \pi]^k\right)$ such that $f(x) := g(x_\lambda), \forall \, x \in [0, \pi]^k$, where $x = (x_1, \ldots, x_\lambda, \ldots, x_k)$ and $g \in C_+([0, \pi])$, we get that

$$M_{\overrightarrow{N}}^{(M)}(f)(x) = M_{N_\lambda}^{(M)}(g)(x_\lambda),$$

(9.110)

by the maximum multiplicative principle (9.27) and simplification of (9.107).

We present

Theorem 9.35 *Let $x \in [0, \pi]^k, k \in \mathbb{N} - \{1\}$, be fixed, and let $f \in C^n\left([0, \pi]^k, \mathbb{R}_+\right)$, $n \in \mathbb{N} - \{1\}$. We assume that $f_\alpha(x) = 0$, for all $\alpha : |\alpha| = 1, \ldots, n$. Then*

$$\left| M_{\overrightarrow{N}}^{(M)}(f)(x) - f(x) \right| \leq \frac{(k\pi)^{n-1}}{2} \left(\max_{\alpha:|\alpha|=n} \omega_1\left(f_\alpha, \frac{1}{(N_{\min})^{\frac{1}{n+1}}}\right) \right).$$

(9.111)

$$\left[\frac{(k\pi)^2}{(n+1)!} \frac{1}{(N_{min})^{\frac{n}{n+1}}} + \frac{k\pi}{2n!N_{min}} + \frac{1}{8\,(n-1)!\,(N_{min})^{\frac{n+2}{n+1}}} \right],$$

$\forall \overrightarrow{N} = (N_1, ..., N_k) \in \mathbb{N}^k$, where $N_{min} := \min\{N_1, ..., N_k\}$.
We have that $\lim\limits_{\overrightarrow{N} \to (\infty,...,\infty)} M_{\overrightarrow{N}}^{(M)}(f)(x) = f(x)$.

Proof By (9.26) we get:

$$\left| M_{\overrightarrow{N}}^{(M)}(f)(x) - f(x) \right| \overset{(9.110)}{\leq} \left(\max_{\alpha:|\alpha|=n} \omega_1(f_\alpha, h) \right) \cdot$$

$$\left[\frac{k^n}{(n+1)!h} \left(\sum_{i=1}^{k} M_{N_i}^{(M)}\left(|t_i - x_i|^{n+1}\right)(x_i) \right) + \frac{k^{n-1}}{2n!} \left(\sum_{i=1}^{k} M_{N_i}^{(M)}\left(|t_i - x_i|^{n}\right)(x_i) \right) \right.$$

$$\tag{9.112}$$

$$\left. + \frac{hk^{n-2}}{8\,(n-1)!} \left(\sum_{i=1}^{k} M_{N_i}^{(M)}\left(|t_i - x_i|^{n-1}\right)(x_i) \right) \right] \overset{(9.105)}{\leq}$$

$$\frac{1}{2N_{min}} \left(\max_{\alpha:|\alpha|=n} \omega_1(f_\alpha, h) \right) \left[\frac{k^{n+1}\pi^{n+1}}{(n+1)!} + \frac{k^n \pi^n}{2n!} + \frac{hk^{n-1}\pi^{n-1}}{8\,(n-1)!} \right] =: (\xi).$$

Above notice that $\sum_{i=1}^{k} M_{N_i}^{(M)}\left(|t_i - x_i|^{n}\right)(x_i) \overset{(9.105)}{\leq} \sum_{i=1}^{k} \frac{\pi^n}{2N_i} \leq \frac{k\pi^n}{2N_{min}}$, etc.

Next we choose $h := \left(\frac{1}{N_{min}} \right)^{\frac{1}{n+1}}$, then $h^n = \left(\frac{1}{N_{min}} \right)^{\frac{n}{n+1}}$ and $h^{n+1} = \frac{1}{N_{min}}$.
We have

$$(\xi) = \frac{(k\pi)^{n-1}}{2} \left(\max_{\alpha:|\alpha|=n} \omega_1 \left(f_\alpha, \frac{1}{(N_{min})^{\frac{1}{n+1}}} \right) \right) \cdot \tag{9.113}$$

$$\left[\frac{(k\pi)^2}{(n+1)!} \frac{1}{(N_{min})^{\frac{n}{n+1}}} + \frac{k\pi}{2n!N_{min}} + \frac{1}{8\,(n-1)!\,(N_{min})^{\frac{n+2}{n+1}}} \right],$$

proving the claim. ∎

We also give

Proposition 9.36 *Let* $x \in [0, \pi]^k, k \in \mathbb{N} - \{1\}$, *be fixed and let* $f \in C^1([0, \pi], \mathbb{R}_+)$. *We assume that* $\frac{\partial f(x)}{\partial x_i} = 0$, *for* $i = 1, ..., k$. *Then*

$$\left| M_{\overrightarrow{N}}^{(M)}(f)(x) - f(x) \right| \leq \frac{1}{2} \left(\max_{i=1,...,k} \omega_1 \left(\frac{\partial f}{\partial x_i}, \frac{1}{\sqrt{N_{min}}} \right) \right) \cdot$$

$$\left[\frac{(k\pi)^2}{2\sqrt{N_{min}}} + \frac{k\pi}{2N_{min}} + \frac{1}{8\left(\sqrt{N_{min}}\right)^3} \right], \tag{9.114}$$

$\forall \overrightarrow{N} \in \mathbb{N}^k$, where $N_{min} := \min\{N_1, ..., N_k\}$.

Also it holds $\lim\limits_{\overrightarrow{N}\to(\infty,...,\infty)} M_{\overrightarrow{N}}^{(M)}(f)(x) = f(x)$.

Proof By (9.31) we get:

$$\left| M_{\overrightarrow{N}}^{(M)}(f)(x) - f(x) \right| \overset{(9.110)}{\leq} \left(\max_{i=1,...,k} \omega_1 \left(\frac{\partial f}{\partial x_i}, h \right) \right) \cdot$$

$$\left[\frac{k}{2h} \left(\sum_{i=1}^{k} M_{N_i}^{(M)}\left((t_i - x_i)^2\right)(x_i) \right) + \frac{1}{2} \left(\sum_{i=1}^{k} M_{N_i}^{(M)}(|t_i - x_i|)(x_i) \right) + \frac{h}{8} \right] \overset{(9.105)}{\leq}$$

$$\tag{9.115}$$

$$\frac{1}{2N_{\min}} \left(\max_{i=1,...,k} \omega_1 \left(\frac{\partial f}{\partial x_i}, h \right) \right) \left[\frac{k^2\pi^2}{2h} + \frac{k\pi}{2} + \frac{h}{8} \right] =: (\psi).$$

Next we choose $h := \left(\frac{1}{N_{\min}} \right)^{\frac{1}{2}}$, then $h^2 = \frac{1}{N_{\min}}$.

We have that

$$(\psi) = \frac{1}{2} \left(\max_{i=1,...,k} \omega_1 \left(\frac{\partial f}{\partial x_i}, \frac{1}{\sqrt{N_{\min}}} \right) \right) \cdot$$

$$\left[\frac{k^2\pi^2}{2\sqrt{N_{\min}}} + \frac{k\pi}{2N_{\min}} + \frac{1}{8\,(N_{\min})^{\frac{3}{2}}} \right],$$

$$\tag{9.116}$$

proving the claim. ∎

It follows

Theorem 9.37 *Let* $f \in C_+\left([0, \pi]^k\right)$, $k \in \mathbb{N} - \{1\}$. *Then*

$$\left| M_{\overrightarrow{N}}^{(M)}(f)(x) - f(x) \right| \leq \left(\frac{k\pi}{2} + 1 \right) \omega_1 \left(f, \frac{1}{N_{\min}} \right),$$

$$\tag{9.117}$$

$\forall\, x \in [0, \pi]^k$, $\forall\, \overrightarrow{N} \in \mathbb{N}^k$, *where* $N_{\min} := \min\{N_1, ..., N_k\}$.

That is

$$\left\| M_{\overrightarrow{N}}^{(M)}(f) - f \right\|_{\infty} \leq \left(\frac{k\pi}{2} + 1 \right) \omega_1 \left(f, \frac{1}{N_{\min}} \right).$$

$$\tag{9.118}$$

It holds $\lim\limits_{\overrightarrow{N}\to(\infty,...,\infty)} M_{\overrightarrow{N}}^{(M)}(f) = f$, *uniformly*.

Proof We get that (use of (9.44))

$$\left| M_{\overrightarrow{N}}^{(M)}(f)(x) - f(x) \right| \overset{(9.110)}{\leq} \omega_1(f, h) \left(1 + \frac{1}{h} \left(\sum_{i=1}^{k} M_{N_i}^{(M)}(|t_i - x_i|)(x_i) \right) \right)$$

$$\overset{(9.104)}{\leq} \omega_1 (f, h) \left(1 + \frac{1}{h} \left(\frac{k\pi}{2N_{\min}}\right)\right) \tag{9.119}$$

(setting $h := \frac{1}{N_{\min}}$)

$$= \omega_1 \left(f, \frac{1}{N_{\min}}\right) \left(\frac{k\pi}{2} + 1\right), \ \forall \, x \in [0, \pi]^k, \ \forall \, \overrightarrow{N} \in \mathbb{N}^k,$$

proving the claim. ∎

We make

Remark 9.38 Let $f \in C_+ ([-1, 1])$. Let the Chebyshev knots of second kind $x_{N,k} = \cos \left(\left(\frac{N-k}{N-1}\right) \pi\right) \in [-1, 1], k = 1, ..., N, N \in \mathbb{N} - \{1\}$, which are the roots of $\omega_N (x) = \sin (N - 1) t \sin t, x = \cos t \in [-1, 1]$. Notice that $x_{N,1} = -1$ and $x_{N,N} = 1$.
Define

$$l_{N,k} (x) := \frac{(-1)^{k-1} \omega_N (x)}{(1 + \delta_{k,1} + \delta_{k,N}) (N - 1) (x - x_{N,k})}, \tag{9.120}$$

$N \geq 2, k = 1, ..., N$, and $\omega_N (x) = \prod_{k=1}^N (x - x_{N,k})$ and $\delta_{i,j}$ denotes the Kronecker's symbol, that is $\delta_{i,j} = 1$, if $i = j$, and $\delta_{i,j} = 0$, if $i \neq j$.
The Max-product Lagrange interpolation operators on Chebyshev knots of second kind, plus the endpoints ± 1, are defined by ([5], p. 12)

$$L_N^{(M)} (f) (x) = \frac{\bigvee_{k=1}^N l_{N,k} (x) f (x_{N,k})}{\bigvee_{k=1}^N l_{N,k} (x)}, \ x \in [-1, 1]. \tag{9.121}$$

By [5], pp. 297–298 and [3], we get that

$$L_N^{(M)} (|\cdot - x|^m) (x) \leq \frac{2^{m+1} \pi^2}{3 (N - 1)}, \tag{9.122}$$

$\forall \, x \in (-1, 1)$ and $\forall \, m \in \mathbb{N}; \forall \, N \in \mathbb{N}, N \geq 4$.
We see that $L_N^{(M)} (f) (x) \geq 0$ is well defined and continuous for any $x \in [-1, 1]$. Following [5], p. 289, because $\sum_{k=1}^N l_{N,k} (x) = 1, \forall \, x \in [-1, 1]$, for any x there exists $k \in \{1, ..., N\} : l_{N,k} (x) > 0$, hence $\bigvee_{k=1}^N l_{N,k} (x) > 0$. We have that $l_{N,k} (x_{N,k}) = 1$, and $l_{N,k} (x_{N,j}) = 0$, if $k \neq j$. Furthermore it holds $L_N^{(M)} (f) (x_{N,j}) = f (x_{N,j})$, all $j \in \{1, ..., N\}$, and $L_N^{(M)} (1) = 1$.
By [5], pp. 289–290, $L_N^{(M)}$ are positive sublinear operators.

We give

Definition 9.39 Let $f \in C_+ ([-1, 1]^k), \ k \in \mathbb{N} - \{1\}$, and $\overrightarrow{N} = (N_1, ..., N_k) \in (\mathbb{N} - \{1\})^k$. We define the multivariate Max-product Lagrange interpolation operators on Chebyshev knots of second kind, plus the endpoints ± 1, as follows:

$$L_{\overrightarrow{N}}^{(M)}(f)(x) :=$$

$$\frac{\bigvee_{i_1=1}^{N_1} \bigvee_{i_2=1}^{N_2} \cdots \bigvee_{i_k=1}^{N_k} l_{N_1,i_1}(x_1) \, l_{N_2,i_2}(x_2) \ldots l_{N_k,i_k}(x_k) \, f\left(x_{N_1,i_1}, x_{N_2,i_2}, \ldots, x_{N_k,i_k}\right)}{\bigvee_{i_1=1}^{N_1} \bigvee_{i_2=1}^{N_2} \cdots \bigvee_{i_k=1}^{N_k} l_{N_1,i_1}(x_1) \, l_{N_2,i_2}(x_2) \ldots l_{N_k,i_k}(x_k)},$$

(9.123)

$\forall \, x = (x_1, \ldots, x_k) \in [-1, 1]^k$. Call $N_{\min} := \min\{N_1, \ldots, N_k\}$.

The operators $L_{\overrightarrow{N}}^{(M)}(f)(x)$ are positive sublinear mapping $C_+\left([-1, 1]^k\right)$ into itself, and $L_{\overrightarrow{N}}^{(M)}(1) = 1$.

We also have

$$L_{\overrightarrow{N}}^{(M)}(f)(x) :=$$

$$\frac{\bigvee_{i_1=1}^{N_1} \bigvee_{i_2=1}^{N_2} \cdots \bigvee_{i_k=1}^{N_k} l_{N_1,i_1}(x_1) \, l_{N_2,i_2}(x_2) \ldots l_{N_k,i_k}(x_k) \, f\left(x_{N_1,i_1}, x_{N_2,i_2}, \ldots, x_{N_k,i_k}\right)}{\prod_{\lambda=1}^{k}\left(\bigvee_{i_\lambda=1}^{N_\lambda} l_{N_\lambda,i_\lambda}(x_\lambda)\right)},$$

(9.124)

$\forall \, x = (x_1, \ldots, x_\lambda, \ldots, x_k) \in [-1, 1]^k$, by the maximum multiplicative principle, see (9.27). Notice that $L_{\overrightarrow{N}}^{(M)}(f)\left(x_{N_1,i_1}, \ldots, x_{N_k,i_k}\right) = f\left(x_{N_1,i_1}, \ldots, x_{N_k,i_k}\right)$. The last is also true if $x_{N_1,i_1}, \ldots, x_{N_k,i_k} \in \{-1, 1\}$.

We make

Remark 9.40 The coordinate Max-product Lagrange interpolation operators on Chebyshev knots of second kind, plus the endpoints ± 1, are defined as follows ($\lambda = 1, \ldots, k$):

$$L_{N_\lambda}^{(M)}(g)(x_\lambda) := \frac{\bigvee_{i_\lambda=1}^{N_\lambda} l_{N_\lambda,i_\lambda}(x_\lambda) \, g\left(x_{N_\lambda,i_\lambda}\right)}{\bigvee_{i_\lambda=1}^{N_\lambda} l_{N_\lambda,i_\lambda}(x_\lambda)},$$

(9.125)

$\forall \, N_\lambda \in \mathbb{N}, \, N_\lambda \geq 2$, and $\forall \, x_\lambda \in [-1, 1], \, \forall \, g \in C_+([-1, 1])$.

Here we have ($\lambda = 1, \ldots, k; \, x_\lambda \in [-1, 1]$)

$$l_{N_\lambda,i_\lambda}(x_\lambda) = \frac{(-1)^{i_\lambda - 1} \, \omega_{N_\lambda}(x_\lambda)}{\left(1 + \delta_{i_\lambda,1} + \delta_{i_\lambda,N_\lambda}\right)(N_\lambda - 1)\left(x_\lambda - x_{N_\lambda,i_\lambda}\right)},$$

(9.126)

$N_\lambda \geq 2, \, i_\lambda = 1, \ldots, N_\lambda$ and $\omega_{N_\lambda}(x_\lambda) = \prod_{i_\lambda=1}^{N_\lambda}\left(x_\lambda - x_{N_\lambda,i_\lambda}\right)$; where $x_{N_\lambda,i_\lambda} = \cos\left(\left(\frac{N_\lambda - i_\lambda}{N_\lambda - 1}\right)\pi\right) \in [-1, 1], \, i_\lambda = 1, \ldots, N_\lambda \, (N_\lambda \geq 2)$ are roots of $\omega_{N_\lambda}(x_\lambda) = \sin(N_\lambda - 1) t_\lambda \sin t_\lambda, \, x_\lambda = \cos t_\lambda$. Notice that $x_{N_\lambda,1} = -1, x_{N_\lambda,N_\lambda} = 1$.

In case of $f \in C_+\left([-1, 1]^k\right)$ such that $f(x) := g(x_\lambda), \, \forall \, x \in [-1, 1]^k$, where $x = (x_1, \ldots, x_\lambda, \ldots, x_k)$ and $g \in C_+([-1, 1])$, we get that

$$L_{\overrightarrow{N}}^{(M)}(f)(x) = L_{N_\lambda}^{(M)}(g)(x_\lambda),$$

(9.127)

by the maximum multiplicative principle (9.27) and simplification of (9.124).

We present

Theorem 9.41 *Let $x \in (-1, 1)^k$, $k \in \mathbb{N} - \{1\}$, be fixed, and let $f \in C^n ([-1, 1]^k, \mathbb{R}_+)$, $n \in \mathbb{N} - \{1\}$. We assume that $f_\alpha (x) = 0$, for all $\alpha : |\alpha| = 1, ..., n$. Then*

$$\left| L_{\vec{N}}^{(M)} (f) (x) - f (x) \right| \leq \frac{(2k)^{n-1} \pi^2}{3} \left(\max_{\alpha : |\alpha| = n} \omega_1 \left(f_\alpha, \frac{1}{\sqrt[n+1]{N_{\min} - 1}} \right) \right) \cdot \quad (9.128)$$

$$\left[\frac{8k^2}{(n + 1)! (N_{\min} - 1)^{\frac{n}{n+1}}} + \frac{2k}{n! (N_{\min} - 1)} + \frac{1}{4 (n - 1)! (N_{\min} - 1)^{\frac{n+2}{n+1}}} \right],$$

$\forall \; \vec{N} = (N_1, ..., N_k) \in \mathbb{N}^k$; $N_i \geq 4$, $i = 1, ..., k$, and $N_{\min} := \min\{N_1, ..., N_k\}$. We have that $\lim\limits_{\vec{N} \to (\infty,...,\infty)} L_{\vec{N}}^{(M)} (f) (x) = f (x)$.

Proof By (9.26) we get:

$$\left| L_{\vec{N}}^{(M)} (f) (x) - f (x) \right| \overset{(9.127)}{\leq} \left(\max_{\alpha : |\alpha| = n} \omega_1 (f_\alpha, h) \right) \cdot$$

$$\left[\frac{k^n}{(n + 1)! h} \left(\sum_{i=1}^{k} L_{N_i}^{(M)} \left(|t_i - x_i|^{n+1} \right) (x_i) \right) + \frac{k^{n-1}}{2n!} \left(\sum_{i=1}^{k} L_{N_i}^{(M)} \left(|t_i - x_i|^{n} \right) (x_i) \right) \right.$$

$$\left. + \frac{h k^{n-2}}{8 (n - 1)!} \left(\sum_{i=1}^{k} L_{N_i}^{(M)} \left(|t_i - x_i|^{n-1} \right) (x_i) \right) \right] \overset{(9.122)}{\leq} \quad (9.129)$$

$$\frac{\pi^2}{3 (N_{\min} - 1)} \left(\max_{\alpha : |\alpha| = n} \omega_1 (f_\alpha, h) \right) \left[\frac{k^{n+1} 2^{n+2}}{(n + 1)! h} + \frac{k^n 2^{n+1}}{2n!} + \frac{h k^{n-1} 2^n}{8 (n - 1)!} \right] =: (\xi).$$

Above we notice that $\sum_{i=1}^{k} L_{N_i}^{(M)} \left(|t_i - x_i|^{n} \right) (x_i) \overset{(9.122)}{\leq} \sum_{i=1}^{k} \frac{2^{n+1} \pi^2}{3(N_i - 1)} \leq \frac{2^{n+1} \pi^2 k}{3(N_{\min} - 1)}$, etc.

Next we choose $h := \left(\frac{1}{N_{\min} - 1} \right)^{\frac{1}{n+1}}$, then $h^n = \left(\frac{1}{N_{\min} - 1} \right)^{\frac{n}{n+1}}$ and $h^{n+1} = \frac{1}{N_{\min} - 1}$.
We have

$$(\xi) = \frac{\pi^2}{3} \left(\max_{\alpha : |\alpha| = n} \omega_1 \left(f_\alpha, \frac{1}{\sqrt[n+1]{N_{\min} - 1}} \right) \right) \cdot \quad (9.130)$$

$$\left[\frac{k^{n+1} 2^{n+2}}{(n + 1)!} \frac{1}{(N_{\min} - 1)^{\frac{n}{n+1}}} + \frac{k^n 2^n}{n! (N_{\min} - 1)} + \frac{k^{n-1} 2^{n-1}}{4 (n - 1)!} \frac{1}{(N_{\min} - 1)^{\frac{n+2}{n+1}}} \right],$$

proving the claim. ∎

We also give

Proposition 9.42 *Let* $x \in (-1, 1)^k$, $k \in \mathbb{N}-\{1\}$, *be fixed, and let* $f \in C^1\left([-1, 1]^k, \mathbb{R}_+\right)$. *We assume that* $\frac{\partial f(x)}{\partial x_i} = 0$, *for* $i = 1, ..., k$. *Then*

$$\left|L_{\overrightarrow{N}}^{(M)}(f)(x) - f(x)\right| \leq \frac{\pi^2}{3}\left(\max_{i=1,...,k}\omega_1\left(\frac{\partial f}{\partial x_i}, \frac{1}{\sqrt{N_{\min}-1}}\right)\right) \cdot \qquad (9.131)$$

$$\left[\frac{(2k)^2}{\sqrt{N_{\min}-1}} + \frac{2k}{(N_{\min}-1)} + \frac{1}{8\left(\sqrt{N_{\min}-1}\right)^3}\right],$$

$\forall \overrightarrow{N} = (N_1, ..., N_k) \in \mathbb{N}^k$; $N_i \geq 4$, $i = 1, ..., k$, *and* $N_{\min} := \min\{N_1, ..., N_k\}$. *We have that* $\lim\limits_{\overrightarrow{N} \to (\infty,...,\infty)} L_{\overrightarrow{N}}^{(M)}(f)(x) = f(x)$.

Proof By (9.31) we get:

$$\left|L_{\overrightarrow{N}}^{(M)}(f)(x) - f(x)\right| \overset{(9.127)}{\leq} \left(\max_{i=1,...,k}\omega_1\left(\frac{\partial f}{\partial x_i}, h\right)\right) \cdot$$

$$\left[\frac{k}{2h}\left(\sum_{i=1}^k L_{N_i}^{(M)}\left((t_i - x_i)^2\right)(x_i)\right) + \frac{1}{2}\left(\sum_{i=1}^k L_{N_i}^{(M)}\left(|t_i - x_i|\right)(x_i)\right) + \frac{h}{8}\right] \overset{(9.122)}{\leq} \qquad (9.132)$$

$$\frac{\pi^2}{3(N_{\min}-1)}\left(\max_{i=1,...,k}\omega_1\left(\frac{\partial f}{\partial x_i}, h\right)\right)\left[\frac{k^2 2^3}{2h} + \frac{k^2 2}{2} + \frac{h}{8}\right] =: (\psi).$$

Next we choose $h := \left(\frac{1}{N_{\min}-1}\right)^{\frac{1}{2}}$, then $h^2 = \frac{1}{N_{\min}-1}$.

We have that

$$(\psi) = \frac{\pi^2}{3}\left(\max_{i=1,...,k}\omega_1\left(\frac{\partial f}{\partial x_i}, \frac{1}{\sqrt{N_{\min}-1}}\right)\right) \cdot$$

$$\left[\frac{k^2 2^2}{\sqrt{N_{\min}-1}} + \frac{2k}{(N_{\min}-1)} + \frac{1}{8\left(N_{\min}-1\right)^{\frac{3}{2}}}\right], \qquad (9.133)$$

proving the claim. ∎

It follows

Theorem 9.43 *Let any* $x \in [-1, 1]^k$, $k \in \mathbb{N} - \{1\}$, *and let* $f \in C_+\left([-1, 1]^k\right)$. *Then*

$$\left|L_{\overrightarrow{N}}^{(M)}(f)(x) - f(x)\right| \leq \left(1 + \frac{4\pi^2 k}{3}\right)\omega_1\left(f, \frac{1}{(N_{\min}-1)}\right), \qquad (9.134)$$

and

$$\left\| L_{\overrightarrow{N}}^{(M)} (f) - f \right\|_{\infty} \leq \left(1 + \frac{4\pi^2 k}{3} \right) \omega_1 \left(f, \frac{1}{(N_{\min} - 1)} \right), \quad (9.135)$$

$\forall\, \overrightarrow{N} = (N_1, ..., N_k) \in \mathbb{N}^k;\ N_i \geq 4,\ i = 1, ..., k,$ *and* $N_{\min} := \min\{N_1, ..., N_k\}.$
We have that $\displaystyle\lim_{\overrightarrow{N} \to (\infty,...,\infty)} L_{\overrightarrow{N}}^{(M)} (f) (x) = f (x),\ \forall\, x := (x_1, ..., x_k) \in [-1, 1]^k,$
uniformly.

Proof We get that (use of (9.44))

$$\left| L_{\overrightarrow{N}}^{(M)} (f) (x) - f (x) \right| \overset{(9.127)}{\leq} \omega_1 (f, h) \left(1 + \frac{1}{h} \left(\sum_{i=1}^{k} L_{N_i}^{(M)} (|t_i - x_i|) (x) \right) \right)$$

$$\overset{(9.122)}{\leq} \omega_1 (f, h) \left(1 + \frac{1}{h} \left(\sum_{i=1}^{k} \frac{2^2 \pi^2}{3 (N_i - 1)} \right) \right) \leq \omega_1 (f, h) \left(1 + \frac{1}{h} \left(\frac{4\pi^2 k}{3 (N_{\min} - 1)} \right) \right)$$

$$(9.136)$$

(setting $h := \frac{1}{N_{\min} - 1}$)

$$= \omega_1 \left(f, \frac{1}{(N_{\min} - 1)} \right) \left(1 + \frac{4\pi^2 k}{3} \right), \quad \forall\, x \in (-1, 1)^k,$$

proving the claim. ∎

We make

Remark 9.44 The Chebyshev knots of first kind $x_{N,k} := \cos \left(\frac{(2(N-k)+1)}{2(N+1)} \pi \right) \in (-1, 1),\ k \in \{0, 1, ..., N\},\ -1 < x_{N,0} < x_{N,1} < ... < x_{N,N} < 1,$ are the roots of the first kind Chebyshev polynomial $T_{N+1} (x) := \cos ((N + 1) \arccos x),\ x \in [-1, 1].$
Define ($x \in [-1, 1]$)

$$h_{N,k} (x) := \left(1 - x \cdot x_{N,k} \right) \left(\frac{T_{N+1} (x)}{(N + 1) (x - x_{N,k})} \right)^2, \quad (9.137)$$

the fundamental interpolation polynomials.
The Max-product interpolation Hermite–Fejér operators on Chebyshev knots of the first kind (seep. 12 of [5]) are defined by

$$H_{2N+1}^{(M)} (f) (x) = \frac{\bigvee_{k=0}^{N} h_{N,k} (x) f (x_{N,k})}{\bigvee_{k=0}^{N} h_{N,k} (x)}, \quad \forall\, N \in \mathbb{N}, \quad (9.138)$$

for $f \in C_+ ([-1, 1]),\ \forall\, x \in [-1, 1].$

By [5], p. 287, we have

$$H_{2N+1}^{(M)} \left(|\cdot - x| \right) (x) \leq \frac{2\pi}{N+1}, \ \forall \, x \in [-1, 1], \forall \, N \in \mathbb{N}. \tag{9.139}$$

And by [3], we get that

$$H_{2N+1}^{(M)} \left(|\cdot - x|^m \right) (x) \leq \frac{2^m \pi}{N+1}, \ \forall \, x \in [-1, 1], \forall \, m, N \in \mathbb{N}. \tag{9.140}$$

Notice $H_{2N+1}^{(M)} (1) = 1$, and $H_{2N+1}^{(M)}$ maps $C_+ ([-1, 1])$ into itself, and it is a positive sublinear operator. Furthermore it holds $\bigvee_{k=0}^{N} h_{N,k} (x) > 0$, $\forall \, x \in [-1, 1]$. We also have $h_{N,k} (x_{N,k}) = 1$, and $h_{N,k} (x_{N,j}) = 0$, if $k \neq j$, and $H_{2N+1}^{(M)} (f) (x_{N,j}) = f (x_{N,j})$, for all $j \in \{0, 1, ..., N\}$, see [5], p. 282.

We need

Definition 9.45 Let $f \in C_+ \left([-1, 1]^k \right)$, $k \in \mathbb{N} - \{1\}$, and $\overrightarrow{N} = (N_1, ..., N_k) \in \mathbb{N}^k$. We define the multivariate Max-product interpolation Hermite-Fejér operators on Chebyshev knots of the first kind, as follows:

$$H_{2\overrightarrow{N}+1}^{(M)} (f) (x) :=$$

$$\frac{\bigvee_{i_1=0}^{N_1} \bigvee_{i_2=0}^{N_2} ... \bigvee_{i_k=0}^{N_k} h_{N_1,i_1} (x_1) h_{N_2,i_2} (x_2) ... h_{N_k,i_k} (x_k) f \left(x_{N_1,i_1}, x_{N_2,i_2}, ..., x_{N_k,i_k} \right)}{\bigvee_{i_1=0}^{N_1} \bigvee_{i_2=0}^{N_2} ... \bigvee_{i_k=0}^{N_k} h_{N_1,i_1} (x_1) h_{N_2,i_2} (x_2) ... h_{N_k,i_k} (x_k)},$$
$$\tag{9.141}$$

$\forall \, x = (x_1, ..., x_k) \in [-1, 1]^k$. Call $N_{\min} := \min\{N_1, ..., N_k\}$.

The operators $H_{2\overrightarrow{N}+1}^{(M)} (f) (x)$ are positive sublinear mapping $C_+ \left([-1, 1]^k \right)$ into itself, and $H_{2\overrightarrow{N}+1}^{(M)} (1) = 1$.

We also have

$$H_{2\overrightarrow{N}+1}^{(M)} (f) (x) :=$$

$$\frac{\bigvee_{i_1=0}^{N_1} \bigvee_{i_2=0}^{N_2} ... \bigvee_{i_k=0}^{N_k} h_{N_1,i_1} (x_1) h_{N_2,i_2} (x_2) ... h_{N_k,i_k} (x_k) f \left(x_{N_1,i_1}, x_{N_2,i_2}, ..., x_{N_k,i_k} \right)}{\prod_{\lambda=1}^{k} \left(\bigvee_{i_\lambda=0}^{N_\lambda} h_{N_\lambda,i_\lambda} (x_\lambda) \right)},$$
$$\tag{9.142}$$

$\forall \, x = (x_1, ..., x_\lambda, ..., x_k) \in [-1, 1]^k$, by the maximum multiplicative principle, see (9.27). Notice that $H_{2\overrightarrow{N}+1}^{(M)} (f) \left(x_{N_1,i_1}, ..., x_{N_k,i_k} \right) = f \left(x_{N_1,i_1}, ..., x_{N_k,i_k} \right)$.

We make

Remark 9.46 The coordinate Max-product interpolation Hermite–Fejér operators on Chebyshev knots of the first kind, are defined as follows ($\lambda = 1, ..., k$):

$$H_{2N_\lambda+1}^{(M)}(g)(x_\lambda) := \frac{\vee_{i_\lambda=0}^{N_\lambda} h_{N_\lambda,i_\lambda}(x_\lambda)\, g\left(x_{N_\lambda,i_\lambda}\right)}{\vee_{i_\lambda=0}^{N_\lambda} h_{N_\lambda,i_\lambda}(x_\lambda)}, \qquad (9.143)$$

$\forall\ N_\lambda \in \mathbb{N}$, and $\forall\ x_\lambda \in [-1, 1]$, $\forall\ g \in C_+([-1, 1])$.
   Here we have $(\lambda = 1, ..., k;\ x_\lambda \in [-1, 1])$

$$h_{N_\lambda,i_\lambda}(x_\lambda) = (1 - x_\lambda \cdot x_{N_\lambda,i_\lambda})\left(\frac{T_{N_\lambda+1}(x_\lambda)}{(N_\lambda+1)\left(x_\lambda - x_{N_\lambda,i_\lambda}\right)}\right)^2, \qquad (9.144)$$

where the Chebyshev knots $x_{N_\lambda,i_\lambda} = \cos\left(\frac{(2(N_\lambda-i_\lambda)+1)}{2(N_\lambda+1)}\pi\right) \in (-1, 1)$, $i_\lambda \in \{0, 1, ..., N_\lambda\}$, $-1 < x_{N_\lambda,0} < x_{N_\lambda,1} < ... < x_{N_\lambda,N_\lambda} < 1$ are the roots of the first kind Chebyshev polynomial $T_{N_\lambda+1}(x_\lambda) = \cos((N_\lambda+1)\arccos x_\lambda)$, $x_\lambda \in [-1, 1]$.
   In case of $f \in C_+\left([-1, 1]^k\right)$ such that $f(x) := g(x_\lambda)$, $\forall\ x \in [-1, 1]^k$ and $g \in C_+([-1, 1])$, we get that

$$H_{2\vec{N}+1}^{(M)}(f)(x) = H_{2N_\lambda+1}^{(M)}(g)(x_\lambda), \qquad (9.145)$$

by the maximum multiplicative principle (9.27) and simplification of (9.142).

   We present

Theorem 9.47 *Let $x \in [-1, 1]^k$, $k \in \mathbb{N}-\{1\}$, be fixed, and let $f \in C^n\left([-1, 1]^k, \mathbb{R}_+\right)$, $n \in \mathbb{N}-\{1\}$. We assume that $f_\alpha(x) = 0$, for all $\alpha : |\alpha| = 1, ..., n$. Then*

$$\left|H_{2\vec{N}+1}^{(M)}(f)(x) - f(x)\right| \le 2^{n-2}k^{n-1}\pi\left(\max_{\alpha:|\alpha|=n}\omega_1\left(f_\alpha, \frac{1}{\sqrt[n+1]{N_{\min}+1}}\right)\right)\cdot$$
$$(9.146)$$
$$\left[\frac{8k^2}{(n+1)!\,(N_{\min}+1)^{\frac{n}{n+1}}} + \frac{2k}{n!\,(N_{\min}+1)} + \frac{1}{4(n-1)!\,(N_{\min}+1)^{\frac{n+2}{n+1}}}\right],$$

$\forall\ \vec{N} = (N_1, ..., N_k) \in \mathbb{N}^k$, and $N_{\min} := \min\{N_1, ..., N_k\}$.
   We have that $\lim_{\vec{N}\to(\infty,...,\infty)} H_{2\vec{N}+1}^{(M)}(f)(x) = f(x)$.

Proof By (9.26) we get:

$$\left|H_{2\vec{N}+1}^{(M)}(f)(x) - f(x)\right| \overset{(9.145)}{\le} \left(\max_{\alpha:|\alpha|=n}\omega_1(f_\alpha, h)\right)\cdot$$

$$\left[\frac{k^n}{(n+1)!h}\left(\sum_{i=1}^{k} H_{2N_i+1}^{(M)}\left(|t_i - x_i|^{n+1}\right)(x_i)\right) + \frac{k^{n-1}}{2n!}\left(H_{2N_i+1}^{(M)}\left(|t_i - x_i|^n\right)(x_i)\right)\right.$$
$$(9.147)$$
$$\left. + \frac{hk^{n-2}}{8(n-1)!}\left(\sum_{i=1}^{k} H_{2N_i+1}^{(M)}\left(|t_i - x_i|^{n-1}\right)(x_i)\right)\right] \overset{(9.141)}{\le}$$

$$\left(\frac{\pi}{N_{\min}+1}\right)\left(\max_{\alpha:|\alpha|=n}\omega_1\left(f_\alpha,h\right)\right)\left[\frac{k^{n+1}2^{n+1}}{(n+1)!h}+\frac{k^n2^n}{2n!}+\frac{hk^{n-1}2^{n-1}}{8(n-1)!}\right]=:(\xi).$$

Next we choose $h:=\left(\frac{1}{N_{\min}+1}\right)^{\frac{1}{n+1}}$, then $h^n=\left(\frac{1}{N_{\min}+1}\right)^{\frac{n}{n+1}}$ and $h^{n+1}=\frac{1}{N_{\min}+1}$.

We have

$$(\xi)=\pi\left(\max_{\alpha:|\alpha|=n}\omega_1\left(f_\alpha,\frac{1}{\sqrt[n+1]{N_{\min}+1}}\right)\right). \tag{9.148}$$

$$\left[\frac{(2k)^{n+1}}{(n+1)!\,(N_{\min}+1)^{\frac{n}{n+1}}}+\frac{2^{n-1}k^n}{n!\,(N_{\min}+1)}+\frac{2^{n-2}k^{n-1}}{4(n-1)!\,(N_{\min}+1)^{\frac{n+2}{n+1}}}\right],$$

proving the claim. ∎

We also give

Proposition 9.48 *Let* $x\in[-1,1]^k$, $k\in\mathbb{N}-\{1\}$, *be fixed, and let* $f\in C^1\left([-1,1]^k,\mathbb{R}_+\right)$. *We assume that* $\frac{\partial f(x)}{\partial x_i}=0$, *for* $i=1,...,k$. *Then*

$$\left|H_{2\vec{N}+1}^{(M)}(f)(x)-f(x)\right|\le\pi\left(\max_{i=1,...,k}\omega_1\left(\frac{\partial f}{\partial x_i},\frac{1}{\sqrt{N_{\min}+1}}\right)\right). \tag{9.149}$$

$$\left[\frac{2k^2}{\sqrt{N_{\min}+1}}+\frac{k}{(N_{\min}+1)}+\frac{1}{8\left(\sqrt{N_{\min}+1}\right)^3}\right],$$

$\forall\,\vec{N}=(N_1,...,N_k)\in\mathbb{N}^k$, $N_{\min}:=\min\{N_1,...,N_k\}$.

We have that $\displaystyle\lim_{\vec{N}\to(\infty,...,\infty)}H_{2\vec{N}+1}^{(M)}(f)(x)=f(x)$.

Proof By (9.31) we get

$$\left|H_{2\vec{N}+1}^{(M)}(f)(x)-f(x)\right|\overset{(9.145)}{\le}\left(\max_{i=1,...,k}\omega_1\left(\frac{\partial f}{\partial x_i},h\right)\right).$$

$$\left[\frac{k}{2h}\left(\sum_{i=1}^k H_{2N_i+1}^{(M)}\left((t_i-x_i)^2\right)(x_i)\right)+\frac{1}{2}\left(\sum_{i=1}^k H_{2N_i+1}^{(M)}\left(|t_i-x_i|\right)(x_i)\right)+\frac{h}{8}\right]\overset{(9.141)}{\le}$$

$$\tag{9.150}$$

$$\left(\frac{\pi}{N_{\min}+1}\right)\left(\max_{i=1,...,k}\omega_1\left(\frac{\partial f}{\partial x_i},h\right)\right)\left[\frac{k^2 2^2}{2h}+\frac{k2}{2}+\frac{h}{8}\right]=$$

$$\left(\frac{\pi}{N_{\min}+1}\right)\left(\max_{i=1,...,k}\omega_1\left(\frac{\partial f}{\partial x_i},h\right)\right)\left[\frac{2k^2}{h}+k+\frac{h}{8}\right]=:(\psi).$$

Next we choose $h:=\frac{1}{\sqrt{N_{\min}+1}}$, then $h^2=\frac{1}{N_{\min}+1}$.

We have that

$$(\psi) = \pi \left(\max_{i=1,\dots,k} \omega_1 \left(\frac{\partial f}{\partial x_i}, \frac{1}{\sqrt{N_{\min} + 1}} \right) \right) \cdot$$

$$\left[\frac{2k^2}{\sqrt{N_{\min} + 1}} + \frac{k}{(N_{\min} + 1)} + \frac{1}{8 \left(\sqrt{N_{\min} + 1} \right)^3} \right], \qquad (9.151)$$

proving the claim. ∎

It follows

Theorem 9.49 *Let* $f \in C_+ \left([-1, 1]^k \right)$, $k \in \mathbb{N} - \{1\}$. *Then*

$$\left| H^{(M)}_{2\vec{N}+1} (f) (x) - f (x) \right| \le (2k\pi + 1) \omega_1 \left(f, \frac{1}{N_{\min} + 1} \right), \qquad (9.152)$$

$\forall \, x \in [-1, 1]^k$, *and* $\forall \, \vec{N} = (N_1, \dots, N_k) \in \mathbb{N}^k$, *where* $N_{\min} := \min\{N_1, \dots, N_k\}$.
That is

$$\left\| H^{(M)}_{2\vec{N}+1} (f) - f \right\|_\infty \le (2k\pi + 1) \omega_1 \left(f, \frac{1}{N_{\min} + 1} \right), \qquad (9.153)$$

We get that

$$\lim_{\vec{N} \to (\infty,\dots,\infty)} H^{(M)}_{2\vec{N}+1} (f) = f, \qquad (9.154)$$

uniformly.

Proof We get that (use of (9.44))

$$\left| H^{(M)}_{2\vec{N}+1} (f) (x) - f (x) \right| \overset{(9.145)}{\le} \omega_1 (f, h) \left(1 + \frac{1}{h} \left(\sum_{i=1}^{k} H^{(M)}_{2N_i+1} (|t_i - x_i|) (x_i) \right) \right)$$

$$\overset{(9.140)}{\le} \omega_1 (f, h) \left(1 + \frac{k}{h} \left(\frac{2\pi}{(N_{\min} + 1)} \right) \right) \qquad (9.155)$$

(setting $h := \frac{1}{N_{\min}+1}$)

$$= \omega_1 \left(f, \frac{1}{N_{\min} + 1} \right) (1 + 2k\pi), \quad \forall \, x \in [-1, 1]^k,$$

proving the claim. ∎

We make

Remark 9.50 Let $\theta_{\vec{N}}^{(M)}$ denote any of the Max-product multivariate operators studied in this chapter: $B_{\vec{N}}^{(M)}$, $T_{N}^{(M)}$, $U_{\vec{N}}^{(M)}$, $T_{\vec{N}}^{(M)}$, $M_{\vec{N}}^{(M)}$, $L_{\vec{N}}^{(M)}$ and $H_{2\vec{N}+1}^{(M)}$. We observe that an important contraction property holds:

$$\left\| \theta_{\vec{N}}^{(M)} (f) \right\|_\infty \leq \|f\|_\infty, \tag{9.156}$$

and

$$\left\| \theta_{\vec{N}}^{(M)} \left(\theta_{\vec{N}}^{(M)} (f) \right) \right\|_\infty \leq \left\| \theta_{\vec{N}}^{(M)} (f) \right\|_\infty \leq \|f\|_\infty, \tag{9.157}$$

i.e.

$$\left\| \left(\theta_{\vec{N}}^{(M)} \right)^2 (f) \right\|_\infty \leq \|f\|_\infty, \tag{9.158}$$

and in general holds

$$\left\| \left(\theta_{\vec{N}}^{(M)} \right)^n (f) \right\|_\infty \leq \left\| \left(\theta_{\vec{N}}^{(M)} \right)^{n-1} (f) \right\|_\infty \leq \ldots \leq \|f\|_\infty, \ \forall \, n \in \mathbb{N}. \tag{9.159}$$

We need the following Holder's type inequality:

Theorem 9.51 *Let Q, with the l_1-norm $\|\cdot\|$, be a compact and convex subset of $\mathbb{R}^k$, $k \in \mathbb{N} - \{1\}$ and $L : C_+ (Q) \to C_+ (Q)$, be a positive sublinear operator and $f, g \in C_+ (Q)$, furthermore let $p, q > 1 : \frac{1}{p} + \frac{1}{q} = 1$. Assume that $L \left((f (\cdot))^p \right) (s_*)$, $L \left((g (\cdot))^q \right) (s_*) > 0$ for some $s_* \in Q$. Then*

$$L (f (\cdot) g (\cdot)) (s_*) \leq \left(L \left((f (\cdot))^p \right) (s_*) \right)^{\frac{1}{p}} \left(L \left((g (\cdot))^q \right) (s_*) \right)^{\frac{1}{q}}. \tag{9.160}$$

Proof Let $a, b \geq 0$, $p, q > 1 : \frac{1}{p} + \frac{1}{q} = 1$. The Young's inequality says

$$ab \leq \frac{a^p}{p} + \frac{b^q}{q}. \tag{9.161}$$

Then

$$\frac{f (s)}{\left(L \left((f (\cdot))^p \right) (s_*) \right)^{\frac{1}{p}}} \cdot \frac{g (s)}{\left(L \left((g (\cdot))^q \right) (s_*) \right)^{\frac{1}{q}}} \leq$$

$$\frac{(f (s))^p}{p \left(L \left((f (\cdot))^p \right) (s_*) \right)} + \frac{(g (s))^q}{q \left(L \left((g (\cdot))^q \right) (s_*) \right)}, \ \forall \, s \in Q. \tag{9.162}$$

Hence it holds

$$\frac{L (f (\cdot) g (\cdot)) (s_*)}{\left(L \left((f (\cdot))^p \right) (s_*) \right)^{\frac{1}{p}} \left(L \left((g (\cdot))^q \right) (s_*) \right)^{\frac{1}{q}}} \leq \tag{9.163}$$

$$\frac{\left(L\left((f\left(\cdot\right))^{p}\right)\right)(s_{*})}{p\left(L\left((f\left(\cdot\right))^{p}\right)(s_{*})\right)} + \frac{\left(L\left((g\left(\cdot\right))^{q}\right)\right)(s_{*})}{q\left(L\left((g\left(\cdot\right))^{q}\right)(s_{*})\right)} = \frac{1}{p} + \frac{1}{q} = 1, \text{ for } s_{*} \in Q,$$

proving the claim. ∎

By (9.160), under the assumption $L_N\left(\|\cdot - x\|^{n+1}\right)(x) > 0$, and $L_N(1) = 1$, we obtain

$$L_N\left(\|\cdot - x\|^{n}\right)(x) \le \left(L_N\left(\|\cdot - x\|^{n+1}\right)(x)\right)^{\frac{n}{n+1}}, \tag{9.164}$$

in case of $n = 1$ we derive

$$L_N\left(\|\cdot - x\|\right)(x) \le \sqrt{\left(L_N\left(\|\cdot - x\|^{2}\right)(x)\right)}. \tag{9.165}$$

We give

Theorem 9.52 *Let Q with $\|\cdot\|$ the l_1-norm, be a compact and convex subset of $\mathbb{R}^k$, $k \in \mathbb{N} - \{1\}$, and $f \in C_+(Q)$. Let $\{L_N\}_{N \in \mathbb{N}}$ be positive sublinear operators from $C_+(Q)$ into itself, such that $L_N(1) = 1$, $\forall\, N \in \mathbb{N}$. We assume further that $L_N(\|t - x\|)(x) > 0$, $\forall\, N \in \mathbb{N}$. Then*

$$|L_N(f)(x) - f(x)| \le 2\omega_1(f, L_N(\|t - x\|)(x)), \tag{9.166}$$

$\forall\, N \in \mathbb{N}$, $x = (x_1, ..., x_k) \in Q$; $t = (t_1, ..., t_k) \in Q$, where

$$\omega_1(f, h) := \sup_{\substack{x, y \in Q: \\ \|x-y\| \le h}} |f(x) - f(y)|. \tag{9.167}$$

If $L_N(\|t - x\|)(x) \to 0$, then $L_N(f)(x) \to f(x)$, as $N \to +\infty$.

Proof By Theorem 9.13. ∎

We need

Theorem 9.53 *Let $(Q, \|\cdot\|)$, where $\|\cdot\|$ is the l_1-norm, be a compact and convex subset of $\mathbb{R}^k$, $k \in \mathbb{N} - \{1\}$, and let $x \in Q$ ($x = (x_1, ..., x_k)$) be fixed. Let $f \in C^n(Q)$, $n \in \mathbb{N}$, $h > 0$. We assume that $f_\alpha(x) = 0$, for all $\alpha : |\alpha| = 1, ..., n$. Let $\{L_N\}_{N \in \mathbb{N}}$ be positive sublinear operators from $C_+(Q)$ into $C_+(Q)$, such that $L_N(1) = 1$, $\forall\, N \in \mathbb{N}$. Then*

$$|L_N(f)(x) - f(x)| \le \left(\max_{\alpha : |\alpha| = n} \omega_1(f_\alpha, h)\right) \cdot$$

$$\left[\frac{L_N\left(\|\cdot - x\|^{n+1}\right)(x)}{(n+1)!h} + \frac{L_N\left(\|\cdot - x\|^{n}\right)(x)}{2n!} + \frac{h}{8(n-1)!} L_N\left(\|\cdot - x\|^{n-1}\right)(x)\right], \tag{9.168}$$

$\forall\, N \in \mathbb{N}$.

Proof By (9.19) and (9.25). ∎

It follows

Theorem 9.54 *All as in Theorem 9.53. Additionally assume that* $L_N \left(\| \cdot - x \|^{n+1} \right) (x)$
$> 0, \forall N \in \mathbb{N}.$ *Then*

$$|L_N (f) (x) - f (x)| \le \frac{1}{2n!} \left(3 + \frac{n}{4 (n + 1)} \right) \cdot$$

$$\left(\max_{\alpha : |\alpha| = n} \omega_1 \left(f_\alpha, \frac{1}{(n + 1)} \left(L_N \left(\| \cdot - x \|^{n+1} \right) (x) \right)^{\frac{1}{n+1}} \right) \right) \left(L_N \left(\| \cdot - x \|^{n+1} \right) (x) \right)^{\frac{n}{n+1}},$$
$$(9.169)$$

$\forall N \in \mathbb{N}, x = (x_1, ..., x_k) \in Q, \omega_1$ *as in (9.167) for* f_α.
 If $L_N \left(\| \cdot - x \|^{n+1} \right) (x) \to 0$, *then* $L_N (f) (x) \to f (x)$, *as* $N \to +\infty$.

Proof By Theorem 9.51 notice also that

$$L_N \left(\| \cdot - x \|^{n-1} \right) (x) \le \left(L_N \left(\| \cdot - x \|^{n+1} \right) (x) \right)^{\frac{n-1}{n+1}}. \qquad (9.170)$$

We choose

$$h := \frac{1}{(n + 1)} \left(L_N \left(\| \cdot - x \|^{n+1} \right) (x) \right)^{\frac{1}{n+1}} > 0. \qquad (9.171)$$

That is

$$(h (n + 1))^{n+1} = L_N \left(\| \cdot - x \|^{n+1} \right) (x). \qquad (9.172)$$

We apply (9.168) to have (see also (9.164) and (9.170)).

$$|L_N (f) (x) - f (x)| \le \left(\max_{\alpha : |\alpha| = n} \omega_1 (f_\alpha, h) \right) \cdot$$

$$\left[\frac{L_N \left(\| \cdot - x \|^{n+1} \right) (x)}{(n + 1)! h} + \frac{\left(L_N \left(\| \cdot - x \|^{n+1} \right) (x) \right)^{\frac{n}{n+1}}}{2n!} + \qquad (9.173)$$

$$\frac{h}{8 (n - 1)!} L \left(N \left(\| \cdot - x \|^{n+1} \right) (x) \right)^{\frac{n-1}{n+1}} \right] =$$

$$\left(\max_{\alpha : |\alpha| = n} \omega_1 \left(f_\alpha, \frac{1}{(n + 1)} \left(L_N \left(\| \cdot - x \|^{n+1} \right) (x) \right)^{\frac{1}{n+1}} \right) \right) \cdot$$

$$\left[\frac{h^n (n + 1)^{n+1}}{(n + 1)!} + \frac{h^n (n + 1)^n}{2n!} + \frac{h^n (n + 1)^{n-1}}{8 (n - 1)!} \right] =$$

$$\left(\max_{\alpha:|\alpha|=n} \omega_1 \left(f_\alpha, \frac{1}{(n+1)} \left(L_N \left(\|\cdot - x\|^{n+1}\right)(x)\right)^{\frac{1}{n+1}}\right)\right).$$

$$\left[\frac{(n+1)^{n+1}}{(n+1)!} + \frac{(n+1)^n}{2n!} + \frac{(n+1)^{n-1}}{8(n-1)!}\right] \frac{1}{(n+1)^n} \left(L_N \left(\|\cdot - x\|^{n+1}\right)(x)\right)^{\frac{n}{n+1}} =$$

$$\left[\frac{3}{2n!} + \frac{n}{8(n+1)!}\right] \left(\max_{\alpha:|\alpha|=n} \omega_1 \left(f_\alpha, \frac{1}{(n+1)} \left(L_N \left(\|\cdot - x\|^{n+1}\right)(x)\right)^{\frac{1}{n+1}}\right)\right).$$

$$\left(L_N \left(\|\cdot - x\|^{n+1}\right)(x)\right)^{\frac{n}{n+1}}, \tag{9.174}$$

proving the claim. ■

Final application for $n = 1$ follows:

Corollary 9.55 *Let* $(Q, \|\cdot\|)$, *where* $\|\cdot\|$ *is the* l_1*-norm, be a compact and convex subset of* $\mathbb{R}^k$, $k \in \mathbb{N} - \{1\}$, *and let* $x \in Q$ $(x = (x_1, ..., x_k))$ *be fixed. Let* $f \in C^1(Q)$. *We assume that* $\frac{\partial f}{\partial x_i}(x) = 0$, $i = 1, ..., k$. *Let* $\{L_N\}_{N \in \mathbb{N}}$ *be positive sublinear operators from* $C_+(Q)$ *into* $C_+(Q)$, *such that* $L_N(1) = 1$, $\forall N \in \mathbb{N}$. *Assume that* $L_N \left(\|\cdot - x\|^2\right)(x) > 0$, $\forall N \in \mathbb{N}$. *Then*

$$|L_N(f)(x) - f(x)| \le \frac{25}{16} \left(\max_{i=1,...,k} \omega_1 \left(\frac{\partial f}{\partial x_i}, \frac{1}{2} \left(L_N \left(\|\cdot - x\|^2\right)(x)\right)^{\frac{1}{2}}\right)\right).$$

$$\left(L_N \left(\|\cdot - x\|^2\right)(x)\right)^{\frac{1}{2}}, \tag{9.175}$$

$\forall N \in \mathbb{N}$.
If $L_N \left(\|\cdot - x\|^2\right)(x) \to 0$, *then* $L_N(f)(x) \to f(x)$, *as* $N \to +\infty$.

References

1. G. Anastassiou, *Moments in Probability and Approximation Theory*, Pitman Research Notes in Mathematics Series (Longman Group, New York, 1993)
2. G. Anastassiou, *Approximation by Sublinear Operators* (2017, submitted)
3. G. Anastassiou, *Approximation by Max-Product Operators* (2017, submitted)
4. G. Anastassiou, *Approximation by Multivariate Sublinear and Max-Product Operators* (2017, submitted)
5. B. Bede, L. Coroianu, S. Gal, *Approximation by Max-Product Type Operators* (Springer, Heidelberg, 2016)

Chapter 10
High Order Approximation by Sublinear and Max-Product Operators Using Convexity

Here we consider quantitatively using convexity the approximation of a function by general positive sublinear operators with applications to Max-product operators. These are of Bernstein type, of Favard–Szász–Mirakjan type, of Baskakov type, of Meyer–Köning and Zeller type, of sampling type, of Lagrange interpolation type and of Hermite–Fejér interpolation type. Our results are both: under the presence of smoothness and without any smoothness assumption on the function to be approximated which fulfills a convexity property. It follows [6].

10.1 Background

We make

Remark 10.1 Let $f \in C([a, b])$, $x_0 \in (a, b)$, $0 < h \le \min(x_0 - a, b - x_0)$, and $|f(t) - f(x_0)|$ is convex in $t \in [a, b]$.

By Lemma 8.1.1, p. 243 of [1] we have that

$$|f(t) - f(x_0)| \le \frac{\omega_1(f, h)}{h} |t - x_0|, \ \forall t \in [a, b], \tag{10.1}$$

where

$$\omega_1(f, h) := \sup_{\substack{x, y \in [a,b] \\ |x-y| \le h}} |f(x) - f(y)|, \tag{10.2}$$

the first modulus of continuity of f.

© Springer International Publishing AG, part of Springer Nature 2018
G. A. Anastassiou, *Nonlinearity: Ordinary and Fractional Approximations by Sublinear and Max-Product Operators*, Studies in Systems, Decision and Control 147, https://doi.org/10.1007/978-3-319-89509-3_10

We also make

Remark 10.2 Let $f \in C^n ([a, b])$, $n \in \mathbb{N}$, $x_0 \in (a, b)$, $0 < h \leq \min (x_0 - a, b - x_0)$, and $\left| f^{(n)} (t) - f^{(n)} (x_0) \right|$ is convex in $t \in [a, b]$. We have that

$$f (t) = \sum_{k=0}^{n} \frac{f^{(k)} (x_0)}{k!} (t - x_0)^k + I_t, \tag{10.3}$$

where

$$I_t = \int_{x_0}^{t} \left(\int_{x_0}^{t_1} \cdots \left(\int_{x_0}^{t_{n-1}} \left(f^{(n)} (t_n) - f^{(n)} (x_0) \right) dt_n \right) \cdots \right) dt_1. \tag{10.4}$$

Assuming $f^{(k)} (x_0) = 0$, $k = 1, ..., n$, we get

$$f (t) - f (x_0) = I_t. \tag{10.5}$$

By Lemma 8.1.1, p. 243 of [1] we have

$$\left| f^{(n)} (t) - f^{(n)} (x_0) \right| \leq \frac{\omega_1 \left(f^{(n)}, h \right)}{h} |t - x_0|, \ \forall t \in [a, b]. \tag{10.6}$$

Furthermore it holds

$$|I_t| \leq \frac{\omega_1 \left(f^{(n)}, h \right)}{h} \frac{|t - x_0|^{n+1}}{(n + 1)!}, \ \forall t \in [a, b]. \tag{10.7}$$

Hence we derive that

$$|f (t) - f (x_0)| \overset{(10.5)}{\leq} \frac{\omega_1 \left(f^{(n)}, h \right)}{h} \frac{|t - x_0|^{n+1}}{(n + 1)!}, \ \forall t \in [a, b]. \tag{10.8}$$

We have proved the following results:

Theorem 10.3 *Let* $f \in C ([a, b])$, $x \in (a, b)$, $0 < h \leq \min (x - a, b - x)$, *and* $|f (\cdot) - f (x)|$ *is convex over* $[a, b]$. *Then*

$$|f (\cdot) - f (x)| \leq \frac{\omega_1 (f, h)}{h} |\cdot - x|, \ over \ [a, b]. \tag{10.9}$$

Theorem 10.4 *Let* $f \in C^n ([a, b])$, $n \in \mathbb{N}$, $x \in (a, b)$, $0 < h \leq \min (x - a, b - x)$, *and* $\left| f^{(n)} (\cdot) - f^{(n)} (x) \right|$ *is convex over* $[a, b]$. *Assume that* $f^{(k)} (x) = 0$, $k = 1, ..., n$. *Then*

$$|f (\cdot) - f (x)| \leq \frac{\omega_1 \left(f^{(n)}, h \right)}{h} \frac{|\cdot - x|^{n+1}}{(n + 1)!}, \ over \ [a, b]. \tag{10.10}$$

We give

Definition 10.5 Call $C_+([a,b]) := \{f : [a,b] \to \mathbb{R}_+ \text{ and continuous}\}$. Let L_N from $C_+([a,b])$ into $C_+([a,b])$ be a sequence of operators satisfying the following properties (see also [7], p. 17):
   (i) (positive homogeneous)

$$L_N(\alpha f) = \alpha L_N(f), \ \forall \alpha \geq 0, \forall f \in C_+([a,b]), \tag{10.11}$$

   (ii) (Monotonicity)
   if $f, g \in C_+([a,b])$ satisfy $f \leq g$, then

$$L_N(f) \leq L_N(g), \ \forall N \in \mathbb{N}, \tag{10.12}$$

   (iii) (Subadditivity)

$$L_N(f+g) \leq L_N(f) + L_N(g), \ \forall f, g \in C_+([a,b]). \tag{10.13}$$

We call L_N positive sublinear operators.

We make

Remark 10.6 As in [7], p. 17, we get that for $f, g \in C_+([a,b])$

$$|L_N(f)(x) - L_N(g)(x)| \leq L_N(|f-g|)(x), \ \forall x \in [a,b]. \tag{10.14}$$

From now on we assume that $L_N(1) = 1, \ \forall N \in \mathbb{N}$. Hence it holds

$$|L_N(f)(x) - f(x)| \leq L_N(|f(\cdot) - f(x)|)(x), \ \forall x \in [a,b], \forall N \in \mathbb{N}, \tag{10.15}$$

see also [7], p. 17.

We obtain the following results:

Theorem 10.7 *Let* $f \in C_+([a,b])$, $x \in (a,b)$, $0 < h \leq \min(x-a, b-x)$, *and* $|f(\cdot) - f(x)|$ *is a convex function over* $[a,b]$. *Let* $\{L_N\}_{N \in \mathbb{N}}$ *positive sublinear operators from* $C_+([a,b])$ *into itself, such that* $L_N(1) = 1, \ \forall N \in \mathbb{N}$. *Then*

$$|L_N(f)(x) - f(x)| \leq \frac{\omega_1(f,h)}{h} L_N(|\cdot - x|)(x), \ \forall N \in \mathbb{N}. \tag{10.16}$$

Proof By (10.9) and (10.15). ∎

Theorem 10.8 *Let* $f \in C^n([a,b], \mathbb{R}_+)$, $n \in \mathbb{N}$, $x \in (a,b)$, $0 < h \leq \min(x-a, b-x)$, *and* $|f^{(n)}(\cdot) - f^{(n)}(x)|$ *is convex over* $[a,b]$. *Assume that* $f^{(k)}(x) = 0, k = 1, ..., n$. *Let* $\{L_N\}_{N \in \mathbb{N}}$ *positive sublinear operators from* $C_+([a,b])$ *into itself, such that* $L_N(1) = 1, \ \forall N \in \mathbb{N}$. *Then*

$$|L_N (f) (x) - f (x)| \le \frac{\omega_1 \left(f^{(n)}, h\right)}{h (n + 1)!} L_N \left(|\cdot - x|^{n+1}\right) (x) , \forall N \in \mathbb{N}. \qquad (10.17)$$

Proof By (10.10) and (10.15).   ∎

We continue with

Theorem 10.9 *Let* $f \in C_+ ([a, b]), x \in (a, b), 0 < L_N (|\cdot - x|) (x) \le \min (x - a, b - x), \forall N \in \mathbb{N}, and |f (\cdot) - f (x)| is a convex function over [a, b]. Here L_N are positive sublinear operators from $C_+ ([a, b])$ into itself, such that $L_N (1) = 1, \forall N \in \mathbb{N}$. Then*

$$|L_N (f) (x) - f (x)| \le \omega_1 (f, L_N (|\cdot - x|) (x)) , \forall N \in \mathbb{N}. \qquad (10.18)$$

If $L_N (|\cdot - x|) (x) \to 0$, then $L_N (f) (x) \to f (x)$, as $N \to +\infty$.

Proof By (10.16).   ∎

Theorem 10.10 *Let* $f \in C^n ([a, b], \mathbb{R}_+), n \in \mathbb{N}, x \in (a, b), 0 < L_N \left(|\cdot - x|^{n+1}\right) (x) \le \min (x - a, b - x), \forall N \in \mathbb{N}, and \left|f^{(n)} (\cdot) - f^{(n)} (x)\right| is convex over [a, b]. Assume that $f^{(k)} (x) = 0, k = 1, ..., n$. Here $\{L_N\}_{N \in \mathbb{N}}$ are positive sublinear operators from $C_+ ([a, b])$ into itself, such that $L_N (1) = 1, \forall N \in \mathbb{N}$. Then*

$$|L_N (f) (x) - f (x)| \le \frac{\omega_1 \left(f^{(n)}, L_N \left(|\cdot - x|^{n+1}\right) (x)\right)}{(n + 1)!}, \forall N \in \mathbb{N}. \qquad (10.19)$$

If $L_N \left(|\cdot - x|^{n+1}\right) (x) \to 0$, then $L_N (f) (x) \to f (x)$, as $N \to +\infty$.

Proof By (10.17).   ∎

Next we combine both Theorems 10.7, 10.8:

Theorem 10.11 *Let* $f \in C^n ([a, b], \mathbb{R}_+), n \in \mathbb{Z}_+, x \in (a, b), 0 < h \le \min (x - a, b - x), and \left|f^{(n)} (\cdot) - f^{(n)} (x)\right| is convex over [a, b]. Assume that $f^{(k)} (x) = 0, k = 1, ..., n$. Let $\{L_N\}_{N \in \mathbb{N}}$ positive sublinear operators from $C_+ ([a, b])$ into itself, such that $L_N (1) = 1, \forall N \in \mathbb{N}$. Then*

$$|L_N (f) (x) - f (x)| \le \frac{\omega_1 \left(f^{(n)}, h\right)}{h (n + 1)!} L_N \left(|\cdot - x|^{n+1}\right) (x) , \forall N \in \mathbb{N}; n \in \mathbb{Z}_+. \qquad (10.20)$$

The initial conditions $f^{(k)} (x) = 0, k = 1, ..., n$, are void when $n = 0$.

In this chapter we study under convexity quantitatively the approximation properties of Max-product operators to the unit. These are special cases of positive sublinear operators. We present also results regarding the convergence to the unit of general positive sublinear operators under convexity. Special emphasis is given to our study about approximation under the presence of smoothness. This chapter is inspired from [7].

Under our convexity conditions the derived convergence inequalities are elegant and compact with very small constants.

10.2 Main Results

Here we apply Theorem 10.11 to Max-product operators.
We make

Remark 10.12 We start with the Max-product Bernstein operators ([7], p. 10)

$$B_N^{(M)}(f)(x) = \frac{\bigvee_{k=0}^N p_{N,k}(x) f\left(\frac{k}{N}\right)}{\bigvee_{k=0}^N p_{N,k}(x)}, \forall N \in \mathbb{N}, \tag{10.21}$$

$p_{N,k}(x) = \binom{N}{k} x^k (1-x)^{N-1}$, $x \in [0, 1]$, $\bigvee$ stands for maximum, and $f \in C_+([0, 1]) = \{f : [0, 1] \to \mathbb{R}_+ \text{ is continuous}\}$, where $\mathbb{R}_+ := [0, \infty)$.

Clearly $B_N^{(M)}$ is a positive sublinear operators from $C_+([0, 1])$ into itself with $B_N^{(M)}(1) = 1$.

By [7], p. 31, we have

$$B_N^{(M)}(|\cdot - x|)(x) \le \frac{6}{\sqrt{N+1}}, \forall x \in [0, 1], \forall N \in \mathbb{N}. \tag{10.22}$$

And by [2] we get:

$$B_N^{(M)}(|\cdot - x|^m)(x) \le \frac{6}{\sqrt{N+1}}, \forall x \in [0, 1], m, N \in \mathbb{N}. \tag{10.23}$$

Denote by

$$C_+^n([0, 1]) = \{f : [0, 1] \to \mathbb{R}_+, n\text{-times continuously differentiable}\}, n \in \mathbb{Z}_+.$$

We present

Theorem 10.13 *Let* $f \in C_+^n([0, 1])$, $n \in \mathbb{Z}_+$, $x \in (0, 1)$ *and* $N^* \in \mathbb{N} : 0 < \frac{1}{\sqrt{N^*+1}} \le \min(x, 1-x)$, *and* $\left|f^{(n)}(\cdot) - f^{(n)}(x)\right|$ *is convex over* $[0, 1]$. *Assume that* $f^{(k)}(x) = 0$, $k = 1, ..., n$. *Then*

$$\left|B_N^{(M)}(f)(x) - f(x)\right| \le \frac{6\omega_1\left(f^{(n)}, \frac{1}{\sqrt{N+1}}\right)}{(n+1)!}, \forall N \in \mathbb{N} : N \ge N^*. \tag{10.24}$$

It holds $\lim_{N \to +\infty} B_N^{(M)}(f)(x) = f(x)$.

Proof By (10.20) we get

$$\left|B_N^{(M)}(f)(x) - f(x)\right| \le \frac{\omega_1\left(f^{(n)}, h\right)}{h(n+1)!} B_N^{(M)}\left(|\cdot - x|^{n+1}\right)(x) \overset{(10.23)}{\le}$$

$$\frac{\omega_1\left(f^{(n)}, h\right)}{h\,(n+1)!}\,\frac{6}{\sqrt{N+1}} =$$

(setting $h := \frac{1}{\sqrt{N+1}}$)

$$\frac{6\omega_1\left(f^{(n)}, \frac{1}{\sqrt{N+1}}\right)}{(n+1)!}, \tag{10.25}$$

proving the claim.   ∎

We make

Remark 10.14 Here we focus on the truncated Favard–Szász–Mirakjan operators

$$T_N^{(M)}(f)(x) = \frac{\bigvee_{k=0}^{N} s_{N,k}(x)\, f\left(\frac{k}{N}\right)}{\bigvee_{k=0}^{N} s_{N,k}(x)}, \quad x \in [0,1],\ N \in \mathbb{N},\ f \in C_+([0,1]),$$

$$\tag{10.26}$$

$s_{N,k}(x) = \frac{(Nx)^k}{k!}$, see also [7], p. 11.

By [7], p. 178–179 we have

$$T_N^{(M)}(|\cdot - x|)(x) \le \frac{3}{\sqrt{N}}, \quad \forall\, x \in [0,1],\ \forall\, N \in \mathbb{N}. \tag{10.27}$$

And by [2] we get

$$T_N^{(M)}(|\cdot - x|^m)(x) \le \frac{3}{\sqrt{N}}, \quad \forall\, x \in [0,1],\ \forall\, m, N \in \mathbb{N}. \tag{10.28}$$

The operators $T_N^{(M)}$ are positive sublinear from $C_+([0,1])$ into itself with $T_N^{(M)}(1) = 1$, $\forall\, N \in \mathbb{N}$.

We give

Theorem 10.15 *Let $f \in C_+^n([0,1])$, $n \in \mathbb{Z}_+$, $x \in (0,1)$ and $N^* \in \mathbb{N} : 0 < \frac{1}{\sqrt{N^*}} \le \min(x, 1-x)$, and $\left|f^{(n)}(\cdot) - f^{(n)}(x)\right|$ is convex over $[0,1]$. Assume that $f^{(k)}(x) = 0$, $k = 1, ..., n$. Then*

$$\left|T_N^{(M)}(f)(x) - f(x)\right| \le \frac{3\omega_1\left(f^{(n)}, \frac{1}{\sqrt{N}}\right)}{(n+1)!}, \quad \forall\, N \in \mathbb{N} : N \ge N^*. \tag{10.29}$$

It holds $\lim\limits_{N \to +\infty} T_N^{(M)}(f)(x) = f(x)$.

Proof By (10.20) we get

$$\left|T_N^{(M)}(f)(x) - f(x)\right| \le \frac{\omega_1\left(f^{(n)}, h\right)}{h\,(n+1)!}\, T_N^{(M)}\left(|\cdot - x|^{n+1}\right)(x) \overset{(10.28)}{\le}$$

$$\frac{\omega_1 \left(f^{(n)}, h\right)}{h \, (n+1)!} \frac{3}{\sqrt{N}} =$$

(setting $h := \frac{1}{\sqrt{N}}$)

$$\frac{3\omega_1 \left(f^{(n)}, \frac{1}{\sqrt{N}}\right)}{(n+1)!}, \tag{10.30}$$

proving the claim. ■

We make

Remark 10.16 Next we study the truncated Max-product Baskakov operators (see [7], p. 11)

$$U_N^{(M)} (f) (x) = \frac{\bigvee_{k=0}^{N} b_{N,k} (x) \, f \left(\frac{k}{N}\right)}{\bigvee_{k=0}^{N} b_{N,k} (x)}, \quad x \in [0, 1], \ f \in C_+ ([0, 1]), \ N \in \mathbb{N}, \tag{10.31}$$

where

$$b_{N,k} (x) = \binom{N+k-1}{k} \frac{x^k}{(1+x)^{N+k}}.$$

From [7], pp. 217–218, we get ($x \in [0, 1]$)

$$\left(U_N^{(M)} (|\cdot - x|)\right) (x) \leq \frac{12}{\sqrt{N+1}}, \quad N \geq 2, N \in \mathbb{N}. \tag{10.32}$$

And as in [2], we obtain ($m \in \mathbb{N}$)

$$\left(U_N^{(M)} (|\cdot - x|^m)\right) (x) \leq \frac{12}{\sqrt{N+1}}, \quad N \geq 2, N \in \mathbb{N}, \ \forall \, x \in [0, 1]. \tag{10.33}$$

Also it holds $U_N^{(M)} (1) (x) = 1$, and $U_N^{(M)}$ are positive sublinear operators from $C_+ ([0, 1])$ into itself, $\forall \, N \in \mathbb{N}$.

We give

Theorem 10.17 *Let $f \in C_+^n ([0, 1])$, $n \in \mathbb{Z}_+$, $x \in (0, 1)$ and $N^* \in \mathbb{N} - \{1\} : 0 < \frac{1}{\sqrt{N^*+1}} \leq \min (x, 1-x)$, and $\left|f^{(n)} (\cdot) - f^{(n)} (x)\right|$ is convex over $[0, 1]$. Assume that $f^{(k)} (x) = 0$, $k = 1, ..., n$. Then*

$$\left|U_N^{(M)} (f) (x) - f (x)\right| \leq \frac{12\omega_1 \left(f^{(n)}, \frac{1}{\sqrt{N+1}}\right)}{(n+1)!}, \quad \forall \, N \in \mathbb{N} : N \geq N^*. \tag{10.34}$$

It holds $\lim_{N \to +\infty} U_N^{(M)} (f) (x) = f (x)$.

Proof By (10.20) we get

$$\left| U_N^{(M)}\left(f\right)(x) - f(x) \right| \le \frac{\omega_1\left(f^{(n)}, h\right)}{h(n+1)!} U_N^{(M)}\left(|\cdot - x|^{n+1}\right)(x) \overset{(10.33)}{\le}$$

$$\frac{\omega_1\left(f^{(n)}, h\right)}{h(n+1)!} \frac{12}{\sqrt{N+1}} =$$

(setting $h := \frac{1}{\sqrt{N+1}}$)

$$\frac{12\omega_1\left(f^{(n)}, \frac{1}{\sqrt{N+1}}\right)}{(n+1)!}, \tag{10.35}$$

proving the claim. ■

We make

Remark 10.18 Here we study Max-product Meyer–Köning and Zeller operators (see [7], p. 11) defined by

$$Z_N^{(M)}\left(f\right)(x) = \frac{\bigvee_{k=0}^{\infty} s_{N,k}(x)\, f\left(\frac{k}{N+k}\right)}{\bigvee_{k=0}^{\infty} s_{N,k}(x)}, \quad \forall N \in \mathbb{N},\, f \in C_+\left([0,1]\right), \tag{10.36}$$

$$s_{N,k}(x) = \binom{N+k}{k} x^k,\, x \in [0,1].$$

By [7], p. 253, we get that

$$Z_N^{(M)}\left(|\cdot - x|\right)(x) \le \frac{8\left(1+\sqrt{5}\right)}{3} \frac{\sqrt{x}\,(1-x)}{\sqrt{N}}, \quad \forall x \in [0,1],\, N \ge 4. \tag{10.37}$$

And by [2], we derive that

$$Z_N^{(M)}\left(|\cdot - x|^m\right)(x) \le \frac{8\left(1+\sqrt{5}\right)}{3} \frac{\sqrt{x}\,(1-x)}{\sqrt{N}}, \tag{10.38}$$

$\forall\, x \in [0,1],\, N \ge 4,\, \forall\, m \in \mathbb{N}$.

The ceiling $\left\lceil \frac{8\left(1+\sqrt{5}\right)}{3} \right\rceil = 9$, and using a basic calculus technique (see [4]) we get that $g(x) := (1-x)\sqrt{x}$ has an absolute maximum over $(0,1]: g\left(\frac{1}{3}\right) = \frac{2}{3\sqrt{3}}$. That is $(1-x)\sqrt{x} \le \frac{2}{3\sqrt{3}}, \forall\, x \in [0,1]$.

Consequently it holds

$$Z_N^{(M)}\left(|\cdot - x|^m\right)(x) \le \frac{6}{\sqrt{3}\sqrt{N}}, \tag{10.39}$$

$\forall x \in [0, 1], \forall N \in \mathbb{N} : N \geq 4, \forall m \in \mathbb{N}$.

Also it holds $Z_N^{(M)}(1) = 1$, and $Z_N^{(M)}$ are positive sublinear operators from $C_+([0, 1])$ into itself, $\forall N \in \mathbb{N}$.

We give

Theorem 10.19 *Let $f \in C_+^n([0, 1])$, $n \in \mathbb{Z}_+$, $x \in (0, 1)$ and $N^* \in \mathbb{N} : N^* \geq 4$ with $0 < \frac{1}{\sqrt{N^*}} \leq \min(x, 1 - x)$, and $\left| f^{(n)}(\cdot) - f^{(n)}(x) \right|$ is convex over $[0, 1]$. Assume that $f^{(k)}(x) = 0$, $k = 1, ..., n$. Then*

$$\left| Z_N^{(M)}(f)(x) - f(x) \right| \leq \left(\frac{6}{\sqrt{3}(n+1)!} \right) \omega_1 \left(f^{(n)}, \frac{1}{\sqrt{N}} \right), \quad \forall N \in \mathbb{N} : N \geq N^*.$$

(10.40)

It holds $\lim_{N \to +\infty} Z_N^{(M)}(f)(x) = f(x)$.

Proof By (10.20) we get

$$\left| Z_N^{(M)}(f)(x) - f(x) \right| \leq \frac{\omega_1 \left(f^{(n)}, h \right)}{h(n+1)!} Z_N^{(M)} \left(|\cdot - x|^{n+1} \right)(x) \overset{(10.39)}{\leq}$$

$$\frac{\omega_1 \left(f^{(n)}, h \right)}{h(n+1)!} \frac{6}{\sqrt{3}\sqrt{N}} =$$

(setting $h := \frac{1}{\sqrt{N}}$)

$$\left(\frac{6}{\sqrt{3}(n+1)!} \right) \omega_1 \left(f^{(n)}, \frac{1}{\sqrt{N}} \right),$$

(10.41)

proving the claim. ∎

We make

Remark 10.20 Here we mention the Max-product truncated sampling operators (see [7], p. 13) defined by

$$W_N^{(M)}(f)(x) := \frac{\bigvee_{k=0}^{N} \frac{\sin(Nx - k\pi)}{Nx - k\pi} f\left(\frac{k\pi}{N} \right)}{\bigvee_{k=0}^{N} \frac{\sin(Nx - k\pi)}{Nx - k\pi}}, \quad x \in [0, \pi],$$

(10.42)

$f : [0, \pi] \to \mathbb{R}_+$, continuous,
   and

$$K_N^{(M)}(f)(x) := \frac{\bigvee_{k=0}^{N} \frac{\sin^2(Nx - k\pi)}{(Nx - k\pi)^2} f\left(\frac{k\pi}{N} \right)}{\bigvee_{k=0}^{N} \frac{\sin^2(Nx - k\pi)}{(Nx - k\pi)^2}}, \quad x \in [0, \pi],$$

(10.43)

$f : [0, \pi] \to \mathbb{R}_+$, continuous.

By convention we take $\frac{\sin(0)}{0} = 1$, which implies for every $x = \frac{k\pi}{N}, k \in \{0, 1, ..., N\}$
that we have $\frac{\sin(Nx-k\pi)}{Nx-k\pi} = 1$.

We define the Max-product truncated combined sampling operators (see also [5])

$$M_N^{(M)}(f)(x) := \frac{\bigvee_{k=0}^N \rho_{N,k}(x) f\left(\frac{k\pi}{N}\right)}{\bigvee_{k=0}^N \rho_{N,k}(x)}, x \in [0, \pi], \tag{10.44}$$

$f \in C_+([0, \pi])$, where

$$M_N^{(M)}(f)(x) := \begin{cases} W_N^{(M)}(f)(x), & \text{if } \rho_{N,k}(x) := \frac{\sin(Nx-k\pi)}{Nx-k\pi}, \\ K_N^{(M)}(f)(x), & \text{if } \rho_{N,k}(x) := \left(\frac{\sin(Nx-k\pi)}{Nx-k\pi}\right)^2. \end{cases} \tag{10.45}$$

By [7], p. 346 and p. 352 we get

$$\left(M_N^{(M)}(|\cdot - x|)\right)(x) \le \frac{\pi}{2N}, \tag{10.46}$$

and by [3] ($m \in \mathbb{N}$) we have

$$\left(M_N^{(M)}(|\cdot - x|^m)\right)(x) \le \frac{\pi^m}{2N}, \forall x \in [0, \pi], \forall N \in \mathbb{N}. \tag{10.47}$$

Also it holds $M_N^{(M)}(1) = 1$, and $M_N^{(M)}$ are positive sublinear operators from C_+
$([0, \pi])$ into itself, $\forall N \in \mathbb{N}$.

We give

Theorem 10.21 Let $f \in C^n([0, \pi], \mathbb{R}_+)$, $n \in \mathbb{Z}_+$, $x \in (0, \pi)$ and $N^* \in \mathbb{N} : 0 < \frac{1}{N^*} \le \min(x, \pi - x)$, and $\left|f^{(n)}(\cdot) - f^{(n)}(x)\right|$ is convex over $[0, \pi]$. Assume that $f^{(k)}(x) = 0, k = 1, ..., n$. Then

$$\left|M_N^{(M)}(f)(x) - f(x)\right| \le \left(\frac{\pi^{n+1}}{2(n+1)!}\right) \omega_1\left(f^{(n)}, \frac{1}{N}\right), \tag{10.48}$$

$\forall N \in \mathbb{N} : N \ge N^*; n \in \mathbb{Z}_+$.

It holds $\lim_{N \to +\infty} M_N^{(M)}(f)(x) = f(x)$.

Proof By (10.20) we have:

$$\left|M_N^{(M)}(f)(x) - f(x)\right| \le \frac{\omega_1\left(f^{(n)}, h\right)}{h(n+1)!} M_N^{(M)}\left(|\cdot - x|^{n+1}\right)(x) \overset{(10.47)}{\le}$$

$$\frac{\omega_1\left(f^{(n)}, h\right)}{h(n+1)!} \frac{\pi^{n+1}}{2N} =$$

(setting $h := \frac{1}{N}$)

$$\left(\frac{\pi^{n+1}}{2\,(n+1)!}\right)\omega_1\left(f^{(n)},\frac{1}{N}\right), \tag{10.49}$$

proving the claim. ∎

We make

Remark 10.22 The Chebyshev knots of first kind $x_{N,k} := \cos\left(\frac{(2(N-k)+1)}{2(N+1)}\pi\right) \in (-1,1)$, $k \in \{0,1,...,N\}$, $-1 < x_{N,0} < x_{N,1} < ... < x_{N,N} < 1$, are the roots of the first kind Chebyshev polynomial $T_{N+1}(x) := \cos((N+1)\arccos x)$, $x \in [-1,1]$.
Define ($x \in [-1,1]$)

$$h_{N,k}(x) := \left(1 - x \cdot x_{N,k}\right)\left(\frac{T_{N+1}(x)}{(N+1)\left(x - x_{N,k}\right)}\right)^2, \tag{10.50}$$

the fundamental interpolation polynomials.

The Max-product interpolation Hermite–Fejér operators on Chebyshev knots of the first kind (see p. 12 of [7]) are defined by

$$H_{2N+1}^{(M)}(f)(x) = \frac{\bigvee_{k=0}^{N} h_{N,k}(x)\,f\left(x_{N,k}\right)}{\bigvee_{k=0}^{N} h_{N,k}(x)}, \forall N \in \mathbb{N}, \tag{10.51}$$

for $f \in C_+([-1,1])$, $\forall\, x \in [-1,1]$.
By [7], p. 287, we have

$$H_{2N+1}^{(M)}\left(|\cdot - x|\right)(x) \leq \frac{2\pi}{N+1}, \forall x \in [-1,1], \forall N \in \mathbb{N}. \tag{10.52}$$

And by [3], we get that

$$H_{2N+1}^{(M)}\left(|\cdot - x|^m\right)(x) \leq \frac{2^m\pi}{N+1}, \forall x \in [-1,1], \forall m, N \in \mathbb{N}. \tag{10.53}$$

Notice $H_{2N+1}^{(M)}(1) = 1$, and $H_{2N+1}^{(M)}$ maps $C_+([-1,1])$ into itself, and it is a positive sublinear operator. Furthermore it holds $\bigvee_{k=0}^{N} h_{N,k}(x) > 0$, $\forall\, x \in [-1,1]$. We also have $h_{N,k}\left(x_{N,k}\right) = 1$, and $h_{N,k}\left(x_{N,j}\right) = 0$, if $k \neq j$, and $H_{2N+1}^{(M)}(f)\left(x_{N,j}\right) = f\left(x_{N,j}\right)$, for all $j \in \{0,1,...,N\}$, see [7], p. 282.

We give

Theorem 10.23 Let $f \in C^n([-1,1],\mathbb{R}_+)$, $n \in \mathbb{Z}_+$, $x \in (-1,1)$ and let $N^* \in \mathbb{N}$: $0 < \frac{1}{N^*+1} \leq \min(x+1,1-x)$, and $\left|f^{(n)}(\cdot) - f^{(n)}(x)\right|$ is convex over $[-1,1]$. Assume that $f^{(k)}(x) = 0$, $k = 1,...,n$. Then

$$\left| H_{2N+1}^{(M)} \left(f \right) (x) - f (x) \right| \le \left(\frac{2^{n+1} \pi}{(n+1)!} \right) \omega_1 \left(f^{(n)}, \frac{1}{N+1} \right), \qquad (10.54)$$

$\forall\, N \ge N^*,\ N \in \mathbb{N};\ n \in \mathbb{Z}_+.$
 It holds $\displaystyle \lim_{N \to +\infty} H_{2N+1}^{(M)} \left(f \right) (x) = f (x).$

Proof By (10.20) we get

$$\left| H_{2N+1}^{(M)} \left(f \right) (x) - f (x) \right| \le \frac{\omega_1 \left(f^{(n)}, h \right)}{h (n+1)!} H_{2N+1}^{(M)} \left(|\cdot - x|^{n+1} \right) (x) \overset{(10.53)}{\le}$$

$$\frac{\omega_1 \left(f^{(n)}, h \right)}{h (n+1)!} \left(\frac{2^{n+1} \pi}{N+1} \right) =$$

(setting $h := \frac{1}{N+1}$)

$$\left(\frac{2^{n+1} \pi}{(n+1)!} \right) \omega_1 \left(f^{(n)}, \frac{1}{N+1} \right), \qquad (10.55)$$

proving the claim.   ∎

We make

Remark 10.24 Let $f \in C_+ \left([-1, 1] \right)$. Let the Chebyshev knots of second kind $x_{N,k} = \cos \left(\left(\frac{N-k}{N-1} \right) \pi \right) \in [-1, 1]$, $k = 1, ..., N$, $N \in \mathbb{N} - \{1\}$, which are the roots of $\omega_N (x) = \sin (N-1) t \sin t$, $x = \cos t \in [-1, 1]$. Notice that $x_{N,1} = -1$ and $x_{N,N} = 1$.
 Define

$$l_{N,k} (x) := \frac{(-1)^{k-1} \omega_N (x)}{\left(1 + \delta_{k,1} + \delta_{k,N} \right) (N-1) \left(x - x_{N,k} \right)}, \qquad (10.56)$$

$N \ge 2$, $k = 1, ..., N$, and $\omega_N (x) = \prod_{k=1}^{N} \left(x - x_{N,k} \right)$ and $\delta_{i,j}$ denotes the Kronecher's symbol, that is $\delta_{i,j} = 1$, if $i = j$, and $\delta_{i,j} = 0$, if $i \ne j$.
 The Max-product Lagrange interpolation operators on Chebyshev knots of second kind, plus the endpoints ± 1, are defined by ([7], p. 12)

$$L_N^{(M)} \left(f \right) (x) = \frac{\bigvee_{k=1}^{N} l_{N,k} (x)\, f \left(x_{N,k} \right)}{\bigvee_{k=1}^{N} l_{N,k} (x)}, \quad x \in [-1, 1]. \qquad (10.57)$$

By [7], pp. 297–298 and [3], we get that

$$L_N^{(M)} \left(|\cdot - x|^m \right) (x) \le \frac{2^{m+1} \pi^2}{3 (N-1)}, \qquad (10.58)$$

$\forall\, x \in (-1, 1)$ and $\forall\, m \in \mathbb{N};\ \forall\, N \in \mathbb{N},\ N \ge 4.$

We see that $L_N^{(M)}(f)(x) \geq 0$ is well defined and continuous for any $x \in [-1, 1]$. Following [7], p. 289, because $\sum_{k=1}^{N} l_{N,k}(x) = 1$, $\forall\, x \in [-1, 1]$, for any x there exists $k \in \{1, ..., N\} : l_{N,k}(x) > 0$, hence $\bigvee_{k=1}^{N} l_{N,k}(x) > 0$. We have that $l_{N,k}(x_{N,k}) = 1$, and $l_{N,k}(x_{N,j}) = 0$, if $k \neq j$. Furthermore it holds $L_N^{(M)}(f)(x_{N,j}) = f(x_{N,j})$, all $j \in \{1, ..., N\}$, and $L_N^{(M)}(1) = 1$.

By [7], pp. 289–290, $L_N^{(M)}$ are positive sublinear operators.

Finally we present

Theorem 10.25 Let $f \in C^n([-1, 1], \mathbb{R}_+)$, $n \in \mathbb{Z}_+$, $x \in (-1, 1)$ and let $N^* \in \mathbb{N}$: $N^* \geq 4$, with $0 < \frac{1}{N^*-1} \leq \min(x + 1, 1 - x)$, and $\left| f^{(n)}(\cdot) - f^{(n)}(x) \right|$ is convex over $[-1, 1]$. Assume that $f^{(k)}(x) = 0$, $k = 1, ..., n$. Then

$$\left| L_N^{(M)}(f)(x) - f(x) \right| \leq \left(\frac{2^{n+2}\pi^2}{3(n+1)!} \right) \omega_1 \left(f^{(n)}, \frac{1}{N-1} \right), \tag{10.59}$$

$\forall\, N \in \mathbb{N} : N \geq N^* \geq 4; \; n \in \mathbb{Z}_+$.
It holds $\lim_{N \to +\infty} L_N^{(M)}(f)(x) = f(x)$.

Proof Using (10.20) we get:

$$\left| L_N^{(M)}(f)(x) - f(x) \right| \leq \frac{\omega_1\left(f^{(n)}, h\right)}{h(n+1)!} L_N^{(M)}\left(|\cdot - x|^{n+1} \right)(x) \stackrel{(10.58)}{\leq}$$

$$\frac{\omega_1\left(f^{(n)}, h\right)}{h(n+1)!} \left(\frac{2^{n+2}\pi^2}{3(N-1)} \right) =$$

(setting $h := \frac{1}{N-1}$)

$$\left(\frac{2^{n+2}\pi^2}{3(n+1)!} \right) \omega_1 \left(f^{(n)}, \frac{1}{N-1} \right), \tag{10.60}$$

proving the claim. ∎

References

1. G. Anastassiou, *Moments in Probability and Approximation Theory*, Pitman Research Notes in Mathematics Series (Longman Group UK, New York, 1993)
2. G. Anastassiou, *Approximation by Sublinear Operators* (2017, submitted)
3. G. Anastassiou, *Approximation by Max-Product Operators* (2017, submitted)
4. G. Anastassiou, *Approximation of Fuzzy Numbers by Max-Product Operators* (2017, submitted)
5. G. Anastassiou, *Approximated by Multivariate Sublinear and Max-Product Operators Under Convexity* (2017, submitted)
6. G. Anastassiou, *Approximation by Sublinear and Max-product Operators Using Convexity* (2017, submitted)
7. B. Bede, L. Coroianu, S. Gal, *Approximation by Max-Product Type Operators* (Springer, Heidelberg, 2016)

Chapter 11
High Order Conformable Fractional Approximation by Max-Product Operators Using Convexity

Here we consider the approximation of functions by a large variety of Max-Product operators under conformable fractional differentiability and using convexity. These are positive sublinear operators. Our study relies on our general results about positive sublinear operators. We derive Jackson type inequalities under conformable fractional initial conditions and convexity. So our approach is quantitative by obtaining inequalities where their right hand sides involve the modulus of continuity of a high order conformable fractional derivative of the function under approximation. Due to the convexity assumptions our inequalities are compact and elegant with small constants. It follows [5].

11.1 Background

In this chapter we study under convexity quantitatively the conformable fractional approximation properties of Max-product operators to the unit. These are special cases of positive sublinear operators. We first present results regarding the convergence to the unit of general positive sublinear operators under convexity. The focus of our study is approximation under the presence of conformable fractional smoothness.

Under our convexity conditions the derived conformable fractional convergence inequalities are elegant and compact with very small constants.

This chapter is inspired by [7].

We make

Remark 11.1 Let $x, y \in [a, b] \subseteq [0, \infty)$, and $g(x) = x^{\alpha}, 0 < \alpha \le 1$, then $g'(x) = \alpha x^{\alpha-1} = \frac{\alpha}{x^{1-\alpha}}$, for $x \in (0, \infty)$. Since $a \le x \le b$, then $\frac{1}{x} \ge \frac{1}{b} > 0$ and $\frac{\alpha}{x^{1-\alpha}} \ge \frac{\alpha}{b^{1-\alpha}} > 0$.

© Springer International Publishing AG, part of Springer Nature 2018

G. A. Anastassiou, *Nonlinearity: Ordinary and Fractional Approximations by Sublinear and Max-Product Operators*, Studies in Systems, Decision and Control 147, https://doi.org/10.1007/978-3-319-89509-3_11

Assume $y > x$. By the mean value theorem we get

$$y^\alpha - x^\alpha = \frac{\alpha}{\xi^{1-\alpha}} (y - x), \text{ where } \xi \in (x, y).$$ (11.1)

A similar to (11.1) equality when $x > y$ is true.

Then we obtain

$$\frac{\alpha}{b^{1-\alpha}} |y - x| \le |y^\alpha - x^\alpha| = \frac{\alpha}{\xi^{1-\alpha}} |y - x|.$$ (11.2)

Thus, it holds

$$\frac{\alpha}{b^{1-\alpha}} |y - x| \le |y^\alpha - x^\alpha|.$$ (11.3)

Hence we get

$$|y - x| \le \frac{b^{1-\alpha}}{\alpha} |y^\alpha - x^\alpha|,$$ (11.4)

$\forall \, x, y \in [a, b] \subset [0, \infty), \alpha \in (0, 1]$.

We also make

Remark 11.2 For $0 < \alpha \le 1, x, y, t, s \ge 0$, we have

$$2^{\alpha-1} (x^\alpha + y^\alpha) \le (x + y)^\alpha \le x^\alpha + y^\alpha.$$ (11.5)

Assume that $t > s$, then

$$t = t - s + s \Rightarrow t^\alpha = (t - s + s)^\alpha \le (t - s)^\alpha + s^\alpha,$$

hence $t^\alpha - s^\alpha \le (t - s)^\alpha$.

Similarly, when $s > t \Rightarrow s^\alpha - t^\alpha \le (s - t)^\alpha$.

Therefore it holds

$$|t^\alpha - s^\alpha| \le |t - s|^\alpha , \forall t, s \in [0, \infty).$$ (11.6)

We need

Definition 11.3 ([1, 8]) Let $f : [0, \infty) \to \mathbb{R}$. The conformable α-fractional derivative for $\alpha \in (0, 1]$ is given by

$$D_\alpha f (t) := \lim_{\varepsilon \to 0} \frac{f \left(t + \varepsilon t^{1-\alpha}\right) - f (t)}{\varepsilon},$$ (11.7)

$$D_\alpha f (0) = \lim_{t \to 0+} D_\alpha f (t).$$ (11.8)

If f is differentiable, then

$$D_\alpha f(t) = t^{1-\alpha} f'(t),\tag{11.9}$$

where f' is the usual derivative.

We define

$$D_\alpha^n f = D_\alpha^{n-1}(D_\alpha f) \text{ and } D_\alpha^0 f = f.\tag{11.10}$$

If $f : [0, \infty) \to \mathbb{R}$ is α-differentiable at $t_0 > 0$, $\alpha \in (0, 1]$, then f is continuous at t_0, see [8].

We will use

Theorem 11.4 (see *[6]*) (Taylor formula) *Let* $\alpha \in (0, 1]$ *and* $n \in \mathbb{N}$. *Suppose* f *is* $(n + 1)$ *times conformable* α-*fractional differentiable on* $[0, \infty)$, *and* $s, t \in [0, \infty)$, *and* $D_\alpha^{n+1} f$ *is assumed to be continuous on* $[0, \infty)$. *Then we have*

$$f(t) = \sum_{k=0}^{n} \frac{1}{k!} \left(\frac{t^\alpha - s^\alpha}{\alpha} \right)^k D_\alpha^k f(s) + \frac{1}{n!} \int_s^t \left(\frac{t^\alpha - \tau^\alpha}{\alpha} \right)^n D_\alpha^{n+1} f(\tau) \tau^{a-1} d\tau.$$

$$\tag{11.11}$$

The case $n = 0$ follows.

Corollary 11.5 (*[4]*) *Let* $\alpha \in (0, 1]$. *Suppose* f *is* α -*fractional differentiable on* $[0, \infty)$, *and* $s, t \in [0, \infty)$. *Assume that* $D_\alpha f$ *is continuous on* $[0, \infty)$. *Then*

$$f(t) = f(s) + \int_s^t D_\alpha f(\tau) \tau^{a-1} d\tau.\tag{11.12}$$

Note 11.6 *Theorem 11.4 and Corollary 11.5 are also true for* $f : [a, b] \to \mathbb{R}$, $[a, b] \subseteq [0, \infty)$, $s, t \in [a, b]$.

We need

Definition 11.7 Let $f \in C([a, b])$. We define the first modulus of continuity of f as:

$$\omega_1(f, \delta) := \sup_{\substack{x, y \in [a,b]: \\ |x-y| \le \delta}} |f(x) - f(y)|, \delta > 0.\tag{11.13}$$

11.2   Main Results

We give

Theorem 11.8 *Let* $\alpha \in (0, 1]$ *and* $n \in \mathbb{Z}_+$. *Suppose* f *is* $(n + 1)$ *times conformable* α-*fractional differentiable on* $[a, b] \subseteq [0, \infty)$, *and* $t \in [a, b]$, $x_0 \in (a, b)$, *and* $D_\alpha^{n+1} f$ *is continuous on* $[a, b]$. *Let* $0 < h \le \min(x_0 - a, b - x_0)$ *and* *assume* $\left| D_\alpha^{n+1} f \right|$ *is convex over* $[a, b]$. *Furthermore assume that* $D_\alpha^k f(x_0) = 0$, $k = 1, \dots, n+1$. *Then*

$$|f(t) - f(x_0)| \leq \left(\frac{\omega_1 \left(D_\alpha^{n+1} f, h \right) b^{1-\alpha}}{(n+2)! \alpha^{n+2} h} \right) |t - x_0|^{(n+2)\alpha}, \tag{11.14}$$

$\forall\, t \in [a, b]$.

Proof We have that

$$\frac{1}{n!} \int_s^t \left(\frac{t^\alpha - \tau^\alpha}{\alpha} \right)^n D_\alpha^{n+1} f(s) \tau^{\alpha-1} d\tau = \frac{D_\alpha^{n+1} f(s)}{n!} \int_s^t \left(\frac{t^\alpha - \tau^\alpha}{\alpha} \right)^n \tau^{\alpha-1} d\tau \tag{11.15}$$

(by $\frac{d\tau^\alpha}{d\tau} = \alpha \tau^{\alpha-1} \Rightarrow d\tau^\alpha = \alpha \tau^{\alpha-1} d\tau \Rightarrow \frac{1}{\alpha} d\tau^\alpha = \tau^{\alpha-1} d\tau$)

$$= \frac{D_\alpha^{n+1} f(s)}{\alpha^{n+1} n!} \int_s^t (t^\alpha - \tau^\alpha)^n d\tau^\alpha$$

(by $t \leq \tau \leq s \Rightarrow t^\alpha \leq \tau^\alpha (=: z) \leq s^\alpha$)

$$= \frac{D_\alpha^{n+1} f(s)}{\alpha^{n+1} n!} \int_{s^\alpha}^{t^\alpha} (t^\alpha - z)^n \, dz = \frac{D_\alpha^{n+1} f(s)}{\alpha^{n+1} n!} \frac{(t^\alpha - s^\alpha)^{n+1}}{n+1} \tag{11.16}$$

$$= \frac{D_\alpha^{n+1} f(s)}{(n+1)!} \left(\frac{t^\alpha - s^\alpha}{\alpha} \right)^{n+1}.$$

Therefore it holds

$$\frac{1}{n!} \int_s^t \left(\frac{t^\alpha - \tau^\alpha}{\alpha} \right)^n D_\alpha^{n+1} f(s) \tau^{\alpha-1} d\tau = \frac{D_\alpha^{n+1} f(s)}{(n+1)!} \left(\frac{t^\alpha - s^\alpha}{\alpha} \right)^{n+1}. \tag{11.17}$$

By (11.11) and (11.12) we get:

$$f(t) = \sum_{k=0}^{n+1} \frac{1}{k!} \left(\frac{t^\alpha - s^\alpha}{\alpha} \right)^k D_\alpha^k f(s) + \tag{11.18}$$

$$\frac{1}{n!} \int_s^t \left(\frac{t^\alpha - \tau^\alpha}{\alpha} \right)^n \left(D_\alpha^{n+1} f(\tau) - D_\alpha^{n+1} f(s) \right) \tau^{\alpha-1} d\tau.$$

In particular it holds

$$f(t) = \sum_{k=0}^{n+1} \frac{1}{k!} \left(\frac{t^\alpha - x_0^\alpha}{\alpha} \right)^k D_\alpha^k f(x_0) + \tag{11.19}$$

$$\frac{1}{n!}\int_{x_0}^{t}\left(\frac{t^{\alpha}-\tau^{\alpha}}{\alpha}\right)^{n}\left(D_{\alpha}^{n+1}f\left(\tau\right)-D_{\alpha}^{n+1}f\left(x_0\right)\right)\tau^{\alpha-1}d\tau.$$

By the assumption $D_{\alpha}^{k}f\left(x_0\right)=0$, $k=1,\ldots,n+1$, we can write

$$f\left(t\right)-f\left(x_0\right)=\frac{1}{n!}\int_{x_0}^{t}\left(\frac{t^{\alpha}-\tau^{\alpha}}{\alpha}\right)^{n}\left(D_{\alpha}^{n+1}f\left(\tau\right)-D_{\alpha}^{n+1}f\left(x_0\right)\right)\tau^{\alpha-1}d\tau. \quad (11.20)$$

By assumption here $\left|D_{\alpha}^{n+1}f\right|$ is convex over $[a,b]\subseteq[0,\infty)$; $x_0\in(a,b)$.

Let $h:0<h\leq\min\left(x_0-a,b-x_0\right)$, by Lemma 8.1.1, p. 243 of [2] we get that

$$\left|D_{\alpha}^{n+1}f\left(t\right)-D_{\alpha}^{n+1}f\left(x_0\right)\right|\leq\frac{\omega_1\left(D_{\alpha}^{n+1}f,h\right)}{h}\left|t-x_0\right|, \quad (11.21)$$

$\forall\,t\in[a,b]$.

Using (11.4) we obtain

$$\left|D_{\alpha}^{n+1}f\left(t\right)-D_{\alpha}^{n+1}f\left(x_0\right)\right|\leq\frac{\omega_1\left(D_{\alpha}^{n+1}f,h\right)}{h}\frac{b^{1-\alpha}}{\alpha}\left|t^{\alpha}-x_0^{\alpha}\right|, \quad (11.22)$$

$\forall\,t\in[a,b]$.

Next we estimate (11.20).

(1). We observe that $(t\geq x_0)$

$$\left|f\left(t\right)-f\left(x_0\right)\right|\overset{(11.20)}{\leq}\frac{1}{n!}\int_{x_0}^{t}\left(\frac{t^{\alpha}-\tau^{\alpha}}{\alpha}\right)^{n}\left|D_{\alpha}^{n+1}f\left(\tau\right)-D_{\alpha}^{n+1}f\left(x_0\right)\right|\tau^{\alpha-1}d\tau\overset{(11.22)}{\leq}$$

$$\frac{\omega_1\left(D_{\alpha}^{n+1}f,h\right)b^{1-\alpha}}{n!h\alpha^{n+2}}\int_{x_0}^{t}\left(t^{\alpha}-\tau^{\alpha}\right)^{n}\left(\tau^{\alpha}-x_0^{\alpha}\right)d\tau^{\alpha}= \quad (11.23)$$

$$\left(\frac{\omega_1\left(D_{\alpha}^{n+1}f,h\right)b^{1-\alpha}}{n!h\alpha^{n+2}}\right)\int_{x_0^{\alpha}}^{t^{\alpha}}\left(t^{\alpha}-z\right)^{(n+1)-1}\left(z-x_0^{\alpha}\right)^{2-1}dz=$$

$$\left(\frac{\omega_1\left(D_{\alpha}^{n+1}f,h\right)b^{1-\alpha}}{n!h\alpha^{n+2}}\right)\frac{\Gamma\left(n+1\right)\Gamma\left(2\right)}{\Gamma\left(n+3\right)}\left(t^{\alpha}-x_0^{\alpha}\right)^{n+2}=$$

$$\left(\frac{\omega_1\left(D_{\alpha}^{n+1}f,h\right)b^{1-\alpha}}{n!h\alpha^{n+2}}\right)\frac{n!}{(n+2)!}\left(t^{\alpha}-x_0^{\alpha}\right)^{n+2}=$$

$$\left(\frac{\omega_1\left(D_{\alpha}^{n+1}f,h\right)b^{1-\alpha}}{(n+2)!h\alpha^{n+2}}\right)\left(t^{\alpha}-x_0^{\alpha}\right)^{n+2}. \quad (11.24)$$

We have proved that (case of $t \geq x_0$)

$$|f(t) - f(x_0)| \leq \left(\frac{\omega_1\left(D_\alpha^{n+1} f, h\right) b^{1-\alpha}}{(n+2)! h \alpha^{n+2}}\right)\left(t^\alpha - x_0^\alpha\right)^{n+2}. \tag{11.25}$$

(2) We observe that ($t \leq x_0$)

$$|f(t) - f(x_0)| \overset{(11.20)}{=} \frac{1}{n!}\left|\int_{x_0}^t \left(\frac{t^\alpha - \tau^\alpha}{\alpha}\right)^n \left(D_\alpha^{n+1} f(\tau) - D_\alpha^{n+1} f(x_0)\right) \tau^{\alpha-1} d\tau\right| =$$

$$\frac{1}{n! \alpha}\left|\int_t^{x_0} \left(\frac{\tau^\alpha - t^\alpha}{\alpha}\right)^n \left(D_\alpha^{n+1} f(\tau) - D_\alpha^{n+1} f(x_0)\right) d\tau^\alpha\right| \leq \tag{11.26}$$

$$\frac{1}{n! \alpha^{n+1}} \int_t^{x_0} \left(\tau^\alpha - t^\alpha\right)^n \left|D_\alpha^{n+1} f(\tau) - D_\alpha^{n+1} f(x_0)\right| d\tau^\alpha \overset{(11.22)}{\leq}$$

$$\frac{1}{n! \alpha^{n+1}}\left(\frac{\omega_1\left(D_\alpha^{n+1} f, h\right) b^{1-\alpha}}{h}\right) \int_t^{x_0} \left(x_0^\alpha - \tau^\alpha\right)\left(\tau^\alpha - t^\alpha\right)^n d\tau^\alpha =$$

$$\left(\frac{\omega_1\left(D_\alpha^{n+1} f, h\right) b^{1-\alpha}}{n! \alpha^{n+2} h}\right) \int_{t^\alpha}^{x_0^\alpha} \left(x_0^\alpha - z\right)^{2-1}\left(z - t^\alpha\right)^{(n+1)-1} dz = \tag{11.27}$$

$$\left(\frac{\omega_1\left(D_\alpha^{n+1} f, h\right) b^{1-\alpha}}{n! \alpha^{n+2} h}\right) \frac{\Gamma(2) \Gamma(n+1)}{\Gamma(n+3)}\left(x_0^\alpha - t^\alpha\right)^{n+2} =$$

$$\left(\frac{\omega_1\left(D_\alpha^{n+1} f, h\right) b^{1-\alpha}}{n! \alpha^{n+2} h}\right) \frac{n!}{(n+2)!}\left(x_0^\alpha - t^\alpha\right)^{n+2} =$$

$$\left(\frac{\omega_1\left(D_\alpha^{n+1} f, h\right) b^{1-\alpha}}{(n+2)! \alpha^{n+2} h}\right)\left(x_0^\alpha - t^\alpha\right)^{n+2}.$$

We have proved ($t \leq x_0$) that

$$|f(t) - f(x_0)| \leq \left(\frac{\omega_1\left(D_\alpha^{n+1} f, h\right) b^{1-\alpha}}{(n+2)! \alpha^{n+2} h}\right)\left(x_0^\alpha - t^\alpha\right)^{n+2}. \tag{11.28}$$

In conclusion we have established that

$$|f(t) - f(x_0)| \leq \left(\frac{\omega_1\left(D_\alpha^{n+1} f, h\right) b^{1-\alpha}}{(n+2)! \alpha^{n+2} h}\right)\left|t^\alpha - x_0^\alpha\right|^{n+2}, \forall t \in [a, b]. \tag{11.29}$$

By (11.6) we have

$$\left| t^\alpha - x_0^\alpha \right| \le \left| t - x_0 \right|^\alpha . \tag{11.30}$$

Thus by (11.29) and (11.30) the claim is proved. ∎

We rewrite the statement of Theorem 11.8 in a convenient way as follows:

Theorem 11.9 *Let $\alpha \in (0, 1]$ and $n \in \mathbb{N}$. Suppose f is n times conformable α-fractional differentiable on $[a, b] \subseteq [0, \infty)$, and $x \in (a, b)$, and $D_\alpha^n f$ is continuous on $[a, b]$. Let $0 < h \le \min (x - a, b - x)$ and assume $\left| D_\alpha^n f \right|$ is convex over $[a, b]$. Furthermore assume that $D_\alpha^k f (x) = 0$, $k = 1, ..., n$. Then, over $[a, b]$, we have*

$$\left| f (\cdot) - f (x) \right| \le \left(\frac{\omega_1 \left(D_\alpha^n f, h \right) b^{1-\alpha}}{(n + 1)! \alpha^{n+1} h} \right) \left| \cdot - x \right|^{(n+1)\alpha} . \tag{11.31}$$

We need

Definition 11.10 Here $C_+ ([a, b]) := \{ f : [a, b] \subseteq [0, \infty) \to \mathbb{R}_+ $, continuous functions$\}$. Let $L_N : C_+ ([a, b]) \to C_+ ([a, b])$, operators, $\forall N \in \mathbb{N}$, such that
(i)

$$L_N (\alpha f) = \alpha L_N (f) , \quad \forall \alpha \ge 0, \forall f \in C_+ ([a, b]) , \tag{11.32}$$

(ii) if $f, g \in C_+ ([a, b]) : f \le g$, then

$$L_N (f) \le L_N (g) , \quad \forall N \in \mathbb{N}, \tag{11.33}$$

(iii)

$$L_N (f + g) \le L_N (f) + L_N (g) , \quad \forall f, g \in C_+ ([a, b]) . \tag{11.34}$$

We call $\{L_N\}_{N \in \mathbb{N}}$ positive sublinear operators.

We need a Hölder's type inequality, see next:

Theorem 11.11 *(see [3]) Let $L : C_+ ([a, b]) \to C_+ ([a, b])$, be a positive sublinear operator and $f, g \in C_+ ([a, b])$, furthermore let $p, q > 1 : \frac{1}{p} + \frac{1}{q} = 1$. Assume that $L \left((f (\cdot))^p \right) (s_*) , L \left((g (\cdot))^q \right) (s_*) > 0$ for some $s_* \in [a, b]$. Then*

$$L (f (\cdot) g (\cdot)) (s_*) \le \left(L \left((f (\cdot))^p \right) (s_*) \right)^{\frac{1}{p}} \left(L \left((g (\cdot))^q \right) (s_*) \right)^{\frac{1}{q}} . \tag{11.35}$$

We make

Remark 11.12 By [7], p. 17, we get: let $f, g \in C_+ ([a, b])$, then

$$\left| L_N (f) (x) - L_N (g) (x) \right| \le L_N (\left| f - g \right|) (x) , \quad \forall x \in [a, b] \subseteq [0, \infty). \tag{11.36}$$

Furthermore, we also have that

$$|L_N (f) (x) - f (x)| \leq L_N (|f (\cdot) - f (x)|) (x) + |f (x)| |L_N (e_0) (x) - 1|,$$
$$(11.37)$$

$\forall x \in [a, b] \subseteq [0, \infty); e_0 (t) = 1.$

From now on we assume that $L_N (1) = 1$. Hence it holds

$$|L_N (f) (x) - f (x)| \leq L_N (|f (\cdot) - f (x)|) (x), \forall x \in [a, b] \subseteq [0, \infty). \quad (11.38)$$

We give

Theorem 11.13 *Let* $\alpha \in (0, 1]$ *and* $n \in \mathbb{N}$. *Suppose* $f \in C_+ ([a, b])$ *is n times conformable* α *-fractional differentiable on* $[a, b] \subseteq [0, \infty)$, *and* $x \in (a, b)$, *and* $D_\alpha^n f$ *is continuous on* $[a, b]$. *Let* $0 < h \leq \min (x - a, b - x)$ *and assume* $|D_\alpha^n f|$ *is convex over* $[a, b]$. *Furthermore assume that* $D_\alpha^k f (x) = 0, k = 1, ..., n$. *Let* $\{L_N\}_{N \in \mathbb{N}}$ *from* $C_+ ([a, b])$ *into itself, positive sublinear operators such that:* $L_N (1) = 1, \forall N \in \mathbb{N}$. *Then*

$$|L_N (f) (x) - f (x)| \leq \left(\frac{\omega_1 \left(D_\alpha^n f, h \right) b^{1-\alpha}}{(n + 1)! \alpha^{n+1} h} \right) L_N \left(|\cdot - x|^{(n+1)\alpha} \right) (x), \forall N \in \mathbb{N}.$$
$$(11.39)$$

Proof By (11.31) and (11.38). ∎

We give

Theorem 11.14 *All as in Theorem 11.13. Additionally assume that* $L_N \left(|\cdot - x|^{(n+1)(\alpha+1)} \right) (x) > 0, \ \forall N \in \mathbb{N}$. *Then*

$$|L_N (f) (x) - f (x)| \leq \left(\frac{\omega_1 \left(D_\alpha^n f, h \right) b^{1-\alpha}}{(n + 1)! \alpha^{n+1} h} \right) \left(L_N \left(|\cdot - x|^{(n+1)(\alpha+1)} \right) (x) \right)^{\frac{\alpha}{\alpha+1}},$$
$$(11.40)$$

$\forall N \in \mathbb{N}$.

Proof By (11.39) and Theorem 11.11: we have

$$L_N \left(|\cdot - x|^{(n+1)\alpha} \right) (x) \leq \left(L_N \left(|\cdot - x|^{(n+1)(\alpha+1)} \right) (x) \right)^{\frac{\alpha}{\alpha+1}}, \quad (11.41)$$

proving the claim. ∎

We present

Theorem 11.15 *Let* $\{L_N\}_{N \in \mathbb{N}}$ *from* $C_+ ([a, b])$ *into itself, positive sublinear operators, such that:* $L_N (1) = 1, \forall N \in \mathbb{N}$. *Additionally assume that* $L_N \left(|\cdot - x|^{(n+1)(\alpha+1)} \right)$ $(x) > 0, \ \forall N \in \mathbb{N}; x \in (a, b)$. *Here* $\alpha \in (0, 1]$ *and* $n \in \mathbb{N}$. *Suppose* $f \in C_+ ([a, b])$ *is n times conformable* α-*fractional differentiable on* $[a, b] \subseteq [0, \infty)$, *and* $D_\alpha^n f$ *is continuous on* $[a, b]$. *Assume here that* $0 < \left(L_N \left(|\cdot - x|^{(n+1)(\alpha+1)} \right) (x) \right)^{\frac{\alpha}{\alpha+1}} \leq$

$\min(x - a, b - x)$, $\forall N \in \mathbb{N} : N \geq N^* \in \mathbb{N}$, *and assume* $\left| D_\alpha^n f \right|$ *is convex over* $[a, b]$. *Furthermore assume that* $D_\alpha^k f(x) = 0$, $k = 1, ..., n$. *Then*

$$|L_N(f)(x) - f(x)| \leq \frac{b^{1-\alpha} \omega_1 \left(D_\alpha^n f, \left(L_N \left(|\cdot - x|^{(n+1)(\alpha+1)} \right)(x) \right)^{\frac{\alpha}{\alpha+1}} \right)}{(n+1)! \alpha^{n+1}}, \quad (11.42)$$

$\forall N \in \mathbb{N} : N \geq N^* \in \mathbb{N}$.

If $L_N \left(|\cdot - x|^{(n+1)(\alpha+1)} \right)(x) \to 0$, *then* $L_N(f)(x) \to f(x)$, *as* $N \to +\infty$.

Proof By (11.40) and choosing $h := \left(L_N \left(|\cdot - x|^{(n+1)(\alpha+1)} \right)(x) \right)^{\frac{\alpha}{\alpha+1}}$. ∎

We also give

Theorem 11.16 *Let* $\{L_N\}_{N \in \mathbb{N}}$ *from* $C_+([a, b])$ *into itself, positive sublinear operators:* $L_N(1) = 1$, $\forall N \in \mathbb{N}$. *Also* $L_N \left(|\cdot - x|^{(n+1)\alpha} \right)(x) > 0$, $\forall N \in \mathbb{N}$. *Here* $\alpha \in (0, 1]$, $n \in \mathbb{N}$ *and* $x \in (a, b)$; $[a, b] \subseteq [0, \infty)$. *Suppose* $f \in C_+([a, b])$ *is* n *times conformable* α-*fractional differentiable on* $[a, b]$, *and* $D_\alpha^n f$ *is continuous on* $[a, b]$. *Let* $0 < L_N \left(|\cdot - x|^{(n+1)\alpha} \right)(x) \leq \min(x - a, b - x)$, $\forall N \geq N^*$; $N, N^* \in \mathbb{N}$, *and assume* $\left| D_\alpha^n f \right|$ *is convex over* $[a, b]$. *Furthermore assume that* $D_\alpha^k f(x) = 0$, $k = 1, ..., n$. *Then*

$$|L_N(f)(x) - f(x)| \leq \frac{b^{1-\alpha} \omega_1 \left(D_\alpha^n f, L_N \left(|\cdot - x|^{(n+1)\alpha} \right)(x) \right)}{(n+1)! \alpha^{n+1}}, \quad (11.43)$$

$\forall N \geq N^*$, *where* $N, N^* \in \mathbb{N}$.

If $L_N \left(|\cdot - x|^{(n+1)\alpha} \right)(x) \to 0$, *then* $L_N(f)(x) \to f(x)$, *as* $N \to +\infty$.

Proof By (11.39) and choosing $h := L_N \left(|\cdot - x|^{(n+1)\alpha} \right)(x)$. ∎

11.3 Applications

(I) Here we apply Theorem 11.14 to well known Max-product operators.
We make

Remark 11.17 In [7], p. 10, the authors introduced the basic Max-product Bernstein operators

$$B_N^{(M)}(f)(x) = \frac{\bigvee_{k=0}^N p_{N,k}(x) f\left(\frac{k}{N}\right)}{\bigvee_{k=0}^N p_{N,k}(x)}, \quad N \in \mathbb{N}, \quad (11.44)$$

where $\vee$ stands for maximum, and $p_{N,k}(x) = \binom{N}{k} x^k (1 - x)^{N-k}$ and $f : [0, 1] \to \mathbb{R}_+ = [0, \infty)$ is continuous.

These are nonlinear and piecewise rational operators.

We have $B_N^{(M)}(1) = 1$, and

$$B_N^{(M)}(|\cdot - x|)(x) \le \frac{6}{\sqrt{N+1}}, \forall x \in [0, 1], \forall N \in \mathbb{N}, \tag{11.45}$$

see [7], p. 31.

$B_N^{(M)}$ are positive sublinear operators and thus they possess the monotonicity property, also since $|\cdot - x| \le 1$, then $|\cdot - x|^\beta \le 1, \forall\, x \in [0, 1], \forall\, \beta > 0$.

Therefore it holds

$$B_N^{(M)}\left(|\cdot - x|^{1+\beta}\right)(x) \le \frac{6}{\sqrt{N+1}}, \forall x \in [0, 1], \forall N \in \mathbb{N}, \forall \beta > 0. \tag{11.46}$$

Furthermore, clearly it holds that

$$B_N^{(M)}\left(|\cdot - x|^{1+\beta}\right)(x) > 0, \forall N \in \mathbb{N}, \forall \beta \ge 0 \text{ and any } x \in (0, 1). \tag{11.47}$$

The operator $B_N^{(M)}$ maps $C_+([0, 1])$ into itself.

We give

Theorem 11.18 *Let $\alpha \in (0, 1]$ and $n \in \mathbb{N}$. Suppose $f \in C_+([0, 1])$ is n times conformable α-fractional differentiable on $[0, 1]$, $x \in (0, 1)$, and $D_\alpha^n f$ is continuous on $[0, 1]$. Let $N^* \in \mathbb{N}$ such that $\frac{1}{(N^*+1)^{\frac{\alpha}{2(\alpha+1)}}} \le \min(x, 1-x)$ and assume $\left|D_\alpha^n f\right|$ is convex over $[0, 1]$. Furthermore assume that $D_\alpha^k f(x) = 0$, $k = 1, ..., n$. Then*

$$\left|B_N^{(M)}(f)(x) - f(x)\right| \le \left(\frac{6^{\frac{\alpha}{\alpha+1}}}{(n+1)!\alpha^{n+1}}\right) \omega_1\left(D_\alpha^n f, \frac{1}{(N+1)^{\frac{\alpha}{2(\alpha+1)}}}\right), \tag{11.48}$$

$\forall\, N \ge N^*, N \in \mathbb{N}$.

It holds $\lim\limits_{N \to +\infty} B_N^{(M)}(f)(x) = f(x)$.

Proof By (11.47) we get that $B_N^{(M)}\left(|\cdot - x|^{(n+1)(\alpha+1)}\right)(x) > 0, \forall\, N \in \mathbb{N}$.

By (11.40), (11.46), we obtain that

$$\left|B_N^{(M)}(f)(x) - f(x)\right| \le \left(\frac{\omega_1\left(D_\alpha^n f, h\right)}{(n+1)!\alpha^{n+1}h}\right) \left(B_N^{(M)}\left(|\cdot - x|^{(n+1)(\alpha+1)}\right)(x)\right)^{\frac{\alpha}{\alpha+1}} \le$$

$$\left(\frac{\omega_1\left(D_\alpha^n f, h\right)}{(n+1)!\alpha^{n+1}h}\right)\left(\frac{6}{\sqrt{N+1}}\right)^{\frac{\alpha}{\alpha+1}} = \tag{11.49}$$

(setting $h := \left(\frac{1}{\sqrt{N+1}}\right)^{\frac{\alpha}{\alpha+1}}$)

$$\left(\frac{6^{\frac{\alpha}{\alpha+1}}}{(n+1)!\alpha^{n+1}}\right)\omega_1\left(D_\alpha^n f, \frac{1}{(N+1)^{\frac{\alpha}{2(\alpha+1)}}}\right),$$

proving the claim. ∎

We continue with

Remark 11.19 The truncated Favard–Szász–Mirakjan operators are given by

$$T_N^{(M)}(f)(x) = \frac{\bigvee_{k=0}^N s_{N,k}(x) f\left(\frac{k}{N}\right)}{\bigvee_{k=0}^N s_{N,k}(x)}, x \in [0,1], N \in \mathbb{N}, f \in C_+([0,1]),$$

(11.50)

$s_{N,k}(x) = \frac{(Nx)^k}{k!}$, see also [7], p. 11.

By [7], p. 178–179, we get that

$$T_N^{(M)}(|\cdot - x|)(x) \le \frac{3}{\sqrt{N}}, \forall x \in [0,1], \forall N \in \mathbb{N}.$$

(11.51)

Clearly it holds

$$T_N^{(M)}\left(|\cdot - x|^{1+\beta}\right)(x) \le \frac{3}{\sqrt{N}}, \forall x \in [0,1], \forall N \in \mathbb{N}, \forall \beta > 0.$$

(11.52)

The operators $T_N^{(M)}$ are positive sublinear operators mapping $C_+([0,1])$ into itself, with $T_N^{(M)}(1) = 1$.

Furthermore it holds

$$T_N^{(M)}\left(|\cdot - x|^\lambda\right)(x) = \frac{\bigvee_{k=0}^N \frac{(Nx)^k}{k!}\left|\frac{k}{N} - x\right|^\lambda}{\bigvee_{k=0}^N \frac{(Nx)^k}{k!}} > 0, \forall x \in (0,1], \forall \lambda \ge 1, \forall N \in \mathbb{N}.$$

(11.53)

We give

Theorem 11.20 *Let* $\alpha \in (0,1]$ *and* $n \in \mathbb{N}$. *Suppose* $f \in C_+([0,1])$ *is* n *times conformable* α*-fractional differentiable on* $[0,1]$, $x \in (0,1)$, *and* $D_\alpha^n f$ *is continuous on* $[0,1]$. *Let* $N^* \in \mathbb{N}$ *such that* $\frac{1}{(N^*)^{\frac{\alpha}{2(\alpha+1)}}} \le \min(x, 1-x)$ *and assume* $\left|D_\alpha^n f\right|$ *is convex over* $[0,1]$. *Furthermore assume that* $D_\alpha^k f(x) = 0$, $k = 1, ..., n$. *Then*

$$\left|T_N^{(M)}(f)(x) - f(x)\right| \le \left(\frac{3^{\frac{\alpha}{\alpha+1}}}{(n+1)!\alpha^{n+1}}\right)\omega_1\left(D_\alpha^n f, \frac{1}{N^{\frac{\alpha}{2(\alpha+1)}}}\right),$$

(11.54)

$\forall N \ge N^*$, $N \in \mathbb{N}$.

It holds $\lim_{N\to+\infty} T_N^{(M)}(f)(x) = f(x)$.

Proof By (11.53) we have $T_N^{(M)} \left(|\cdot - x|^{(n+1)(\alpha+1)} \right) (x) > 0, \forall N \in \mathbb{N}$.
By (11.40), (11.52), we get that

$$\left| T_N^{(M)} (f) (x) - f (x) \right| \leq \left(\frac{\omega_1 \left(D_\alpha^n f, h \right)}{(n+1)! \alpha^{n+1} h} \right) \left(\frac{3}{\sqrt{N}} \right)^{\frac{\alpha}{\alpha+1}} =$$

(setting $h := \left(\frac{1}{\sqrt{N}} \right)^{\frac{\alpha}{\alpha+1}}$)

$$\left(\frac{3^{\frac{\alpha}{\alpha+1}}}{(n+1)! \alpha^{n+1}} \right) \omega_1 \left(D_\alpha^n f, \frac{1}{N^{\frac{\alpha}{2(\alpha+1)}}} \right), \tag{11.55}$$

proving the claim. ∎

We make

Remark 11.21 Next we study the truncated Max-product Baskakov operators (see [7], p. 11)

$$U_N^{(M)} (f) (x) = \frac{\bigvee_{k=0}^N b_{N,k} (x) f \left(\frac{k}{N} \right)}{\bigvee_{k=0}^N b_{N,k} (x)}, x \in [0, 1], f \in C_+ ([0, 1]), N \in \mathbb{N},$$

$$\tag{11.56}$$

where

$$b_{N,k} (x) = \binom{N+k-1}{k} \frac{x^k}{(1+x)^{N+k}}. \tag{11.57}$$

From [7], pp. 217–218, we get ($x \in [0, 1]$)

$$\left(U_N^{(M)} (|\cdot - x|) \right) (x) \leq \frac{2\sqrt{3} \left(\sqrt{2} + 2 \right)}{\sqrt{N+1}}, N \geq 2, N \in \mathbb{N}. \tag{11.58}$$

Let $\lambda \geq 1$, clearly then it holds

$$\left(U_N^{(M)} (|\cdot - x|^\lambda) \right) (x) \leq \frac{2\sqrt{3} \left(\sqrt{2} + 2 \right)}{\sqrt{N+1}}, \forall N \geq 2, N \in \mathbb{N}. \tag{11.59}$$

Also it holds $U_N^{(M)} (1) = 1$, and $U_N^{(M)}$ are positive sublinear operators from $C_+ ([0, 1])$ into itself. Furthermore it holds

$$U_N^{(M)} (|\cdot - x|^\lambda) (x) > 0, \forall x \in (0, 1], \forall \lambda \geq 1, \forall N \in \mathbb{N}. \tag{11.60}$$

We give

Theorem 11.22 *Let* $\alpha \in (0, 1]$ *and* $n \in \mathbb{N}$. *Suppose* $f \in C_+ ([0, 1])$ *is* n *times conformable* α-*fractional differentiable on* $[0, 1]$, $x \in (0, 1)$, *and* $D_\alpha^n f$ *is continuous on* $[0, 1]$. *Let* $N^* \in \mathbb{N}-\{1\}$ *such that* $\frac{1}{(N^*+1)^{\frac{\alpha}{2(\alpha+1)}}} \leq \min (x, 1 - x)$ *and assume* $\left| D_\alpha^n f \right|$ *is convex over* $[0, 1]$. *Furthermore assume that* $D_\alpha^k f (x) = 0$, $k = 1, ..., n$. *Then*

$$\left| U_N^{(M)} (f) (x) - f (x) \right| \leq \left(\frac{\left(2\sqrt{3} \left(\sqrt{2} + 2 \right) \right)^{\frac{\alpha}{\alpha+1}}}{(n + 1)! \alpha^{n+1}} \right) \omega_1 \left(D_\alpha^n f, \frac{1}{(N + 1)^{\frac{\alpha}{2(\alpha+1)}}} \right),$$

(11.61)

$\forall N \in \mathbb{N} : N \geq N^* \geq 2$.
   It holds $\lim\limits_{N \to +\infty} U_N^{(M)} (f) (x) = f (x)$.

Proof By (11.60) we have that $U_N^{(M)} \left(| \cdot - x |^{(n+1)(\alpha+1)} \right) (x) > 0$, $\forall N \in \mathbb{N}$.
   By (11.40) and (11.59) we get

$$\left| U_N^{(M)} (f) (x) - f (x) \right| \leq \left(\frac{\omega_1 \left(D_\alpha^n f, h \right)}{(n + 1)! \alpha^{n+1} h} \right) \left(\frac{2\sqrt{3} \left(\sqrt{2} + 2 \right)}{\sqrt{N + 1}} \right)^{\frac{\alpha}{\alpha+1}} = \quad (11.62)$$

(setting $h := \left(\frac{1}{\sqrt{N+1}} \right)^{\frac{\alpha}{\alpha+1}}$)

$$\left(\frac{\left(2\sqrt{3} \left(\sqrt{2} + 2 \right) \right)^{\frac{\alpha}{\alpha+1}}}{(n + 1)! \alpha^{n+1}} \right) \omega_1 \left(D_\alpha^n f, \frac{1}{(N + 1)^{\frac{\alpha}{2(\alpha+1)}}} \right),$$

proving the claim. ■

   We continue with

Remark 11.23 Here we study the Max-product Meyer-Köning and Zeller operators (see [7], p. 11) defined by

$$Z_N^{(M)} (f) (x) = \frac{\bigvee_{k=0}^{\infty} s_{N,k} (x) f \left(\frac{k}{N+k} \right)}{\bigvee_{k=0}^{\infty} s_{N,k} (x)}, \forall N \in \mathbb{N}, f \in C_+ ([0, 1]), \quad (11.63)$$

$$s_{N,k} (x) = \binom{N + k}{k} x^k, x \in [0, 1].$$
   By [7], p. 253, we get that

$$Z_N^{(M)} (| \cdot - x |) (x) \leq \frac{8 \left(1 + \sqrt{5} \right)}{3} \frac{\sqrt{x} (1 - x)}{\sqrt{N}}, \forall x \in [0, 1], \forall N \geq 4, N \in \mathbb{N}.$$

(11.64)

As before we get that (for $\lambda \geq 1$)

$$Z_N^{(M)} \left(|\cdot - x|^\lambda \right)(x) \leq \frac{8\left(1+\sqrt{5}\right)}{3} \frac{\sqrt{x}\,(1-x)}{\sqrt{N}}, \tag{11.65}$$

$\forall\, x \in [0, 1]$, $N \geq 4$, $N \in \mathbb{N}$.

Also it holds $Z_N^{(M)}(1) = 1$, and $Z_N^{(M)}$ are positive sublinear operators from $C_+([0, 1])$ into itself. Also it holds

$$Z_N^{(M)} \left(|\cdot - x|^\lambda \right)(x) > 0,\ \forall x \in (0, 1),\ \forall \lambda \geq 1,\ \forall N \in \mathbb{N}. \tag{11.66}$$

We give

Theorem 11.24 *Let $\alpha \in (0, 1]$ and $n \in \mathbb{N}$. Suppose $f \in C_+([0, 1])$ is n times conformable α-fractional differentiable on $[0, 1]$, $x \in (0, 1)$, and $D_\alpha^n f$ is continuous on $[0, 1]$. Let $N^* \in \mathbb{N}$, $N^* \geq 4$, such that $\frac{1}{(N^*)^{\frac{\alpha}{2(\alpha+1)}}} \leq \min(x, 1-x)$ and assume $\left|D_\alpha^n f\right|$ is convex over $[0, 1]$. Furthermore assume that $D_\alpha^k f(x) = 0$, $k = 1, ..., n$. Then*

$$\left| Z_N^{(M)}(f)(x) - f(x) \right| \leq \left(\frac{8\left(1+\sqrt{5}\right)}{3}\sqrt{x}\,(1-x) \right)^{\frac{\alpha}{\alpha+1}} \left(\frac{\omega_1\left(D_\alpha^n f, \frac{1}{N^{\frac{\alpha}{2(\alpha+1)}}}\right)}{(n+1)!\,\alpha^{n+1}} \right), \tag{11.67}$$

$\forall\, N \geq N^* \geq 4$, $N \in \mathbb{N}$.

It holds $\displaystyle \lim_{N \to +\infty} Z_N^{(M)}(f)(x) = f(x)$.

Proof By (11.66) we get that $Z_N^{(M)} \left(|\cdot - x|^{(n+1)(\alpha+1)} \right)(x) > 0$, $\forall\, N \in \mathbb{N}$.

By (11.40) and (11.65) we obtain

$$\left| Z_N^{(M)}(f)(x) - f(x) \right| \leq \tag{11.68}$$

$$\left(\frac{\omega_1\left(D_\alpha^n f, h\right)}{(n+1)!\,\alpha^{n+1}h} \right) \left(\frac{8\left(1+\sqrt{5}\right)}{3}\sqrt{x}\,(1-x) \right)^{\frac{\alpha}{\alpha+1}} \left(\frac{1}{\sqrt{N}} \right)^{\frac{\alpha}{\alpha+1}} =$$

(setting $h := \left(\frac{1}{\sqrt{N}} \right)^{\frac{\alpha}{\alpha+1}}$)

$$\left(\frac{8\left(1+\sqrt{5}\right)}{3}\sqrt{x}\,(1-x) \right)^{\frac{\alpha}{\alpha+1}} \left(\frac{\omega_1\left(D_\alpha^n f, \frac{1}{N^{\frac{\alpha}{2(\alpha+1)}}}\right)}{(n+1)!\,\alpha^{n+1}} \right),$$

proving the claim. ∎

We make

Remark 11.25 Here we deal with the Max-product truncated sampling operators (see [7], p. 13) defined by

$$W_N^{(M)}(f)(x) = \frac{\bigvee_{k=0}^{N} \frac{\sin(Nx-k\pi)}{Nx-k\pi} f\left(\frac{k\pi}{N}\right)}{\bigvee_{k=0}^{N} \frac{\sin(Nx-k\pi)}{Nx-k\pi}}, \tag{11.69}$$

and

$$K_N^{(M)}(f)(x) = \frac{\bigvee_{k=0}^{N} \frac{\sin^2(Nx-k\pi)}{(Nx-k\pi)^2} f\left(\frac{k\pi}{N}\right)}{\bigvee_{k=0}^{N} \frac{\sin^2(Nx-k\pi)}{(Nx-k\pi)^2}}, \tag{11.70}$$

$\forall\, x \in [0, \pi]$, $f : [0, \pi] \to \mathbb{R}_+$ a continuous function.

Following [7], p. 343, and making the convention $\frac{\sin(0)}{0} = 1$ and denoting $s_{N,k}(x) = \frac{\sin(Nx-k\pi)}{Nx-k\pi}$, we get that $s_{N,k}\left(\frac{k\pi}{N}\right) = 1$, and $s_{N,k}\left(\frac{j\pi}{N}\right) = 0$, if $k \neq j$, furthermore $W_N^{(M)}(f)\left(\frac{j\pi}{N}\right) = f\left(\frac{j\pi}{N}\right)$, for all $j \in \{0, ..., N\}$.

Clearly $W_N^{(M)}(f)$ is a well-defined function for all $x \in [0, \pi]$, and it is continuous on $[0, \pi]$, also $W_N^{(M)}(1) = 1$.

By [7], p. 344, $W_N^{(M)}$ are positive sublinear operators.

Call $I_N^+(x) = \{k \in \{0, 1, ..., N\}; s_{N,k}(x) > 0\}$, and set $x_{N,k} := \frac{k\pi}{N}$, $k \in \{0, 1, ..., N\}$.

We see that

$$W_N^{(M)}(f)(x) = \frac{\bigvee_{k\in I_N^+(x)} s_{N,k}(x) f\left(x_{N,k}\right)}{\bigvee_{k\in I_N^+(x)} s_{N,k}(x)}. \tag{11.71}$$

By [7], p. 346, we have

$$W_N^{(M)}(|\cdot - x|)(x) \leq \frac{\pi}{2N}, \forall N \in \mathbb{N}, \forall x \in [0, \pi]. \tag{11.72}$$

Notice also $|x_{N,k} - x| \leq \pi, \forall\, x \in [0, \pi]$.

Therefore ($\lambda \geq 1$) it holds

$$W_N^{(M)}(|\cdot - x|^\lambda)(x) \leq \frac{\pi^{\lambda-1}\pi}{2N} = \frac{\pi^\lambda}{2N}, \forall x \in [0, \pi], \forall N \in \mathbb{N}. \tag{11.73}$$

If $x \in \left(\frac{j\pi}{N}, \frac{(j+1)\pi}{N}\right)$, with $j \in \{0, 1, ..., N\}$, we obtain $nx - j\pi \in (0, \pi)$ and thus $s_{N,j}(x) = \frac{\sin(Nx-j\pi)}{Nx-j\pi} > 0$, see [7], pp. 343-344.

Consequently it holds ($\lambda \geq 1$)

$$W_N^{(M)}(|\cdot - x|^\lambda)(x) = \frac{\bigvee_{k\in I_N^+(x)} s_{N,k}(x) |x_{N,k} - x|^\lambda}{\bigvee_{k\in I_N^+(x)} s_{N,k}(x)} > 0, \forall x \in [0, \pi], \tag{11.74}$$

such that $x \neq x_{N,k}$, for any $k \in \{0, 1, ..., N\}$.

We give

Theorem 11.26 *Let $\alpha \in (0, 1]$ and $n \in \mathbb{N}$. Suppose $f \in C_+([0, \pi])$ is n times conformable α-fractional differentiable on $[0, \pi]$, and $x \in (0, \pi)$, such that $x \neq \frac{k\pi}{N}$, $k \in \{0, 1, ..., N\}$, $\forall N \in \mathbb{N}$, and $D_\alpha^n f$ is continuous on $[0, \pi]$. Let $N^* \in \mathbb{N}$ such that $\frac{1}{(N^*)^{\frac{\alpha}{\alpha+1}}} \leq \min(x, \pi - x)$ and assume $\left| D_\alpha^n f \right|$ is convex over $[0, \pi]$. Furthermore assume that $D_\alpha^k f(x) = 0$, $k = 1, ..., n$. Then*

$$\left| W_N^{(M)}(f)(x) - f(x) \right| \leq \left(\frac{\pi^{n\alpha+1}}{2^{\frac{\alpha}{\alpha+1}}(n+1)!\alpha^{n+1}} \right) \omega_1 \left(D_\alpha^n f, \frac{1}{N^{\frac{\alpha}{\alpha+1}}} \right), \quad (11.75)$$

$\forall N \in \mathbb{N} : N \geq N^*$.
    It holds $\lim\limits_{N \to +\infty} W_N^{(M)}(f)(x) = f(x)$.

Proof By (11.74) we have $W_N^{(M)} \left(|\cdot - x|^{(n+1)(\alpha+1)} \right)(x) > 0$, $\forall N \in \mathbb{N}$.
    By (11.40) and (11.73), we obtain

$$\left| W_N^{(M)}(f)(x) - f(x) \right| \leq \left(\frac{\omega_1 \left(D_\alpha^n f, h \right) \pi^{1-\alpha}}{(n+1)!\alpha^{n+1}h} \right) \left(\frac{\pi^{(n+1)(\alpha+1)}}{2N} \right)^{\frac{\alpha}{\alpha+1}} = \quad (11.76)$$

$$\left(\frac{\pi^{n\alpha+1}\omega_1 \left(D_\alpha^n f, h \right)}{(n+1)!\alpha^{n+1}h} \right) \left(\frac{1}{2N} \right)^{\frac{\alpha}{\alpha+1}} =$$

(setting $h := \left(\frac{1}{N} \right)^{\frac{\alpha}{\alpha+1}}$)

$$\left(\frac{\pi^{n\alpha+1}}{2^{\frac{\alpha}{\alpha+1}}(n+1)!\alpha^{n+1}} \right) \omega_1 \left(D_\alpha^n f, \frac{1}{N^{\frac{\alpha}{\alpha+1}}} \right),$$

proving the claim.   ∎

We make

Remark 11.27 Here we continue with the Max-product truncated sampling operators (see [7], p. 13) defined by

$$K_N^{(M)}(f)(x) = \frac{\bigvee_{k=0}^{N} \frac{\sin^2(Nx-k\pi)}{(Nx-k\pi)^2} f\left(\frac{k\pi}{N} \right)}{\bigvee_{k=0}^{N} \frac{\sin^2(Nx-k\pi)}{(Nx-k\pi)^2}}, \quad (11.77)$$

$\forall x \in [0, \pi]$, $f : [0, \pi] \to \mathbb{R}_+$ a continuous function.

Following [7], p. 350, and making the convention $\frac{\sin(0)}{0} = 1$ and denoting $s_{N,k}(x) = \frac{\sin^2(Nx-k\pi)}{(Nx-k\pi)^2}$, we get that $s_{N,k}\left(\frac{k\pi}{N}\right) = 1$, and $s_{N,k}\left(\frac{j\pi}{N}\right) = 0$, if $k \neq j$, furthermore $K_N^{(M)}(f)\left(\frac{j\pi}{N}\right) = f\left(\frac{j\pi}{N}\right)$, for all $j \in \{0, ..., N\}$.

Since $s_{N,j}\left(\frac{j\pi}{N}\right) = 1$ it follows that $\bigvee_{k=0}^{N} s_{N,k}\left(\frac{j\pi}{N}\right) \geq 1 > 0$, for all $j \in \{0, 1, ..., N\}$. Hence $K_N^{(M)}(f)$ is well-defined function for all $x \in [0, \pi]$, and it is continuous on $[0, \pi]$, also $K_N^{(M)}(1) = 1$. By [7], p. 350, $K_N^{(M)}$ are positive sublinear operators.

Denote $x_{N,k} := \frac{k\pi}{N}, k \in \{0, 1, ..., N\}$.

By [7], p. 352, we have

$$K_N^{(M)}(|\cdot - x|)(x) \leq \frac{\pi}{2N}, \forall N \in \mathbb{N}, \forall x \in [0, \pi]. \tag{11.78}$$

Notice also $|x_{N,k} - x| \leq \pi, \forall x \in [0, \pi]$.

Therefore ($\lambda \geq 1$) it holds

$$K_N^{(M)}\left(|\cdot - x|^\lambda\right)(x) \leq \frac{\pi^{\lambda-1}\pi}{2N} = \frac{\pi^\lambda}{2N}, \forall x \in [0, \pi], \forall N \in \mathbb{N}. \tag{11.79}$$

If $x \in \left(\frac{j\pi}{N}, \frac{(j+1)\pi}{N}\right)$, with $j \in \{0, 1, ..., N\}$, we obtain $nx - j\pi \in (0, \pi)$ and thus $s_{N,j}(x) = \frac{\sin^2(Nx-j\pi)}{(Nx-j\pi)^2} > 0$, see [7], pp. 350.

Consequently it holds ($\lambda \geq 1$)

$$K_N^{(M)}\left(|\cdot - x|^\lambda\right)(x) = \frac{\bigvee_{k=0}^{N} s_{N,k}(x)|x_{N,k} - x|^\lambda}{\bigvee_{k=0}^{N} s_{N,k}(x)} > 0, \forall x \in [0, \pi], \tag{11.80}$$

such that $x \neq x_{N,k}$, for any $k \in \{0, 1, ..., N\}$.

We give

Theorem 11.28 *Let $\alpha \in (0, 1]$ and $n \in \mathbb{N}$. Suppose $f \in C_+([0, \pi])$ is n times conformable α-fractional differentiable on $[0, \pi]$, and $x \in (0, \pi)$, such that $x \neq \frac{k\pi}{N}$, $k \in \{0, 1, ..., N\}$, $\forall N \in \mathbb{N}$, and $D_\alpha^n f$ is continuous on $[0, \pi]$. Let $N^* \in \mathbb{N}$ such that $\frac{1}{(N^*)^{\frac{\alpha}{\alpha+1}}} \leq \min(x, \pi - x)$ and assume $|D_\alpha^n f|$ is convex over $[0, \pi]$. Furthermore assume that $D_\alpha^k f(x) = 0, k = 1, ..., n$. Then*

$$\left|K_N^{(M)}(f)(x) - f(x)\right| \leq \left(\frac{\pi^{n\alpha+1}}{2^{\frac{\alpha}{\alpha+1}}(n+1)!\alpha^{n+1}}\right)\omega_1\left(D_\alpha^n f, \frac{1}{N^{\frac{\alpha}{\alpha+1}}}\right), \tag{11.81}$$

$\forall N \in \mathbb{N}: N \geq N^*$.

It holds $\lim_{N \to +\infty} K_N^{(M)}(f)(x) = f(x)$.

Proof By (11.80) we have $K_N^{(M)} \left(|\cdot - x|^{(n+1)(\alpha+1)} \right) (x) > 0, \forall\, N \in \mathbb{N}$.
By (11.40) and (11.79), we obtain

$$\left| K_N^{(M)} (f) (x) - f (x) \right| \le \left(\frac{\omega_1 \left(D_\alpha^n f, h \right) \pi^{1-\alpha}}{(n+1)!\alpha^{n+1}h} \right) \left(\frac{\pi^{(n+1)(\alpha+1)}}{2N} \right)^{\frac{\alpha}{\alpha+1}} = \quad (11.82)$$

$$\left(\frac{\pi^{n\alpha+1} \omega_1 \left(D_\alpha^n f, h \right)}{(n+1)!\alpha^{n+1}h} \right) \left(\frac{1}{2N} \right)^{\frac{\alpha}{\alpha+1}} =$$

(setting $h := \left(\frac{1}{N} \right)^{\frac{\alpha}{\alpha+1}}$)

$$\left(\frac{\pi^{n\alpha+1}}{2^{\frac{\alpha}{\alpha+1}} (n+1)!\alpha^{n+1}} \right) \omega_1 \left(D_\alpha^n f, \frac{1}{N^{\frac{\alpha}{\alpha+1}}} \right),$$

proving the claim. ∎

(II) Here we apply Theorem 11.13 to well known Max-product operators in the case of $(n+1)\alpha \ge 1$, that is when $\frac{1}{n+1} \le \alpha \le 1$, where $n \in \mathbb{N}$.
We give

Theorem 11.29 *Let $\alpha \in (0, 1]$ and $n \in \mathbb{N}$. Suppose $f \in C_+ ([0, 1])$ is n times conformable α-fractional differentiable on $[0, 1]$, $x \in (0, 1)$, and $D_\alpha^n f$ is continuous on $[0, 1]$. Let $N^* \in \mathbb{N}$ such that $\frac{1}{\sqrt{N^*+1}} \le \min (x, 1-x)$ and assume $\left| D_\alpha^n f \right|$ is convex over $[0, 1]$. Furthermore assume that $D_\alpha^k f (x) = 0, k = 1, ..., n$. Then*

$$\left| B_N^{(M)} (f) (x) - f (x) \right| \le \left(\frac{6}{(n+1)!\alpha^{n+1}} \right) \omega_1 \left(D_\alpha^n f, \frac{1}{\sqrt{N+1}} \right), \quad (11.83)$$

$\forall\, N \ge N^*, N \in \mathbb{N}$.
It holds $\lim\limits_{N\to+\infty} B_N^{(M)} (f) (x) = f (x)$.

Proof By (11.39), (11.46), we obtain that

$$\left| B_N^{(M)} (f) (x) - f (x) \right| \le \left(\frac{\omega_1 \left(D_\alpha^n f, h \right)}{(n+1)!\alpha^{n+1}h} \right) B_N^{(M)} \left(|\cdot - x|^{(n+1)\alpha} \right) (x) \le$$

$$\left(\frac{\omega_1 \left(D_\alpha^n f, h \right)}{(n+1)!\alpha^{n+1}h} \right) \frac{6}{\sqrt{N+1}} = \quad (11.84)$$

(setting $h := \frac{1}{\sqrt{N+1}}$)

$$\left(\frac{6}{(n+1)!\alpha^{n+1}} \right) \omega_1 \left(D_\alpha^n f, \frac{1}{\sqrt{N+1}} \right),$$

proving the claim. ■

Theorem 11.30 *Let* $\alpha \in (0, 1]$ *and* $n \in \mathbb{N}$. *Suppose* $f \in C_+ ([0, 1])$ *is* n *times conformable* α *-fractional differentiable on* $[0, 1]$, $x \in (0, 1)$, *and* $D_\alpha^n f$ *is continuous on* $[0, 1]$. *Let* $N^* \in \mathbb{N}$ *such that* $\frac{1}{\sqrt{N^*}} \leq \min (x, 1 - x)$ *and assume* $\left| D_\alpha^n f \right|$ *is convex over* $[0, 1]$. *Furthermore assume that* $D_\alpha^k f (x) = 0$, $k = 1, ..., n$. *Then*

$$\left| T_N^{(M)} (f) (x) - f (x) \right| \leq \left(\frac{3}{(n + 1)! \alpha^{n+1}} \right) \omega_1 \left(D_\alpha^n f, \frac{1}{\sqrt{N}} \right), \qquad (11.85)$$

$\forall \, N \geq N^*, \, N \in \mathbb{N}$.
   It holds $\lim\limits_{N \to +\infty} T_N^{(M)} (f) (x) = f (x)$.

Proof By (11.39), (11.52), we get that

$$\left| T_N^{(M)} (f) (x) - f (x) \right| \leq \left(\frac{\omega_1 \left(D_\alpha^n f, h \right)}{(n + 1)! \alpha^{n+1} h} \right) \frac{3}{\sqrt{N}} =$$

(setting $h := \frac{1}{\sqrt{N}}$)

$$\left(\frac{3}{(n + 1)! \alpha^{n+1}} \right) \omega_1 \left(D_\alpha^n f, \frac{1}{\sqrt{N}} \right), \qquad (11.86)$$

proving the claim. ■

Theorem 11.31 *Let* $\alpha \in (0, 1]$ *and* $n \in \mathbb{N}$. *Suppose* $f \in C_+ ([0, 1])$ *is* n *times conformable* α-*fractional differentiable on* $[0, 1]$, $x \in (0, 1)$, *and* $D_\alpha^n f$ *is continuous on* $[0, 1]$. *Let* $N^* \in \mathbb{N} - \{1\}$ *such that* $\frac{1}{\sqrt{N^*+1}} \leq \min (x, 1 - x)$ *and assume* $\left| D_\alpha^n f \right|$ *is convex over* $[0, 1]$. *Furthermore assume that* $D_\alpha^k f (x) = 0$, $k = 1, ..., n$. *Then*

$$\left| U_N^{(M)} (f) (x) - f (x) \right| \leq \frac{\left(2\sqrt{3} \left(\sqrt{2} + 2 \right) \right)}{(n + 1)! \alpha^{n+1}} \omega_1 \left(D_\alpha^n f, \frac{1}{\sqrt{N + 1}} \right), \qquad (11.87)$$

$\forall \, N \in \mathbb{N} : N \geq N^* \geq 2$.
   It holds $\lim\limits_{N \to +\infty} U_N^{(M)} (f) (x) = f (x)$.

Proof By (11.39), (11.59), we get

$$\left| U_N^{(M)} (f) (x) - f (x) \right| \leq \left(\frac{\omega_1 \left(D_\alpha^n f, h \right)}{(n + 1)! \alpha^{n+1} h} \right) \left(\frac{2\sqrt{3} \left(\sqrt{2} + 2 \right)}{\sqrt{N + 1}} \right) = \qquad (11.88)$$

(setting $h := \frac{1}{\sqrt{N+1}}$)

$$\frac{\left(2\sqrt{3}\left(\sqrt{2}+2\right)\right)}{(n+1)!\alpha^{n+1}}\omega_1\left(D_\alpha^n f,\frac{1}{\sqrt{N+1}}\right),$$

proving the claim. ∎

Theorem 11.32 *Let $\alpha \in (0, 1]$ and $n \in \mathbb{N}$. Suppose $f \in C_+([0, 1])$ is n times conformable α-fractional differentiable on $[0, 1]$, $x \in (0, 1)$, and $D_\alpha^n f$ is continuous on $[0, 1]$. Let $N^* \in \mathbb{N}$, $N^* \geq 4$, such that $\frac{1}{\sqrt{N^*}} \leq \min(x, 1 - x)$ and assume $\left|D_\alpha^n f\right|$ is convex over $[0, 1]$. Furthermore assume that $D_\alpha^k f(x) = 0$, $k = 1, ..., n$. Then*

$$\left|Z_N^{(M)}(f)(x) - f(x)\right| \leq \left(\frac{8\left(1+\sqrt{5}\right)}{3}\sqrt{x}(1-x)\right)\left(\frac{\omega_1\left(D_\alpha^n f, \frac{1}{\sqrt{N}}\right)}{(n+1)!\alpha^{n+1}}\right),$$
(11.89)

$\forall\, N \geq N^* \geq 4$, $N \in \mathbb{N}$.
  It holds $\lim\limits_{N\to+\infty} Z_N^{(M)}(f)(x) = f(x)$.

Proof By (11.39) and (11.65), we obtain

$$\left|Z_N^{(M)}(f)(x) - f(x)\right| \leq \left(\frac{\omega_1\left(D_\alpha^n f, h\right)}{(n+1)!\alpha^{n+1}h}\right)\left(\frac{8\left(1+\sqrt{5}\right)}{3}\sqrt{x}(1-x)\right)\frac{1}{\sqrt{N}} =$$
(11.90)

(setting $h := \frac{1}{\sqrt{N}}$)

$$\frac{8\left(1+\sqrt{5}\right)}{3}\sqrt{x}(1-x)\left(\frac{\omega_1\left(D_\alpha^n f, \frac{1}{\sqrt{N}}\right)}{(n+1)!\alpha^{n+1}}\right),$$

proving the claim. ∎

Theorem 11.33 *Let $\alpha \in (0, 1]$ and $n \in \mathbb{N}$. Suppose $f \in C_+([0, \pi])$ is n times conformable α-fractional differentiable on $[0, \pi]$, and $x \in (0, \pi)$, such that $x \neq \frac{k\pi}{N}$, $k \in \{0, 1, ..., N\}$, $\forall\, N \in \mathbb{N}$, and $D_\alpha^n f$ is continuous on $[0, \pi]$. Let $N^* \in \mathbb{N}$ such that $\frac{1}{N^*} \leq \min(x, \pi - x)$ and assume $\left|D_\alpha^n f\right|$ is convex over $[0, \pi]$. Furthermore assume that $D_\alpha^k f(x) = 0$, $k = 1, ..., n$. Then*

$$\left|W_N^{(M)}(f)(x) - f(x)\right| \leq \left(\frac{\pi^{n\alpha+1}}{2(n+1)!\alpha^{n+1}}\right)\omega_1\left(D_\alpha^n f, \frac{1}{N}\right),$$
(11.91)

$\forall\, N \in \mathbb{N}: N \geq N^*$.
  It holds $\lim\limits_{N\to+\infty} W_N^{(M)}(f)(x) = f(x)$.

Proof By (11.39) and (11.73), we obtain

$$\left| W_N^{(M)} \left(f \right) (x) - f \left(x \right) \right| \le \left(\frac{\omega_1 \left(D_\alpha^n f, h \right) \pi^{1-\alpha}}{(n+1)! \alpha^{n+1} h} \right) \frac{\pi^{(n+1)\alpha}}{2N} = \tag{11.92}$$

$$\left(\frac{\pi^{n\alpha+1} \omega_1 \left(D_\alpha^n f, h \right)}{2(n+1)! \alpha^{n+1} h} \right) \frac{1}{N} =$$

(setting $h := \frac{1}{N}$)

$$\left(\frac{\pi^{n\alpha+1}}{2(n+1)! \alpha^{n+1}} \right) \omega_1 \left(D_\alpha^n f, \frac{1}{N} \right),$$

proving the claim. ∎

Theorem 11.34 *Let* $\alpha \in (0, 1]$ *and* $n \in \mathbb{N}$. *Suppose* $f \in C_+ ([0, \pi])$ *is n times conformable* α *-fractional differentiable on* $[0, \pi]$, *and* $x \in (0, \pi)$, *such that* $x \ne \frac{k\pi}{N}$, $k \in \{0, 1, ..., N\}$, $\forall N \in \mathbb{N}$, *and* $D_\alpha^n f$ *is continuous on* $[0, \pi]$. *Let* $N^* \in \mathbb{N}$ *such that* $\frac{1}{N^*} \le \min (x, \pi - x)$ *and assume* $\left| D_\alpha^n f \right|$ *is convex over* $[0, \pi]$. *Furthermore assume that* $D_\alpha^k f (x) = 0$, $k = 1, ..., n$. *Then*

$$\left| K_N^{(M)} \left(f \right) (x) - f \left(x \right) \right| \le \left(\frac{\pi^{n\alpha+1}}{2(n+1)! \alpha^{n+1}} \right) \omega_1 \left(D_\alpha^n f, \frac{1}{N} \right), \tag{11.93}$$

$\forall N \in \mathbb{N} : N \ge N^*$.
It holds $\lim\limits_{N \to +\infty} K_N^{(M)} \left(f \right) (x) = f \left(x \right)$.

Proof By (11.39) and (11.79), we obtain

$$\left| K_N^{(M)} \left(f \right) (x) - f \left(x \right) \right| \le \left(\frac{\omega_1 \left(D_\alpha^n f, h \right) \pi^{1-\alpha}}{(n+1)! \alpha^{n+1} h} \right) \frac{\pi^{(n+1)\alpha}}{2N} = \tag{11.94}$$

$$\left(\frac{\pi^{n\alpha+1} \omega_1 \left(D_\alpha^n f, h \right)}{2(n+1)! \alpha^{n+1} h} \right) \frac{1}{N} =$$

(setting $h := \frac{1}{N}$)

$$\left(\frac{\pi^{n\alpha+1}}{2(n+1)! \alpha^{n+1}} \right) \omega_1 \left(D_\alpha^n f, \frac{1}{N} \right),$$

proving the claim. ∎

References

1. M. Abu Hammad, R. Khalil, Abel's formula and Wronskian for conformable fractional differential equations. Int. J. Differ. Equ. Appl. **13**(3), 177–183 (2014)
2. G. Anastassiou, *Moments in Probability and Approximation Theory*. Pitman Research Notes in Mathematics Series (Longman Group UK, New York, 1993)
3. G. Anastassiou, *Approximation by Sublinear Operators* (2017, submitted)
4. G. Anastassiou, *Conformable Fractional Approximation by Max-Product Operators* (2017, submitted)
5. G. Anastassiou, *Conformable Fractional Approximations by Max-Product Operators using Convexity* (2017, submitted)
6. D. Anderson, Taylor's formula and integral inequalities for conformable fractional derivatives, in *Contributions in Mathematics and Engineering, in Honor of Constantin Carathéodory* (Springer, Berlin, 2016), pp. 25–43
7. B. Bede, L. Coroianu, S. Gal, *Approximation by Max-Product type Operators* (Springer, New York, 2016)
8. R. Khalil, M. Al Horani, A. Yousef, M. Sababheh, A new definition of fractional derivative. J. Comput. Appl. Math. **264**, 65–70 (2014)

Chapter 12
High Order Approximation by Multivariate Sublinear and Max-Product Operators Under Convexity

Here we search quantitatively under convexity the approximation of multivariate function by general multivariate positive sublinear operators with applications to multivariate Max-product operators. These are of Bernstein type, of Favard-Szász-Mirakjan type, of Baskakov type, of sampling type, of Lagrange interpolation type and of Hermite-Fejér interpolation type. Our results are both: under the presence of smoothness and without any smoothness assumption on the function to be approximated which fulfills a convexity assumption. It follows [4].

12.1 Background

In this chapter we study under convexity quantitatively the approximation properties of multivariate Max-product operators to the unit. These are special cases of multivariate positive sublinear operators. We give also general results regarding the convergence to the unit of multivariate positive sublinear operators under convexity. Special emphasis is given to our study about approximation also under the presence of smoothness. This chapter is inspired from [5].

Let Q be a compact and convex subset of $\mathbb{R}^k$, $k \in \mathbb{N} - \{1\}$ and let $x_0 := (x_{01}, ..., x_{0k}) \in Q^o$ be fixed. Let $f \in C^n(Q)$ and suppose that each nth partial derivative $f_\alpha = \frac{\partial^\alpha f}{\partial x^\alpha}$, where $\alpha := (\alpha_1, ..., \alpha_k), \alpha_i \in \mathbb{Z}^+, i = 1, ..., k$, and $|\alpha| := \sum_{i=1}^k \alpha_i = n$, has relative to Q and the l_1-norm $\|\cdot\|$, a modulus of continuity $\omega_1(f_\alpha, h) \leq w$, where h and w are fixed positive numbers. Here

$$\omega_1(f_\alpha, h) := \sup_{\substack{x, y \in Q \\ \|x-y\|_{l_1} \leq h}} |f_\alpha(x) - f_\alpha(y)|. \tag{12.1}$$

© Springer International Publishing AG, part of Springer Nature 2018
G. A. Anastassiou, *Nonlinearity: Ordinary and Fractional Approximations by Sublinear and Max-Product Operators*, Studies in Systems, Decision and Control 147, https://doi.org/10.1007/978-3-319-89509-3_12

We assume that the ball $B(x_0, h) \subset Q$. We also assume that $f_\alpha(x_0) = 0$, all α : $|\alpha| = 1, ..., n$, and $|f_\alpha(x)|$ is convex in x, all $\alpha : |\alpha| = n$.

The jth derivative of $g_z(t) = f(x_0 + t(z - x_0))$, $(z = (z_1, ..., z_k) \in Q)$ is given by

$$g_z^{(j)}(t) = \left[\left(\sum_{i=1}^{k} (z_i - x_{0i}) \frac{\partial}{\partial x_i} \right)^j f \right] (x_{01} + t(z_1 - x_{01}), ..., x_{0k} + t(z_k - x_{0k})).$$

(12.2)

Consequently it holds

$$f(z_1, ..., z_k) = g_z(1) = \sum_{j=0}^{n} \frac{g_z^{(j)}(0)}{j!} + R_n(z, 0),$$

(12.3)

where

$$R_n(z, 0) := \int_0^1 \left(\int_0^{t_1} \cdots \left(\int_0^{t_{n-1}} \left(g_z^{(n)}(t_n) - g_z^{(n)}(0) \right) dt_n \right) \cdots \right) dt_1.$$

(12.4)

We apply Lemma 8.1.1, [1], p. 243, to $(f_\alpha(x_0 + t(z - x_0)) - f_\alpha(x_0))$ as a function of z, when $\omega_1(f_\alpha, h) \leq w$.

$$|f_\alpha(x_0 + t(z - x_0)) - f_\alpha(x_0)| \leq w \frac{t \|z - x_0\|}{h},$$

(12.5)

all $t \geq 0$.

Let Q be a compact and convex subset of $\mathbb{R}^k$, $k \in \mathbb{N} - \{1\}$, $x_0 \in Q$ fixed, $f \in C^n(Q)$. Then for $j = 1, ..., n$, we have

$$g_z^{(j)}(0) = \sum_{\substack{\alpha:=(\alpha_1, ..., \alpha_k), \, \alpha_i \in \mathbb{Z}^+, \\ i=1, ..., k, \, |\alpha|:=\sum_{i=1}^k \alpha_i = j}} \left(\frac{j!}{\prod_{i=1}^k \alpha_i!} \right) \left(\prod_{i=1}^k (z_i - x_{0i})^{\alpha_i} \right) f_\alpha(x_0).$$

(12.6)

If $f_\alpha(x_0) = 0$, for all $\alpha : |\alpha| = 1, ..., n$, then $g_z^{(j)}(0) = 0$, $j = 1, ..., n$, and by (12.3) we get

$$f(z) - f(x_0) = R_n(z, 0).$$

(12.7)

It follows from (12.2) that

$$|R_n(z, 0)| \leq$$

$$\int_0^1 \left[\int_0^{t_1} \cdots \left[\int_0^{t_{n-1}} \left(\sum_{|\alpha|=n} \frac{n! \prod_{i=1}^k |z_i - x_{0i}|^{\alpha_i}}{\alpha_1! ... \alpha_k!} \frac{w}{h} \|z - x_0\| t_n \right) dt_n \right] \cdots \right] dt_1$$

(12.8)

$$= \frac{w}{h} \frac{\|z - x_0\|^{n+1}}{(n+1)!}.$$

Therefore it holds

$$|R_n(z, 0)| \le \frac{w}{h} \frac{\|z - x_0\|^{n+1}}{(n+1)!}, \quad \text{for all } z \in Q. \tag{12.9}$$

Note that $g_z(0) = f(z_0)$.

That is

$$|f(z) - f(x_0)| \le \frac{w}{h} \frac{\|z - x_0\|^{n+1}}{(n+1)!}, \quad \forall z \in Q. \tag{12.10}$$

We have proved the following fundamental result:

Theorem 12.1 *Let Q with the l_1-norm $\|\cdot\|$, be a compact and convex subset of $\mathbb{R}^k$, $k \ge 1$, let $x_0 = (x_{01}, ..., x_{0k}) \in Q^o$ be fixed. Let $f \in C^n(Q)$, $n \in \mathbb{N}$ and suppose that each nth partial derivative $f_\alpha = \frac{\partial^\alpha f}{\partial x^\alpha}$, where $\alpha = (\alpha_1, ..., \alpha_k)$, $\alpha_i \ge 0$, $i = 1, ..., k$, and $|\alpha| = \sum_{i=1}^{k} \alpha_i = n$ has, relative to Q and the l_1-norm a modulus of continuity $\omega_1(f_\alpha, h)$, and that each $|f_\alpha(x)|$ is a convex function of $x \in Q$, all $\alpha : |\alpha| = n$. Assume further that $f_\alpha(x_0) = 0$, for all $\alpha : |\alpha| = 1, ..., n$; and $h > 0$ such that the ball in $\mathbb{R}^k : B(x_0, h)$ is contained in Q. Then*

$$\|f(z) - f(x_0)\| \le \frac{\left(\max_{\alpha : |\alpha| = n} \omega_1(f_\alpha, h) \right)}{h} \frac{\|z - x_0\|^{n+1}}{(n+1)!}, \quad \forall z \in Q. \tag{12.11}$$

In conclusion we have

Theorem 12.2 *Let Q with the l_1-norm $\|\cdot\|$, be a compact and convex subset of $\mathbb{R}^k$, $k \ge 1$, let $x = (x_1, ..., x_k) \in Q^o$ be fixed. Let $f \in C^n(Q)$, $n \in \mathbb{N}$ and suppose that each nth partial derivative $f_\alpha = \frac{\partial^\alpha f}{\partial x^\alpha}$, where $\alpha = (\alpha_1, ..., \alpha_k)$, $\alpha_i \ge 0$, $i = 1, ..., k$, and $|\alpha| = \sum_{i=1}^{k} \alpha_i = n$, has relative to Q and the l_1-norm a modulus of continuity $\omega_1(f_\alpha, h)$, and that each $|f_\alpha(t)|$ is a convex function of $t \in Q$, all $\alpha : |\alpha| = n$. Assume further that $f_\alpha(x) = 0$, for all $\alpha : |\alpha| = 1, ..., n$; and $h > 0$ such that the ball in $\mathbb{R}^k : B(x, h) \subset Q$. Then*

$$\|f(t) - f(x)\| \le \frac{\left(\max_{\alpha : |\alpha| = n} \omega_1(f_\alpha, h) \right)}{h} \frac{\|t - x\|^{n+1}}{(n+1)!} \le \tag{12.12}$$

$$\frac{\left(\max_{\alpha : |\alpha| = n} \omega_1(f_\alpha, h) \right)}{h(n+1)!} k^n \left(\sum_{i=1}^{k} |t_i - x_i|^{n+1} \right), \tag{12.13}$$

$\forall t \in Q$, where $t = (t_1, ..., t_k)$.

Proof By Theorem 12.1 and a convexity argument.   ∎

We need

Definition 12.3 Let Q be a compact and convex subset of $\mathbb{R}^k$, $k \in \mathbb{N} - \{1\}$. Here we denote

$$C_+ (Q) = \{f : Q \to \mathbb{R}_+ \text{ and continuous}\}.$$

Let $L_N : C_+ (Q) \to C_+ (Q)$, $N \in \mathbb{N}$, be a sequence of operators satisfying the following properties:
   (i) (positive homogeneous)

$$L_N (\alpha f) = \alpha L_N (f), \ \forall \, \alpha \geq 0, \, f \in C_+ (Q); \qquad (12.14)$$

   (ii) (monotonicity)
   if $f, g \in C_+ (Q)$ satisfy $f \leq g$, then

$$L_N (f) \leq L_N (g), \ \forall \, N \in \mathbb{N}, \qquad (12.15)$$

and
   (iii) (subadditivity)

$$L_N (f + g) \leq L_N (f) + L_N (g), \ \forall \, f, g \in C_+ (Q). \qquad (12.16)$$

We call L_N positive sublinear operators.

Remark 12.4 (to Definition 12.3) Let $f, g \in C_+ (Q)$. We see that $f = f - g + g \leq |f - g| + g$. Then $L_N (f) \leq L_N (|f - g|) + L_N (g)$, and $L_N (f) - L_N (g) \leq L_N (|f - g|)$.
   Similarly $g = g - f + f \leq |g - f| + f$, hence $L_N (g) \leq L_N (|f - g|) + L_N (f)$, and $L_N (g) - L_N (f) \leq L_N (|f - g|)$.
   Consequently it holds

$$|L_N (f) (x) - L_N (g) (x)| \leq L_N (|f - g|) (x), \ \forall \, x \in Q. \qquad (12.17)$$

In this chapter we treat $L_N : L_N (1) = 1$.
   We observe that

$$|L_N (f) (x) - f (x)| = |L_N (f) (x) - L_N (f (x)) (x)| \overset{(12.17)}{\leq}$$

$$L_N (|f (\cdot) - f (x)|) (x), \ \forall \, x \in Q. \qquad (12.18)$$

We give

Theorem 12.5 *All as in Theorem 12.2, $f \in C^n (Q, \mathbb{R}_+)$. Let $\{L_N\}_{N \in \mathbb{N}}$ be positive sublinear operators mapping $C_+ (Q)$ into itself, such that $L_N (1) = 1$. Then*

$$|L_N(f)(x) - f(x)| \le \frac{\left(\max\limits_{\alpha:|\alpha|=n} \omega_1(f_\alpha, h)\right) k^n}{h(n+1)!} \left(\sum_{i=1}^{k} L_N\left(|t_i - x_i|^{n+1}\right)(x)\right),$$
$$\tag{12.19}$$

$\forall\, N \in \mathbb{N}$.

Proof By Theorem 12.2, see Definition 12.3, and by (12.18). ∎

We need

The Maximum Multiplicative Principle 12.6 *Here* $\vee$ *stands for maximum. Let* $\alpha_i > 0$, $i = 1, ..., n$; $\beta_j > 0$, $j = 1, ..., m$. *Then*

$$\vee_{i=1}^{n} \vee_{j=1}^{m} \alpha_i \beta_j = \left(\vee_{i=1}^{n} \alpha_i\right)\left(\vee_{j=1}^{m} \beta_j\right). \tag{12.20}$$

Proof Obvious. ∎

We make

Remark 12.7 In [5], p. 10, the authors introduced the basic Max-product Bernstein operators

$$B_N^{(M)}(f)(x) = \frac{\bigvee_{k=0}^{N} p_{N,k}(x) f\left(\frac{k}{N}\right)}{\bigvee_{k=0}^{N} p_{N,k}(x)}, \quad N \in \mathbb{N}, \tag{12.21}$$

where $p_{N,k}(x) = \binom{N}{k} x^k (1-x)^{N-k}$, $x \in [0, 1]$, and $f : [0, 1] \to \mathbb{R}_+$ is continuous.

In [5], p. 31, they proved that

$$B_N^{(M)}(|\cdot - x|)(x) \le \frac{6}{\sqrt{N+1}}, \quad \forall\, x \in [0, 1], \forall\, N \in \mathbb{N}. \tag{12.22}$$

And in [2] was proved that

$$B_N^{(M)}(|\cdot - x|^m)(x) \le \frac{6}{\sqrt{N+1}}, \quad \forall\, x \in [0, 1], \forall\, m, N \in \mathbb{N}. \tag{12.23}$$

Under our convexity conditions the derived convergence inequalities are very elegant and compact.

12.2 Main Results

From now on $Q = [0, 1]^k$, $k \in \mathbb{N} - \{1\}$, except otherwise specified.
We mention

Definition 12.8 Let $f \in C_+\left([0, 1]^k\right)$, and $\overrightarrow{N} = (N_1, ..., N_k) \in \mathbb{N}^k$. We define the multivariate Max-product Bernstein operators as follows:

$$B_{\overrightarrow{N}}^{(M)}(f)(x) :=$$

$$\frac{\bigvee_{i_1=0}^{N_1} \bigvee_{i_2=0}^{N_2} \cdots \bigvee_{i_k=0}^{N_k} p_{N_1,i_1}(x_1) \, p_{N_2,i_2}(x_2) \cdots p_{N_k,i_k}(x_k) \, f\left(\frac{i_1}{N_1}, ..., \frac{i_k}{N_k}\right)}{\bigvee_{i_1=0}^{N_1} \bigvee_{i_2=0}^{N_2} \cdots \bigvee_{i_k=0}^{N_k} p_{N_1,i_1}(x_1) \, p_{N_2,i_2}(x_2) \cdots p_{N_k,i_k}(x_k)}, \quad (12.24)$$

$\forall \, x = (x_1, ..., x_k) \in [0, 1]^k$. Call $N_{\min} := \min\{N_1, ..., N_k\}$.

The operators $B_{\overrightarrow{N}}^{(M)}(f)(x)$ are positive sublinear and they map $C_+\left([0, 1]^k\right)$ into itself, and $B_{\overrightarrow{N}}^{(M)}(1) = 1$.

See also [5], p. 123 the bivariate case. We also have

$$B_{\overrightarrow{N}}^{(M)}(f)(x) :=$$

$$\frac{\bigvee_{i_1=0}^{N_1} \bigvee_{i_2=0}^{N_2} \cdots \bigvee_{i_k=0}^{N_k} p_{N_1,i_1}(x_1) \, p_{N_2,i_2}(x_2) \cdots p_{N_k,i_k}(x_k) \, f\left(\frac{i_1}{N_1}, ..., \frac{i_k}{N_k}\right)}{\prod_{\lambda=1}^{k} \left(\bigvee_{i_\lambda=0}^{N_\lambda} p_{N_\lambda,i_\lambda}(x_\lambda)\right)}, \quad (12.25)$$

$\forall \, x \in [0, 1]^k$, by the maximum multiplicative principle, see (12.20).

We make

Remark 12.9 The coordinate Max-product Bernstein operators are defined as follows ($\lambda = 1, ..., k$):

$$B_{N_\lambda}^{(M)}(g)(x_\lambda) := \frac{\bigvee_{i_\lambda=0}^{N_\lambda} p_{N_\lambda,i_\lambda}(x_\lambda) \, g\left(\frac{i_\lambda}{N_\lambda}\right)}{\bigvee_{i_\lambda=0}^{N_\lambda} p_{N_\lambda,i_\lambda}(x_\lambda)}, \quad (12.26)$$

$\forall \, N_\lambda \in \mathbb{N}$, and $\forall \, x_\lambda \in [0, 1]$, $\forall \, g \in C_+([0, 1]) := \{g : [0, 1] \to \mathbb{R}_+ \text{ continuous}\}$.

Here we have

$$p_{N_\lambda,i_\lambda}(x_\lambda) = \binom{N_\lambda}{i_\lambda} x_\lambda^{i_\lambda} (1 - x_\lambda)^{N_\lambda - i_\lambda}, \quad \text{for all } \lambda = 1, ..., k; \ x_\lambda \in [0, 1]. \quad (12.27)$$

In case of $f \in C_+\left([0, 1]^k\right)$ is such that $f(x) := g(x_\lambda)$, $\forall \, x \in [0, 1]^k$, where $x = (x_1, ..., x_\lambda, ..., x_k)$ and $g \in C_+([0, 1])$, we get that

$$B_{\overrightarrow{N}}^{(M)}(f)(x) = B_{N_\lambda}^{(M)}(g)(x_\lambda), \quad (12.28)$$

by the maximum multiplicative principle (12.20) and simplification of (12.25).

Clearly it holds that

$$B_{\overrightarrow{N}}^{(M)}(f)(x) = f(x), \ \forall \, x = (x_1, ..., x_k) \in [0, 1]^k : x_\lambda \in \{0, 1\}, \ \lambda = 1, ..., k.$$

$$(12.29)$$

We present

Theorem 12.10 *Let* $[0, 1]^k$, $k \in \mathbb{N} - \{1\}$ *with the* l_1*-norm and let* $x = (x_1, ..., x_k) \in (0, 1)^k$ *be fixed. Let* $f \in C^n([0, 1]^k, \mathbb{R}_+)$, $n \in \mathbb{N}$, *and suppose that each* n*th partial derivative* $f_\alpha = \frac{\partial^\alpha f}{\partial x^\alpha}$, *where* $\alpha = (\alpha_1, ..., \alpha_k)$, $\alpha_i \geq 0$, $i = 1, ..., k$, *and* $|\alpha| = \sum_{i=1}^k \alpha_i = n$ *has, relative to* $[0, 1]^k$ *and the* l_1*-norm a modulus of continuity* $\omega_1(f_\alpha, \cdot)$, *and that each* $|f_\alpha(t)|$ *is a convex function of* $t \in [0, 1]^k$, *all* $\alpha : |\alpha| = n$. *Assume further that* $f_\alpha(x) = 0$, *for all* $\alpha : |\alpha| = 1, ..., n$; *and the ball in* $\mathbb{R}^k$:
$$B\left(x, \frac{1}{\sqrt{N_{\min}^* + 1}}\right) \subset [0, 1]^k, \text{ for a sufficiently large } N_{\min}^* := \min\{N_1^*, ..., N_k^*\}; \text{ where}$$
$\overrightarrow{N}^* := (N_1^*, ..., N_k^*) \in \mathbb{N}^k$. *Then*

$$\left| B_{\overrightarrow{N}}^{(M)}(f)(x) - f(x) \right| \leq \left(\frac{6k^{n+1}}{(n+1)!} \right) \left(\max_{\alpha : |\alpha| = n} \omega_1 \left(f_\alpha, \frac{1}{\sqrt{N_{\min} + 1}} \right) \right), \quad (12.30)$$

$\forall \, \overrightarrow{N} \in \mathbb{N}^k$, *where* $\overrightarrow{N} := (N_1, ..., N_k) \in \mathbb{N}^k$ *and* $N_{\min} := \min\{N_1, ..., N_k\}$, *with* $N_{\min} \geq N_{\min}^*$.

It holds $\lim_{\overrightarrow{N} \to (\infty, ..., \infty)} B_{\overrightarrow{N}}^{(M)}(f)(x) = f(x)$.

Proof By (12.19) we get:

$$\left| B_{\overrightarrow{N}}^{(M)}(f)(x) - f(x) \right| \overset{(12.28)}{\leq} \frac{\left(\max_{\alpha : |\alpha| = n} \omega_1(f_\alpha, h) \right) k^n}{h(n+1)!} \cdot$$

$$\left(\sum_{i=1}^k B_{N_i}^{(M)} \left(|t_i - x_i|^{n+1} \right)(x_i) \right) \overset{(12.23)}{\leq} \quad (12.31)$$

$$\frac{\left(\max_{\alpha : |\alpha| = n} \omega_1(f_\alpha, h) \right) k^n}{h(n+1)!} \left(\sum_{i=1}^k \frac{6}{\sqrt{N_i + 1}} \right) \leq$$

$$\frac{\left(\max_{\alpha : |\alpha| = n} \omega_1(f_\alpha, h) \right) k^{n+1}}{h(n+1)!} \left(\frac{6}{\sqrt{N_{\min} + 1}} \right) =$$

$$\left(\text{setting } h := \frac{1}{\sqrt{N_{\min} + 1}} \right) \quad (12.32)$$

$$\left(\frac{6k^{n+1}}{(n+1)!} \right) \left(\max_{\alpha:|\alpha|=n} \omega_1 \left(f_\alpha, \frac{1}{\sqrt{N_{\min}+1}} \right) \right),$$

proving the claim.   ∎

We need

Theorem 12.11 *Let Q with the l_1-norm $\|\cdot\|$, be a compact and convex subset of $\mathbb{R}^k$, $k \in \mathbb{N} - \{1\}$, and $f \in C_+(Q) : |f(t) - f(x)|$ is a convex function in $t := (t_1, ..., t_k) \in Q$ for a fixed $x = (x_1, ..., x_k) \in Q^o$. We denote $\omega_1(f, h) := \sup_{\substack{x,y \in Q: \\ \|x-y\| \le h}}$ $|f(x) - f(y)|$, $h > 0$, the first modulus of continuity of f. We assume that the ball in $\mathbb{R}^k : B(x, h) \subset Q$. Let $\{L_N\}_{N \in \mathbb{N}}$ be positive sublinear operators from $C_+(Q)$ into $C_+(Q)$, $L_N(1) = 1$, $\forall N \in \mathbb{N}$. Then*

$$|L_N(f)(x) - f(x)| \le \frac{\omega_1(f, h)}{h} L_N(\|t - x\|)(x) \le$$

$$\frac{\omega_1(f, h)}{h} \left(\sum_{i=1}^{k} L_N(|t_i - x_i|)(x) \right), \tag{12.33}$$

$\forall N \in \mathbb{N}$.

Proof By Lemma 8.1.1, p. 243 of [1] we get that:

$$|f(t) - f(x)| \le \frac{\omega_1(f, h)}{h} \|t - x\|, \ \forall t \in Q. \tag{12.34}$$

By (12.18) we have

$$|L_N(f)(x) - f(x)| \le L_N(|f(t) - f(x)|)(x) \le \tag{12.35}$$

$$\frac{\omega_1(f, h)}{h} L_N(\|t - x\|)(x) \le \frac{\omega_1(f, h)}{h} \left(\sum_{i=1}^{k} L_N(|t_i - x_i|)(x) \right),$$

proving the claim.   ∎

We give

Theorem 12.12 *Let $[0, 1]^k$, $k \in \mathbb{N} - \{1\}$, with the l_1-norm $\|\cdot\|$ and $f \in C_+([0, 1]^k)$: $|f(t) - f(x)|$ is a convex function in $t \in [0, 1]^k$ for a fixed $x \in (0, 1)^k$. Assume that the ball in $\mathbb{R}^k : B\left(x, \frac{1}{\sqrt{N_{\min}^*+1}} \right) \subset [0, 1]^k$, for a sufficiently large $N_{\min}^* := \min\{N_1^*, ..., N_k^*\}$; where $\vec{N}^* := (N_1^*, ..., N_k^*) \in \mathbb{N}^k$. Then*

$$\left| B_{\overrightarrow{N}}^{(M)} (f) (x) - f (x) \right| \le 6k\omega_1 \left(f, \frac{1}{\sqrt{N_{\min} + 1}} \right), \tag{12.36}$$

$\forall \overrightarrow{N} \in \mathbb{N}^k$, where $\overrightarrow{N} := (N_1, ..., N_k) \in \mathbb{N}^k$ and $N_{\min} := \min\{N_1, ..., N_k\}$, with $N_{\min} \ge N_{\min}^*$.

It holds $\lim_{\overrightarrow{N} \to (\infty,...,\infty)} B_{\overrightarrow{N}}^{(M)} (f) (x) = f (x)$.

Proof By (12.33) we have that

$$\left| B_{\overrightarrow{N}}^{(M)} (f) (x) - f (x) \right| \overset{(12.28)}{\le} \frac{\omega_1 (f, h)}{h} \left(\sum_{i=1}^{k} B_{N_i}^{(M)} (|t_i - x_i|) (x_i) \right)$$

$$\overset{(12.22)}{\le} \frac{\omega_1 (f, h)}{h} \left(\sum_{i=1}^{k} \frac{6}{\sqrt{N_i + 1}} \right) \le \frac{\omega_1 (f, h)}{h} \left(\frac{6k}{\sqrt{N_{\min} + 1}} \right)$$

(choosing $h := \frac{1}{\sqrt{N_{\min}+1}}$)

$$= 6k\omega_1 \left(f, \frac{1}{\sqrt{N_{\min} + 1}} \right), \tag{12.37}$$

proving the claim. ∎

We continue with

Definition 12.13 ([5], p. 123) We define the bivariate Max-product Bernstein type operators:

$$\overline{T}_N^{(M)} (f) (x, y) := \frac{\bigvee_{i=0}^{N} \bigvee_{j=0}^{N-i} \binom{N}{i} \binom{N-i}{j} x^i y^j (1 - x - y)^{N-i-j} f \left(\frac{i}{N}, \frac{j}{N} \right)}{\bigvee_{i=0}^{N} \bigvee_{j=0}^{N-i} \binom{N}{i} \binom{N-i}{j} x^i y^j (1 - x - y)^{N-i-j}}, \tag{12.38}$$

$\forall (x, y) \in \Delta := \{(x, y) : x \ge 0, y \ge 0, x + y \le 1\}$, $\forall N \in \mathbb{N}$, and $\forall f \in C_+ (\Delta)$.

Remark 12.14 By [5], p. 137, Theorem 2.7.5 there, $\overline{T}_N^{(M)}$ is a positive sublinear operator mapping $C_+ (\Delta)$ into itself and $\overline{T}_N^{(M)} (1) = 1$, furthermore it holds

$$\left| \overline{T}_N^{(M)} (f) - \overline{T}_N^{(M)} (g) \right| \le \overline{T}_N^{(M)} (|f - g|), \ \forall f, g \in C_+ (\Delta), \forall N \in \mathbb{N}. \tag{12.39}$$

By [5], p. 125 we get that $\overline{T}_N^{(M)} (f) (1, 0) = f (1, 0), \overline{T}_N^{(M)} (f) (0, 1) = f (0, 1)$, and $\overline{T}_N^{(M)} (f) (0, 0) = f (0, 0)$.

By [5], p. 139, we have that $((x, y) \in \Delta)$:

$$\overline{T}_N^{(M)} \left(|\cdot - x| \right) (x, y) = B_N^{(M)} \left(|\cdot - x| \right) (x), \tag{12.40}$$

and

$$\overline{T}_N^{(M)} \left(|\cdot - y| \right) (x, y) = B_N^{(M)} \left(|\cdot - y| \right) (y). \tag{12.41}$$

Working exactly the same way as (12.40), (12.41) are proved we also derive ($m \in \mathbb{N}$, $(x, y) \in \Delta$):

$$\overline{T}_N^{(M)} \left(|\cdot - x|^m \right) (x, y) = B_N^{(M)} \left(|\cdot - x|^m \right) (x), \tag{12.42}$$

and

$$\overline{T}_N^{(M)} \left(|\cdot - y|^m \right) (x, y) = B_N^{(M)} \left(|\cdot - y|^m \right) (y). \tag{12.43}$$

We present

Theorem 12.15 *Let $\Delta \subset \mathbb{R}^2$ is endowed with the l_1-norm. Let $x := (x_1, x_2) \in \Delta^o$ be fixed, and $f \in C^n (\Delta, \mathbb{R}_+)$, $n \in \mathbb{N}$. We assume that $f_\alpha (x) = 0$, for all $\alpha : |\alpha| = 1, ..., n$, and $|f_\alpha|$ is a convex function for all $\alpha : |\alpha| = n$. For a sufficiently large $N^* \in \mathbb{N}$ we have that the disc in $\mathbb{R}^2 : D \left(x, \frac{1}{\sqrt{N^*+1}} \right) \subset \Delta$. Then*

$$\left| \overline{T}_N^{(M)} (f) (x_1, x_2) - f (x_1, x_2) \right| \le \frac{3 \cdot 2^{n+2} \left(\max_{\alpha:|\alpha|=n} \omega_1 \left(f_\alpha, \frac{1}{\sqrt{N+1}} \right) \right)}{(n + 1)!}, \tag{12.44}$$

$\forall N \ge N^*, N \in \mathbb{N}$.

It holds $\lim_{N \to \infty} \overline{T}_N^{(M)} (f) (x_1, x_2) = f (x_1, x_2)$.

Proof By (12.19) we get (here $x := (x_1, x_2) \in \Delta$):

$$\left| \overline{T}_N^{(M)} (f) (x_1, x_2) - f (x_1, x_2) \right| \le \frac{\left(\max_{\alpha:|\alpha|=n} \omega_1 (f_\alpha, h) \right) 2^n}{h (n + 1)!} \cdot$$

$$\left(\sum_{i=1}^{2} \overline{T}_N^{(M)} \left(|t_i - x_i|^{n+1} \right) (x) \right) \overset{\text{(by (12.42), (12.43))}}{=}$$

$$\frac{\left(\max_{\alpha:|\alpha|=n} \omega_1 (f_\alpha, h) \right) 2^n}{h (n + 1)!} \left(\sum_{i=1}^{2} B_N^{(M)} \left(|t_i - x_i|^{n+1} \right) (x_i) \right) \overset{(12.23)}{\le} \tag{12.45}$$

$$\frac{\left(\max_{\alpha:|\alpha|=n} \omega_1 (f_\alpha, h) \right) 2^{n+1}}{h (n + 1)!} \frac{6}{\sqrt{N + 1}} = \frac{3 \cdot 2^{n+2} \left(\max_{\alpha:|\alpha|=n} \omega_1 (f_\alpha, h) \right)}{(n + 1)! h \sqrt{N + 1}} =$$

$$\text{(setting } h := \frac{1}{\sqrt{N+1}})\qquad(12.46)$$

$$\frac{3 \cdot 2^{n+2} \left(\max_{\alpha:|\alpha|=n} \omega_1 \left(f_\alpha, \frac{1}{\sqrt{N+1}} \right) \right)}{(n+1)!},$$

proving the claim. ∎

It follows:

Theorem 12.16 *Let* $\Delta \subset \mathbb{R}^2$ *is endowed with the* l_1 *-norm. Let* $f \in C_+(\Delta)$*. We assume that* $|f(t) - f(x)|$ *is convex in* $t \in \Delta$ *for a fixed* $x \in \Delta^o$*. For a sufficiently large* $N^* \in \mathbb{N}$ *we have that the disc in* $\mathbb{R}^2$: $D\left(x, \frac{1}{\sqrt{N^*+1}}\right) \subset \Delta$. *Then*

$$\left| \overline{T}_N^{(M)}(f)(x_1, x_2) - f(x_1, x_2) \right| \le 12\omega_1 \left(f, \frac{1}{\sqrt{N+1}} \right),\qquad(12.47)$$

$\forall\, N \ge N^*, N \in \mathbb{N}.$
  It holds $\lim_{N \to \infty} \overline{T}_N^{(M)}(f)(x_1, x_2) = f(x_1, x_2).$

Proof By (12.33) we have

$$\left| \overline{T}_N^{(M)}(f)(x_1, x_2) - f(x_1, x_2) \right| \le \frac{\omega_1(f, h)}{h} \left(\sum_{i=1}^{2} \overline{T}_N^{(M)} (|t_i - x_i|)(x) \right)$$

$$\stackrel{\text{(by (12.40), (12.41))}}{=} \frac{\omega_1(f, h)}{h} \left(\sum_{i=1}^{2} B_N^{(M)} (|t_i - x_i|)(x_i) \right) \stackrel{(12.22)}{\le}\qquad(12.48)$$

$$\frac{2\omega_1(f, h)}{h} \frac{6}{\sqrt{N+1}} = \frac{12\omega_1(f, h)}{h\sqrt{N+1}} =$$

(setting $h := \frac{1}{\sqrt{N+1}}$)

$$12\omega_1 \left(f, \frac{1}{\sqrt{N+1}} \right),\qquad(12.49)$$

proving the claim. ∎

We make

Remark 12.17 The Max-product truncated Favard-Szász-Mirakjan operators

$$T_N^{(M)}(f)(x) = \frac{\bigvee_{k=0}^{N} s_{N,k}(x) f\left(\frac{k}{N}\right)}{\bigvee_{k=0}^{N} s_{N,k}(x)}, \quad x \in [0, 1], \ N \in \mathbb{N}, \ f \in C_+([0, 1]),$$

$$(12.50)$$

$s_{N,k}(x) = \frac{(Nx)^k}{k!}$, see also [5], p. 11.
By [5], pp. 178–179, we get that

$$T_N^{(M)}\left(|\cdot - x|\right)(x) \le \frac{3}{\sqrt{N}}, \ \forall\, x \in [0,1],\ \forall\, N \in \mathbb{N}. \tag{12.51}$$

And from [2] we have

$$T_N^{(M)}\left(|\cdot - x|^m\right)(x) \le \frac{3}{\sqrt{N}}, \ \forall\, x \in [0,1],\ \forall\, N, m \in \mathbb{N}. \tag{12.52}$$

We make

Definition 12.18 Let $f \in C_+\left([0,1]^k\right)$, $k \in \mathbb{N} - \{1\}$, and $\overrightarrow{N} = (N_1, ..., N_k) \in \mathbb{N}^k$. We define the multivariate Max-product truncated Favard-Sz ász-Mirakjan operators as follows:

$$T_{\overrightarrow{N}}^{(M)}(f)(x) :=$$

$$\frac{\bigvee_{i_1=0}^{N_1}\bigvee_{i_2=0}^{N_2}\cdots\bigvee_{i_k=0}^{N_k} s_{N_1,i_1}(x_1)\,s_{N_2,i_2}(x_2)\cdots s_{N_k,i_k}(x_k)\,f\left(\frac{i_1}{N_1},...,\frac{i_k}{N_k}\right)}{\bigvee_{i_1=0}^{N_1}\bigvee_{i_2=0}^{N_2}\cdots\bigvee_{i_k=0}^{N_k} s_{N_1,i_1}(x_1)\,s_{N_2,i_2}(x_2)\cdots s_{N_k,i_k}(x_k)}, \tag{12.53}$$

$\forall\, x = (x_1, ..., x_k) \in [0,1]^k$. Call $N_{\min} := \min\{N_1, ..., N_k\}$.
The operators $T_{\overrightarrow{N}}^{(M)}(f)(x)$ are positive sublinear mapping $C_+\left([0,1]^k\right)$ into itself, and $T_{\overrightarrow{N}}^{(M)}(1) = 1$.
We also have

$$T_{\overrightarrow{N}}^{(M)}(f)(x) :=$$

$$\frac{\bigvee_{i_1=0}^{N_1}\bigvee_{i_2=0}^{N_2}\cdots\bigvee_{i_k=0}^{N_k} s_{N_1,i_1}(x_1)\,s_{N_2,i_2}(x_2)\cdots s_{N_k,i_k}(x_k)\,f\left(\frac{i_1}{N_1},...,\frac{i_k}{N_k}\right)}{\prod_{\lambda=1}^{k}\left(\bigvee_{i_\lambda=0}^{N_\lambda} s_{N_\lambda,i_\lambda}(x_\lambda)\right)}, \tag{12.54}$$

$\forall\, x \in [0,1]^k$, by the maximum multiplicative principle, see (12.20).

We make

Remark 12.19 The coordinate Max-product truncated Favard-Szász-Mirakjan operators are defined as follows ($\lambda = 1, ..., k$):

$$T_{N_\lambda}^{(M)}(g)(x_\lambda) := \frac{\bigvee_{i_\lambda=0}^{N_\lambda} s_{N_\lambda,i_\lambda}(x_\lambda)\,g\left(\frac{i_\lambda}{N_\lambda}\right)}{\bigvee_{i_\lambda=0}^{N_\lambda} s_{N_\lambda,i_\lambda}(x_\lambda)}, \tag{12.55}$$

$\forall\, N_\lambda \in \mathbb{N}$, and $\forall\, x_\lambda \in [0,1], \forall\, g \in C_+([0,1])$.
Here we have

$$s_{N_\lambda, i_\lambda}(x_\lambda) = \frac{(N_\lambda x_\lambda)^{i_\lambda}}{i_\lambda!}, \quad \lambda = 1, ..., k; \ x_\lambda \in [0, 1]. \tag{12.56}$$

In case of $f \in C_+\left([0, 1]^k\right)$ such that $f(x) := g(x_\lambda)$, $\forall \ x \in [0, 1]^k$, where $x = (x_1, ..., x_\lambda, ..., x_k)$ and $g \in C_+([0, 1])$, we get that

$$T_{\vec{N}}^{(M)}(f)(x) = T_{N_\lambda}^{(M)}(g)(x_\lambda), \tag{12.57}$$

by the maximum multiplicative principle (12.20) and simplification of (12.54).

We present

Theorem 12.20 Let $[0, 1]^k$, $k \in \mathbb{N} - \{1\}$ with the l_1-norm and let $x = (x_1, ..., x_k) \in (0, 1)^k$ be fixed. Let $f \in C^n\left([0, 1]^k, \mathbb{R}_+\right)$, $n \in \mathbb{N}$, and suppose that each nth partial derivative $f_\alpha = \frac{\partial^\alpha f}{\partial x^\alpha}$, where $\alpha = (\alpha_1, ..., \alpha_k)$, $\alpha_i \geq 0$, $i = 1, ..., k$, and $|\alpha| = \sum_{i=1}^k \alpha_i = n$, has relative to $[0, 1]^k$ and the l_1-norm a modulus of continuity $\omega_1(f_\alpha, \cdot)$, and that each $|f_\alpha(t)|$ is a convex function of $t \in [0, 1]^k$, all $\alpha : |\alpha| = n$. Assume further that $f_\alpha(x) = 0$, for all $\alpha : |\alpha| = 1, ..., n$; and the ball in $\mathbb{R}^k$: $B\left(x, \frac{1}{\sqrt{N_{\min}^*}}\right) \subset [0, 1]^k$, for a sufficiently large $N_{\min}^* := \min\{N_1^*, ..., N_k^*\}$; where $\vec{N}^* := \left(N_1^*, ..., N_k^*\right) \in \mathbb{N}^k$. Then

$$\left| T_{\vec{N}}^{(M)}(f)(x) - f(x) \right| \leq \left(\frac{3k^{n+1}}{(n+1)!}\right)\left(\max_{\alpha:|\alpha|=n} \omega_1\left(f_\alpha, \frac{1}{\sqrt{N_{\min}}}\right)\right), \tag{12.58}$$

$\forall \ \vec{N} \in \mathbb{N}^k$, where $\vec{N} := (N_1, ..., N_k) \in \mathbb{N}^k$ and $N_{\min} := \min\{N_1, ..., N_k\}$, with $N_{\min} \geq N_{\min}^*$.

It holds $\lim_{\vec{N} \to (\infty,...,\infty)} T_{\vec{N}}^{(M)}(f)(x) = f(x)$.

Proof By (12.19) we get:

$$\left| T_{\vec{N}}^{(M)}(f)(x) - f(x) \right| \overset{(12.57)}{\leq} \frac{\left(\max_{\alpha:|\alpha|=n} \omega_1(f_\alpha, h)\right) k^n}{h(n+1)!} \left(\sum_{i=1}^k T_{N_i}^{(M)}\left(|t_i - x_i|^{n+1}\right)(x_i)\right)$$

$$\overset{(12.52)}{\leq} \frac{\left(\max_{\alpha:|\alpha|=n} \omega_1(f_\alpha, h)\right) k^n}{h(n+1)!} \left(\sum_{i=1}^k \frac{3}{\sqrt{N_i}}\right) \leq \frac{3k^{n+1}\left(\max_{\alpha:|\alpha|=n} \omega_1(f_\alpha, h)\right)}{h(n+1)!\sqrt{N_{\min}}} = \tag{12.59}$$

(setting $h := \frac{1}{\sqrt{N_{\min}}}$)

$$\frac{3k^{n+1}\left(\max_{\alpha:|\alpha|=n} \omega_1\left(f_\alpha, \frac{1}{\sqrt{N_{\min}}}\right)\right)}{(n+1)!}, \tag{12.60}$$

proving the claim. ∎

It follows

Theorem 12.21 *Let* $[0, 1]^k, k \in \mathbb{N} - \{1\}$ *with the* l_1-*norm* $\|\cdot\|$ *and* $f \in C_+ \left([0, 1]^k\right)$: $|f(t) - f(x)|$ *is a convex function in* $t \in [0, 1]^k$ *for a fixed* $x \in (0, 1)^k$. *Assume that the ball in* $\mathbb{R}^k : B\left(x, \frac{1}{\sqrt{N_{\min}^*}}\right) \subset [0, 1]^k$, *for a sufficiently large* $N_{\min}^* :=$ $\min\{N_1^*, ..., N_k^*\}$; *where* $\overrightarrow{N}^* := \left(N_1^*, ..., N_k^*\right) \in \mathbb{N}^k$. *Then*

$$\left|T_{\overrightarrow{N}}^{(M)} (f) (x) - f(x)\right| \leq 3k\omega_1 \left(f, \frac{1}{\sqrt{N_{\min}}}\right), \tag{12.61}$$

$\forall \overrightarrow{N} \in \mathbb{N}^k$, *where* $\overrightarrow{N} := (N_1, ..., N_k) \in \mathbb{N}^k$ *and* $N_{\min} := \min\{N_1, ..., N_k\}$, *with* $N_{\min}$ $\geq N_{\min}^*$.

It holds $\lim\limits_{\overrightarrow{N} \to (\infty,...,\infty)} T_{\overrightarrow{N}}^{(M)} (f) (x) = f(x)$.

Proof By (12.33) we have that

$$\left|T_{\overrightarrow{N}}^{(M)} (f) (x) - f(x)\right| \overset{(12.57)}{\leq} \frac{\omega_1 (f, h)}{h} \left(\sum_{i=1}^{k} T_{N_i}^{(M)} (|t_i - x_i|) (x_i)\right)$$

$$\overset{(12.52)}{\leq} \frac{\omega_1 (f, h)}{h} \left(\sum_{i=1}^{k} \frac{3}{\sqrt{N_i}}\right) \leq \frac{\omega_1 (f, h)}{h} \frac{3k}{\sqrt{N_{\min}}}$$

(choosing $h := \frac{1}{\sqrt{N_{\min}}}$)

$$= 3k\omega_1 \left(f, \frac{1}{\sqrt{N_{\min}}}\right), \tag{12.62}$$

proving the claim. ∎

We make

Remark 12.22 We mention the truncated Max-product Baskakov operator (see [5], p. 11)

$$U_N^{(M)} (f) (x) = \frac{\bigvee_{k=0}^{N} b_{N,k} (x) f\left(\frac{k}{N}\right)}{\bigvee_{k=0}^{N} b_{N,k} (x)}, \quad x \in [0, 1], \ f \in C_+ ([0, 1]) , \ \forall N \in \mathbb{N},$$

$$\tag{12.63}$$

where

$$b_{N,k} (x) = \binom{N + k - 1}{k} \frac{x^k}{(1 + x)^{N+k}}. \tag{12.64}$$

From [5], pp. 217–218, we get ($x \in [0, 1]$)

$$\left(U_N^{(M)}\left(|\cdot - x|\right)\right)(x) \le \frac{12}{\sqrt{N+1}}, \quad N \ge 2, N \in \mathbb{N}. \tag{12.65}$$

And as in [2], we obtain $(m \in \mathbb{N})$

$$\left(U_N^{(M)}\left(|\cdot - x|^m\right)\right)(x) \le \frac{12}{\sqrt{N+1}}, \quad N \ge 2, N \in \mathbb{N}, \ \forall\, x \in [0, 1]. \tag{12.66}$$

Definition 12.23 Let $f \in C_+\left([0, 1]^k\right)$, $k \in \mathbb{N} - \{1\}$, and $\overrightarrow{N} = (N_1, ..., N_k) \in \mathbb{N}^k$. We define the multivariate Max-product truncated Baskakov operators as follows:

$$U_{\overrightarrow{N}}^{(M)}(f)(x) :=$$

$$\frac{\bigvee_{i_1=0}^{N_1} \bigvee_{i_2=0}^{N_2} \cdots \bigvee_{i_k=0}^{N_k} b_{N_1,i_1}(x_1)\, b_{N_2,i_2}(x_2) ... b_{N_k,i_k}(x_k)\, f\left(\frac{i_1}{N_1}, ..., \frac{i_k}{N_k}\right)}{\bigvee_{i_1=0}^{N_1} \bigvee_{i_2=0}^{N_2} \cdots \bigvee_{i_k=0}^{N_k} b_{N_1,i_1}(x_1)\, b_{N_2,i_2}(x_2) ... b_{N_k,i_k}(x_k)}, \tag{12.67}$$

$\forall\, x = (x_1, ..., x_k) \in [0, 1]^k$. Call $N_{\min} := \min\{N_1, ..., N_k\}$.

The operators $U_{\overrightarrow{N}}^{(M)}(f)(x)$ are positive sublinear mapping $C_+\left([0, 1]^k\right)$ into itself, and $U_{\overrightarrow{N}}^{(M)}(1) = 1$.

We also have

$$U_{\overrightarrow{N}}^{(M)}(f)(x) :=$$

$$\frac{\bigvee_{i_1=0}^{N_1} \bigvee_{i_2=0}^{N_2} \cdots \bigvee_{i_k=0}^{N_k} b_{N_1,i_1}(x_1)\, b_{N_2,i_2}(x_2) ... b_{N_k,i_k}(x_k)\, f\left(\frac{i_1}{N_1}, ..., \frac{i_k}{N_k}\right)}{\prod_{\lambda=1}^{k}\left(\bigvee_{i_\lambda=0}^{N_\lambda} b_{N_\lambda,i_\lambda}(x_\lambda)\right)}, \tag{12.68}$$

$\forall\, x \in [0, 1]^k$, by the maximum multiplicative principle, see (12.20).

We make

Remark 12.24 The coordinate Max-product truncated Baskakov operators are defined as follows $(\lambda = 1, ..., k)$:

$$U_{N_\lambda}^{(M)}(g)(x_\lambda) := \frac{\bigvee_{i_\lambda=0}^{N_\lambda} b_{N_\lambda,i_\lambda}(x_\lambda)\, g\left(\frac{i_\lambda}{N_\lambda}\right)}{\bigvee_{i_\lambda=0}^{N_\lambda} b_{N_\lambda,i_\lambda}(x_\lambda)}, \tag{12.69}$$

$\forall\, N_\lambda \in \mathbb{N}$, and $\forall\, x_\lambda \in [0, 1]$, $\forall\, g \in C_+([0, 1])$.

Here we have

$$b_{N_\lambda,i_\lambda}(x_\lambda) = \binom{N_\lambda + i_\lambda - 1}{i_\lambda} \frac{x_\lambda^{i_\lambda}}{(1 + x_\lambda)^{N+i_\lambda}}, \quad \lambda = 1, ..., k; \ x_\lambda \in [0, 1]. \tag{12.70}$$

In case of $f \in C_+\left([0, 1]^k\right)$ such that $f(x) := g(x_\lambda)$, $\forall\, x \in [0, 1]^k$, where $x = (x_1, ..., x_\lambda, ..., x_k)$ and $g \in C_+([0, 1])$, we get that

$$U_{\vec{N}}^{(M)}(f)(x) = U_{N_\lambda}^{(M)}(g)(x_\lambda), \tag{12.71}$$

by the maximum multiplicative principle (12.20) and simplification of (12.68).

We present

Theorem 12.25 *Let* $[0, 1]^k$, $k \in \mathbb{N} - \{1\}$ *with the* l_1*-norm and let* $x = (x_1, ..., x_k) \in (0, 1)^k$ *be fixed. Let* $f \in C^n\left([0, 1]^k, \mathbb{R}_+\right)$, $n \in \mathbb{N}$, *and suppose that each nth partial derivative* $f_\alpha = \frac{\partial^\alpha f}{\partial x^\alpha}$, *where* $\alpha = (\alpha_1, ..., \alpha_k)$, $\alpha_i \geq 0$, $i = 1, ..., k$, *and* $|\alpha| = \sum_{i=1}^k \alpha_i = n$ *has, relative to* $[0, 1]^k$ *and the* l_1*-norm a modulus of continuity* $\omega_1(f_\alpha, \cdot)$, *and that each* $|f_\alpha(t)|$ *is a convex function of* $t \in [0, 1]^k$, *all* $\alpha : |\alpha| = n$. *Assume further that* $f_\alpha(x) = 0$, *for all* $\alpha : |\alpha| = 1, ..., n$; *and the ball in* $\mathbb{R}^k$:

$$B\left(x, \frac{1}{\sqrt{N_{\min}^* + 1}}\right) \subset [0, 1]^k, \text{ for a sufficiently large } N_{\min}^* := \min\{N_1^*, ..., N_k^*\}; \text{ where}$$

$\vec{N}^* := \left(N_1^*, ..., N_k^*\right) \in \mathbb{N}^k$. *Then*

$$\left|U_{\vec{N}}^{(M)}(f)(x) - f(x)\right| \leq \left(\frac{12k^{n+1}}{(n+1)!}\right)\left(\max_{\alpha:|\alpha|=n} \omega_1\left(f_\alpha, \frac{1}{\sqrt{N_{\min}+1}}\right)\right), \tag{12.72}$$

$\forall\, \vec{N} \in (\mathbb{N} - \{1\})^k$, *where* $\vec{N} := (N_1, ..., N_k) \in \mathbb{N}^k$ *and* $N_{\min} := \min\{N_1, ..., N_k\}$, *with* $N_{\min} \geq N_{\min}^*$.

It holds $\lim_{\vec{N}\to(\infty,...,\infty)} U_{\vec{N}}^{(M)}(f)(x) = f(x)$.

Proof By (12.19) we get:

$$\left|U_{\vec{N}}^{(M)}(f)(x) - f(x)\right| \overset{(12.71)}{\leq} \frac{\left(\max\limits_{\alpha:|\alpha|=n} \omega_1(f_\alpha, h)\right) k^n}{h(n+1)!} \cdot$$

$$\left(\sum_{i=1}^k U_{N_i}^{(M)}\left(|t_i - x_i|^{n+1}\right)(x_i)\right) \overset{(12.66)}{\leq} \tag{12.73}$$

$$\frac{\left(\max\limits_{\alpha:|\alpha|=n} \omega_1(f_\alpha, h)\right) k^n}{h(n+1)!}\left(\sum_{i=1}^k \frac{12}{\sqrt{N_i+1}}\right) \leq \frac{\left(\max\limits_{\alpha:|\alpha|=n} \omega_1(f_\alpha, h)\right) k^{n+1}}{h(n+1)!}\frac{12}{\sqrt{N_{\min}+1}} =$$

(setting $h := \frac{1}{\sqrt{N_{\min}+1}}$)

$$\left(\frac{12k^{n+1}}{(n+1)!}\right)\left(\max_{\alpha:|\alpha|=n} \omega_1\left(f_\alpha, \frac{1}{\sqrt{N_{\min}+1}}\right)\right), \tag{12.74}$$

proving the claim. ∎

It follows

Theorem 12.26 *Let $[0, 1]^k, k \in \mathbb{N} - \{1\}$, with the l_1-norm $\|\cdot\|$ and $f \in C_+ ([0, 1]^k)$: $|f (t) - f (x)|$ is a convex function in $t \in [0, 1]^k$ for a fixed $x \in (0, 1)^k$. Assume that the ball in $\mathbb{R}^k$: $B \left(x, \frac{1}{\sqrt{N_{\min}^* + 1}} \right) \subset [0, 1]^k$, for a sufficiently large $N_{\min}^* :=$ $\min\{N_1^*, ..., N_k^*\}$; where $\overrightarrow{N}^* := \left(N_1^*, ..., N_k^* \right) \in \mathbb{N}^k$. Then*

$$\left| U_{\overrightarrow{N}}^{(M)} (f) (x) - f (x) \right| \leq 12 k \omega_1 \left(f, \frac{1}{\sqrt{N_{\min} + 1}} \right), \tag{12.75}$$

$\forall \; \overrightarrow{N} \in (\mathbb{N} - \{1\})^k$, where $\overrightarrow{N} := (N_1, ..., N_k) \in \mathbb{N}^k$ and $N_{\min} := \min\{N_1, ..., N_k\}$, with $N_{\min} \geq N_{\min}^*$.
    It holds $\displaystyle\lim_{\overrightarrow{N} \to (\infty, ..., \infty)} U_{\overrightarrow{N}}^{(M)} (f) (x) = f (x)$.

Proof By (12.33) we have that

$$\left| U_{\overrightarrow{N}}^{(M)} (f) (x) - f (x) \right| \overset{(12.71)}{\leq} \frac{\omega_1 (f, h)}{h} \left(\sum_{i=1}^{k} U_{N_i}^{(M)} (|t_i - x_i|) (x_i) \right)$$

$$\overset{(12.65)}{\leq} \frac{\omega_1 (f, h)}{h} \left(\sum_{i=1}^{k} \frac{12}{\sqrt{N_i + 1}} \right) \leq \frac{\omega_1 (f, h)}{h} \left(\frac{12 k}{\sqrt{N_{\min} + 1}} \right)$$

(choosing $h := \frac{1}{\sqrt{N_{\min}+1}}$)

$$= 12 k \omega_1 \left(f, \frac{1}{\sqrt{N_{\min} + 1}} \right),$$

proving the claim. ∎

We make

Remark 12.27 Here we mention the Max-product truncated sampling operators (see [5], p. 13) defined by

$$W_N^{(M)} (f) (x) := \frac{\bigvee_{k=0}^{N} \frac{\sin(Nx - k\pi)}{Nx - k\pi} f \left(\frac{k\pi}{N} \right)}{\bigvee_{k=0}^{N} \frac{\sin(Nx - k\pi)}{Nx - k\pi}}, \quad x \in [0, \pi], \tag{12.77}$$

$f : [0, \pi] \to \mathbb{R}_+$, continuous,
    and

$$K_N^{(M)}(f)(x) := \frac{\bigvee_{k=0}^{N} \frac{\sin^2(Nx-k\pi)}{(Nx-k\pi)^2} f\left(\frac{k\pi}{N}\right)}{\bigvee_{k=0}^{N} \frac{\sin^2(Nx-k\pi)}{(Nx-k\pi)^2}}, \quad x \in [0, \pi], \tag{12.78}$$

$f : [0, \pi] \to \mathbb{R}_+$, continuous.

By convention we take $\frac{\sin(0)}{0} = 1$, which implies for every $x = \frac{k\pi}{N}, k \in \{0, 1, ..., N\}$ that we have $\frac{\sin(Nx-k\pi)}{Nx-k\pi} = 1$.

We define the Max-product truncated combined sampling operators

$$M_N^{(M)}(f)(x) := \frac{\bigvee_{k=0}^{N} \rho_{N,k}(x) f\left(\frac{k\pi}{N}\right)}{\bigvee_{k=0}^{N} \rho_{N,k}(x)}, \quad x \in [0, \pi], \tag{12.79}$$

$f \in C_+([0, \pi])$, where

$$M_N^{(M)}(f)(x) := \begin{cases} W_N^{(M)}(f)(x), \text{ if } \rho_{N,k}(x) := \frac{\sin(Nx-k\pi)}{Nx-k\pi}, \\ K_N^{(M)}(f)(x), \text{ if } \rho_{N,k}(x) := \left(\frac{\sin(Nx-k\pi)}{Nx-k\pi}\right)^2. \end{cases} \tag{12.80}$$

By [5], p. 346 and p. 352 we get

$$\left(M_N^{(M)}(|\cdot - x|)\right)(x) \leq \frac{\pi}{2N}, \tag{12.81}$$

and by [3] $(m \in \mathbb{N})$ we have

$$\left(M_N^{(M)}(|\cdot - x|^m)\right)(x) \leq \frac{\pi^m}{2N}, \quad \forall x \in [0, \pi], \forall N \in \mathbb{N}. \tag{12.82}$$

We give

Definition 12.28 Let $f \in C_+\left([0, \pi]^k\right), k \in \mathbb{N} - \{1\}$, and $\overrightarrow{N} = (N_1, ..., N_k) \in \mathbb{N}^k$. We define the multivariate Max-product truncated combined sampling operators as follows:

$$M_{\overrightarrow{N}}^{(M)}(f)(x) :=$$

$$\frac{\bigvee_{i_1=0}^{N_1} \bigvee_{i_2=0}^{N_2} \cdots \bigvee_{i_k=0}^{N_k} \rho_{N_1,i_1}(x_1) \rho_{N_2,i_2}(x_2) ... \rho_{N_k,i_k}(x_k) f\left(\frac{i_1\pi}{N_1}, \frac{i_2\pi}{N_2}, ..., \frac{i_k\pi}{N_k}\right)}{\bigvee_{i_1=0}^{N_1} \bigvee_{i_2=0}^{N_2} \cdots \bigvee_{i_k=0}^{N_k} \rho_{N_1,i_1}(x_1) \rho_{N_2,i_2}(x_2) ... \rho_{N_k,i_k}(x_k)}, \tag{12.83}$$

$\forall x = (x_1, ..., x_k) \in [0, \pi]^k$. Call $N_{\min} := \min\{N_1, ..., N_k\}$.

The operators $M_{\overrightarrow{N}}^{(M)}(f)(x)$ are positive sublinear mapping $C_+\left([0, \pi]^k\right)$ into itself, and $M_{\overrightarrow{N}}^{(M)}(1) = 1$.

We also have

$$M_{\overrightarrow{N}}^{(M)}(f)(x) :=$$

$$\frac{\vee_{i_1=0}^{N_1} \vee_{i_2=0}^{N_2} \cdots \vee_{i_k=0}^{N_k} \rho_{N_1,i_1}(x_1)\, \rho_{N_2,i_2}(x_2) \cdots \rho_{N_k,i_k}(x_k)\, f\left(\frac{i_1\pi}{N_1}, \frac{i_2\pi}{N_2}, \ldots, \frac{i_k\pi}{N_k}\right)}{\prod_{\lambda=1}^{k}\left(\vee_{i_\lambda=0}^{N_\lambda} \rho_{N_\lambda,i_\lambda}(x_\lambda)\right)},$$

(12.84)

$\forall\, x \in [0, \pi]^k$, by the maximum multiplicative principle, see (12.20).

We make

Remark 12.29 The coordinate Max-product truncated combined sampling operators are defined as follows ($\lambda = 1, \ldots, k$):

$$M_{N_\lambda}^{(M)}(g)(x_\lambda) := \frac{\vee_{i_\lambda=0}^{N_\lambda} \rho_{N_\lambda,i_\lambda}(x_\lambda)\, g\left(\frac{i_\lambda\pi}{N_\lambda}\right)}{\vee_{i_\lambda=0}^{N_\lambda} \rho_{N_\lambda,i_\lambda}(x_\lambda)},$$

(12.85)

$\forall\, N_\lambda \in \mathbb{N}$, and $\forall\, x_\lambda \in [0, \pi]$, $\forall\, g \in C_+([0, \pi])$.
 Here we have ($\lambda = 1, \ldots, k$; $x_\lambda \in [0, \pi]$)

$$\rho_{N_\lambda,i_\lambda}(x_\lambda) = \begin{cases} \frac{\sin(N_\lambda x_\lambda - i_\lambda \pi)}{N_\lambda x_\lambda - i_\lambda \pi}, & \text{if } M_{N_\lambda}^{(M)} = W_{N_\lambda}^{(M)}, \\ \left(\frac{\sin(N_\lambda x_\lambda - i_\lambda \pi)}{N_\lambda x_\lambda - i_\lambda \pi}\right)^2, & \text{if } M_{N_\lambda}^{(M)} = K_{N_\lambda}^{(M)}. \end{cases}$$

(12.86)

In case of $f \in C_+\left([0, \pi]^k\right)$ such that $f(x) := g(x_\lambda)$, $\forall\, x \in [0, \pi]^k$, where $x = (x_1, \ldots, x_\lambda, \ldots, x_k)$ and $g \in C_+([0, \pi])$, we get that

$$M_{\vec{N}}^{(M)}(f)(x) = M_{N_\lambda}^{(M)}(g)(x_\lambda),$$

(12.87)

by the maximum multiplicative principle (12.20) and simplification of (12.84).

We present

Theorem 12.30 *Let $[0, \pi]^k$, $k \in \mathbb{N} - \{1\}$ with the l_1-norm and let $x \in (0, \pi)^k$ be fixed. Let $f \in C^n\left([0, \pi]^k, \mathbb{R}_+\right)$, $n \in \mathbb{N}$, and suppose that each nth partial derivative $f_\alpha = \frac{\partial^\alpha f}{\partial x^\alpha}$, where $\alpha = (\alpha_1, \ldots, \alpha_k)$, $\alpha_i \geq 0$, $i = 1, \ldots, k$, and $|\alpha| = \sum_{i=1}^k \alpha_i = n$, has relative to $[0, \pi]^k$ and the l_1-norm a modulus of continuity $\omega_1(f_\alpha, \cdot)$, and that each $|f_\alpha(t)|$ is a convex function of $t \in [0, \pi]^k$, all $\alpha : |\alpha| = n$. Assume further that $f_\alpha(x) = 0$, for all $\alpha : |\alpha| = 1, \ldots, n$; and the ball in $\mathbb{R}^k : B\left(x, \frac{1}{N_{min}^*}\right) \subset [0, \pi]^k$, for a sufficiently large $N_{min}^* := \min\{N_1^*, \ldots, N_k^*\}$; where $\vec{N}^* := \left(N_1^*, \ldots, N_k^*\right) \in \mathbb{N}^k$. Then*

$$\left|M_{\vec{N}}^{(M)}(f)(x) - f(x)\right| \leq \frac{(k\pi)^{n+1}}{2(n+1)!}\left(\max_{\alpha:|\alpha|=n} \omega_1\left(f_\alpha, \frac{1}{N_{min}}\right)\right),$$

(12.88)

$\forall\, \vec{N} \in \mathbb{N}^k$, where $\vec{N} := (N_1, \ldots, N_k) \in \mathbb{N}^k$ and $N_{min} := \min\{N_1, \ldots, N_k\}$, with $N_{min} \geq N_{min}^*$.
 It holds $\lim_{\vec{N} \to (\infty, \ldots, \infty)} M_{\vec{N}}^{(M)}(f)(x) = f(x)$.

Proof By (12.19) we get:

$$\left| M_{\overrightarrow{N}}^{(M)}(f)(x) - f(x) \right| \overset{(12.87)}{\leq} \frac{\left(\max_{\alpha:|\alpha|=n} \omega_1(f_\alpha, h) \right) k^n}{h(n+1)!} \cdot$$

$$\left(\sum_{i=1}^{k} M_{N_i}^{(M)}\left(|t_i - x_i|^{n+1} \right)(x_i) \right) \overset{(12.82)}{\leq}$$

$$\frac{\left(\max_{\alpha:|\alpha|=n} \omega_1(f_\alpha, h) \right) k^n}{h(n+1)!} \left(\sum_{i=1}^{k} \frac{\pi^{n+1}}{2N_i} \right) \leq \frac{\left(\max_{\alpha:|\alpha|=n} \omega_1(f_\alpha, h) \right) k^{n+1}\pi^{n+1}}{2(n+1)!hN_{\min}} = \quad (12.89)$$

(setting $h := \frac{1}{N_{\min}}$)

$$\frac{\left(\max_{\alpha:|\alpha|=n} \omega_1\left(f_\alpha, \frac{1}{N_{\min}} \right) \right)(k\pi)^{n+1}}{2(n+1)!},$$

proving the claim. ∎

We continue with

Theorem 12.31 *Let* $[0, \pi]^k$, $k \in \mathbb{N} - \{1\}$, *with the* l_1 *-norm and* $f \in C_+\left([0, \pi]^k \right)$: $|f(t) - f(x)|$ *is a convex function in* $t \in [0, \pi]^k$ *for a fixed* $x \in (0, \pi)^k$. *Assume that the ball in* $\mathbb{R}^k : B\left(x, \frac{1}{N_{\min}^*} \right) \subset [0, \pi]^k$, *for a sufficiently large* $N_{\min}^* := \min\{N_1^*, ..., N_k^*\}$; *where* $\overrightarrow{N}^* := (N_1^*, ..., N_k^*) \in \mathbb{N}^k$. *Then*

$$\left| M_{\overrightarrow{N}}^{(M)}(f)(x) - f(x) \right| \leq \left(\frac{k\pi}{2} \right) \omega_1\left(f, \frac{1}{N_{\min}} \right), \quad (12.90)$$

$\forall \overrightarrow{N} \in \mathbb{N}^k$, *where* $\overrightarrow{N} := (N_1, ..., N_k) \in \mathbb{N}^k$ *and* $N_{\min} := \min\{N_1, ..., N_k\}$, *with* $N_{\min} \geq N_{\min}^*$.
It holds $\lim_{\overrightarrow{N} \to (\infty,...,\infty)} M_{\overrightarrow{N}}^{(M)}(f)(x) = f(x)$.

Proof By (12.33) we have that

$$\left| M_{\overrightarrow{N}}^{(M)}(f)(x) - f(x) \right| \overset{(12.87)}{\leq} \frac{\omega_1(f, h)}{h} \left(\sum_{i=1}^{k} M_{N_i}^{(M)}\left(|t_i - x_i| \right)(x_i) \right)$$

$$\overset{(12.81)}{\leq} \frac{\omega_1(f, h)}{h} \left(\sum_{i=1}^{k} \frac{\pi}{2N_i} \right) \leq \frac{\omega_1(f, h)}{h} \frac{(k\pi)}{2N_{\min}} = \quad (12.91)$$

(setting $h := \frac{1}{N_{\min}}$)

$$\left(\frac{k\pi}{2}\right) \omega_1 \left(f, \frac{1}{N_{\min}}\right),$$

proving the claim. ∎

We make

Remark 12.32 Let $f \in C_+ ([-1, 1])$. Let the Chebyshev knots of second kind $x_{N,k} = \cos \left(\left(\frac{N-k}{N-1}\right) \pi\right) \in [-1, 1]$, $k = 1, ..., N$, $N \in \mathbb{N} - \{1\}$, which are the roots of $\omega_N (x) = \sin (N-1) t \sin t$, $x = \cos t \in [-1, 1]$. Notice that $x_{N,1} = -1$ and $x_{N,N} = 1$.

Define

$$l_{N,k}(x) := \frac{(-1)^{k-1} \omega_N(x)}{(1 + \delta_{k,1} + \delta_{k,N})(N-1)(x - x_{N,k})}, \qquad (12.92)$$

$N \geq 2$, $k = 1, ..., N$, and $\omega_N(x) = \prod_{k=1}^N (x - x_{N,k})$ and $\delta_{i,j}$ denotes the Kronecker's symbol, that is $\delta_{i,j} = 1$, if $i = j$, and $\delta_{i,j} = 0$, if $i \neq j$.

The Max-product Lagrange interpolation operators on Chebyshev knots of second kind, plus the endpoints ± 1, are defined by ([5], p. 12)

$$L_N^{(M)} (f)(x) = \frac{\bigvee_{k=1}^N l_{N,k}(x) f(x_{N,k})}{\bigvee_{k=1}^N l_{N,k}(x)}, \quad x \in [-1, 1]. \qquad (12.93)$$

By [5], pp. 297–298 and [3], we get that

$$L_N^{(M)} (|\cdot - x|^m)(x) \leq \frac{2^{m+1} \pi^2}{3(N-1)}, \qquad (12.94)$$

$\forall x \in (-1, 1)$ and $\forall m \in \mathbb{N}$; $\forall N \in \mathbb{N}$, $N \geq 4$.

We see that $L_N^{(M)} (f)(x) \geq 0$ is well defined and continuous for any $x \in [-1, 1]$. Following [5], p. 289, because $\sum_{k=1}^N l_{N,k}(x) = 1$, $\forall x \in [-1, 1]$, for any x there exists $k \in \{1, ..., N\} : l_{N,k}(x) > 0$, hence $\bigvee_{k=1}^N l_{N,k}(x) > 0$. We have that $l_{N,k}(x_{N,k}) = 1$, and $l_{N,k}(x_{N,j}) = 0$, if $k \neq j$. Furthermore it holds $L_N^{(M)} (f)(x_{N,j}) = f(x_{N,j})$, all $j \in \{1, ..., N\}$, and $L_N^{(M)} (1) = 1$.

By [5], pp. 289–290, $L_N^{(M)}$ are positive sublinear operators.

We give

Definition 12.33 Let $f \in C_+ ([-1, 1]^k)$, $k \in \mathbb{N} - \{1\}$, and $\overrightarrow{N} = (N_1, ..., N_k) \in (\mathbb{N} - \{1\})^k$. We define the multivariate Max-product Lagrange interpolation operators on Chebyshev knots of second kind, plus the endpoints ± 1, as follows:

$$L_{\overrightarrow{N}}^{(M)} (f)(x) :=$$

$$\frac{\bigvee_{i_1=1}^{N_1} \bigvee_{i_2=1}^{N_2} \cdots \bigvee_{i_k=1}^{N_k} l_{N_1,i_1}(x_1) l_{N_2,i_2}(x_2) \cdots l_{N_k,i_k}(x_k) f\left(x_{N_1,i_1}, x_{N_2,i_2}, \ldots, x_{N_k,i_k}\right)}{\bigvee_{i_1=1}^{N_1} \bigvee_{i_2=1}^{N_2} \cdots \bigvee_{i_k=1}^{N_k} l_{N_1,i_1}(x_1) l_{N_2,i_2}(x_2) \cdots l_{N_k,i_k}(x_k)},$$

$$(12.95)$$

$\forall\, x = (x_1, \ldots, x_k) \in [-1,1]^k$. Call $N_{\min} := \min\{N_1, \ldots, N_k\}$.

The operators $L_{\vec{N}}^{(M)}(f)(x)$ are positive sublinear mapping $C_+\left([-1,1]^k\right)$ into itself, and $L_{\vec{N}}^{(M)}(1) = 1$.

We also have

$$L_{\vec{N}}^{(M)}(f)(x) :=$$

$$\frac{\bigvee_{i_1=1}^{N_1} \bigvee_{i_2=1}^{N_2} \cdots \bigvee_{i_k=1}^{N_k} l_{N_1,i_1}(x_1) l_{N_2,i_2}(x_2) \cdots l_{N_k,i_k}(x_k) f\left(x_{N_1,i_1}, x_{N_2,i_2}, \ldots, x_{N_k,i_k}\right)}{\prod_{\lambda=1}^{k}\left(\bigvee_{i_\lambda=1}^{N_\lambda} l_{N_\lambda,i_\lambda}(x_\lambda)\right)},$$

$$(12.96)$$

$\forall\, x = (x_1, \ldots, x_\lambda, \ldots, x_k) \in [-1,1]^k$, by the maximum multiplicative principle, see (12.20). Notice that $L_{\vec{N}}^{(M)}(f)\left(x_{N_1,i_1}, \ldots, x_{N_k,i_k}\right) = f\left(x_{N_1,i_1}, \ldots, x_{N_k,i_k}\right)$. The last is also true if $x_{N_1,i_1}, \ldots, x_{N_k,i_k} \in \{-1,1\}$.

We make

Remark 12.34 The coordinate Max-product Lagrange interpolation operators on Chebyshev knots of second kind, plus the endpoints ± 1, are defined as follows $(\lambda = 1, \ldots, k)$:

$$L_{N_\lambda}^{(M)}(g)(x_\lambda) := \frac{\bigvee_{i_\lambda=1}^{N_\lambda} l_{N_\lambda,i_\lambda}(x_\lambda)\, g\left(x_{N_\lambda,i_\lambda}\right)}{\bigvee_{i_\lambda=1}^{N_\lambda} l_{N_\lambda,i_\lambda}(x_\lambda)}, \qquad (12.97)$$

$\forall\, N_\lambda \in \mathbb{N},\, N_\lambda \geq 2$, and $\forall\, x_\lambda \in [-1,1],\, \forall\, g \in C_+([-1,1])$.

Here we have $(\lambda = 1, \ldots, k;\, x_\lambda \in [-1,1])$

$$l_{N_\lambda,i_\lambda}(x_\lambda) = \frac{(-1)^{i_\lambda-1} \omega_{N_\lambda}(x_\lambda)}{\left(1 + \delta_{i_\lambda,1} + \delta_{i_\lambda,N_\lambda}\right)(N_\lambda - 1)\left(x_\lambda - x_{N_\lambda,i_\lambda}\right)}, \qquad (12.98)$$

$N_\lambda \geq 2,\, i_\lambda = 1, \ldots, N_\lambda$ and $\omega_{N_\lambda}(x_\lambda) = \prod_{i_\lambda=1}^{N_\lambda}\left(x_\lambda - x_{N_\lambda,i_\lambda}\right)$; where $x_{N_\lambda,i_\lambda} = \cos\left(\left(\frac{N_\lambda-i_\lambda}{N_\lambda-1}\right)\pi\right) \in [-1,1], i_\lambda = 1, \ldots, N_\lambda\, (N_\lambda \geq 2)$ are roots of $\omega_{N_\lambda}(x_\lambda) = \sin(N_\lambda - 1)$ $t_\lambda \sin t_\lambda,\, x_\lambda = \cos t_\lambda$. Notice that $x_{N_\lambda,1} = -1, x_{N_\lambda,N_\lambda} = 1$.

In case of $f \in C_+\left([-1,1]^k\right)$ such that $f(x) := g(x_\lambda),\, \forall\, x \in [-1,1]^k$, where $x = (x_1, \ldots, x_\lambda, \ldots, x_k)$ and $g \in C_+([-1,1])$, we get that

$$L_{\vec{N}}^{(M)}(f)(x) = L_{N_\lambda}^{(M)}(g)(x_\lambda), \qquad (12.99)$$

by the maximum multiplicative principle (12.20) and simplification of (12.96).

We present

Theorem 12.35 *Let $x \in (-1, 1)^k$, $k \in \mathbb{N} - \{1\}$, be fixed, and let $f \in C^n$ $\left([-1, 1]^k, \mathbb{R}_+\right)$, $n \in \mathbb{N}$. We assume that $f_\alpha(x) = 0$, for all $\alpha : |\alpha| = 1, ..., n$. Here $|f_\alpha|$ is assumed to be convex over $[-1, 1]^k$, all $\alpha : |\alpha| = n$. The set $[-1, 1]^k$, $k \in \mathbb{N} - \{1\}$ is endowed with the l_1-norm and $\omega_1(f_\alpha, \cdot)$, all $\alpha : |\alpha| = n$ is with respect to l_1-norm. For a sufficiently large $N_{\min}^* := \min\{N_1^*, ..., N_k^*\}$, where $\overrightarrow{N}^* :=$ $\left(N_1^*, ..., N_k^*\right) \in \mathbb{N}^k; N_i^* \geq 4, i = 1, ..., k$, the ball in $\mathbb{R}^k : B\left(x, \frac{1}{(N_{\min}^* - 1)}\right) \subset [-1, 1]^k$.*
Then

$$\left|L_{\overrightarrow{N}}^{(M)}(f)(x) - f(x)\right| \leq \left(\frac{2^{n+2} k^{n+1} \pi^2}{3(n+1)!}\right) \left(\max_{\alpha:|\alpha|=n} \omega_1\left(f_\alpha, \frac{1}{(N_{\min} - 1)}\right)\right),$$

$$(12.100)$$

$\forall \overrightarrow{N} \in \mathbb{N}^k$, where $\overrightarrow{N} := (N_1, ..., N_k) \in \mathbb{N}^k$ and $N_{\min} := \min\{N_1, ..., N_k\}$, with $N_{\min} \geq N_{\min}^*$.

It holds $\lim\limits_{\overrightarrow{N} \to (\infty, ..., \infty)} L_{\overrightarrow{N}}^{(M)}(f)(x) = f(x)$.

Proof By (12.19) we get:

$$\left|L_{\overrightarrow{N}}^{(M)}(f)(x) - f(x)\right| \overset{(12.99)}{\leq} \frac{\left(\max\limits_{\alpha:|\alpha|=n} \omega_1(f_\alpha, h)\right) k^n}{h(n+1)!} \cdot$$

$$\left(\sum_{i=1}^{k} L_{N_i}^{(M)}\left(|t_i - x_i|^{n+1}\right)(x_i)\right) \overset{(12.94)}{\leq}$$

$$\frac{\left(\max\limits_{\alpha:|\alpha|=n} \omega_1(f_\alpha, h)\right) k^n}{h(n+1)!} \left(\sum_{i=1}^{k} \frac{2^{n+2} \pi^2}{3(N_i - 1)}\right) \leq \qquad (12.101)$$

$$\frac{\left(\max\limits_{\alpha:|\alpha|=n} \omega_1(f_\alpha, h)\right) k^{n+1}}{h(n+1)!} \left(\frac{2^{n+2} \pi^2}{3}\right) \frac{1}{(N_{\min} - 1)} =$$

(setting $h := \frac{1}{(N_{\min} - 1)}$)

$$\left(\max_{\alpha:|\alpha|=n} \omega_1\left(f_\alpha, \frac{1}{(N_{\min} - 1)}\right)\right) \left(\frac{2^{n+2} k^{n+1} \pi^2}{3(n+1)!}\right),$$

proving the claim. ∎

We continue with

Theorem 12.36 *Let $[-1, 1]^k$, $k \in \mathbb{N} - \{1\}$, with the l_1 -norm $\|\cdot\|$ and $f \in C_+$ $\left([-1, 1]^k\right) : |f(t) - f(x)|$ is a convex function in $t \in [-1, 1]^k$ for a fixed $x \in$*

$(-1, 1)^k$. *For a sufficiently large* $N^*_{\min} := \min\{N^*_1, ..., N^*_k\}$, *where* $\overrightarrow{N}^* := \left(N^*_1, ..., N^*_k\right) \in \mathbb{N}^k; N^*_i \geq 4, i = 1, ..., k$, *the ball in* $\mathbb{R}^k : B\left(x, \frac{1}{(N^*_{\min}-1)}\right) \subset [-1, 1]^k$. *Then*

$$\left|L^{(M)}_{\overrightarrow{N}}(f)(x) - f(x)\right| \leq \left(\frac{4\pi^2 k}{3}\right) \omega_1\left(f, \frac{1}{(N_{\min}-1)}\right), \tag{12.102}$$

$\forall \overrightarrow{N} \in \mathbb{N}^k$, *where* $\overrightarrow{N} := (N_1, ..., N_k) \in \mathbb{N}^k$ *and* $N_{\min} := \min\{N_1, ..., N_k\}$, *with* $N_{\min} \geq N^*_{\min}$.

It holds $\lim\limits_{\overrightarrow{N} \to (\infty,...,\infty)} L^{(M)}_{\overrightarrow{N}}(f)(x) = f(x)$.

Proof By (12.33) we have that

$$\left|L^{(M)}_{\overrightarrow{N}}(f)(x) - f(x)\right| \overset{(12.99)}{\leq} \frac{\omega_1(f, h)}{h} \left(\sum_{i=1}^k L^{(M)}_{N_i}(|t_i - x_i|)(x_i)\right)$$

$$\overset{(12.94)}{\leq} \frac{\omega_1(f, h)}{h} \left(\sum_{i=1}^k \frac{4\pi^2}{3(N_i - 1)}\right) \leq \frac{\omega_1(f, h)}{h}\left(\frac{4\pi^2 k}{3}\right) \frac{1}{(N_{\min}-1)} = \tag{12.103}$$

(setting $h := \frac{1}{(N_{\min}-1)}$)

$$= \omega_1\left(f, \frac{1}{(N_{\min}-1)}\right)\left(\frac{4\pi^2 k}{3}\right),$$

proving the claim. ∎

We make

Remark 12.37 The Chebyshev knots of first kind $x_{N,k} := \cos\left(\frac{(2(N-k)+1)}{2(N+1)}\pi\right) \in (-1, 1), k \in \{0, 1, ..., N\}, -1 < x_{N,0} < x_{N,1} < ... < x_{N,N} < 1$, are the roots of the first kind Chebyshev polynomial $T_{N+1}(x) := \cos((N + 1) \arccos x), x \in [-1, 1]$.
Define $(x \in [-1, 1])$

$$h_{N,k}(x) := (1 - x \cdot x_{N,k}) \left(\frac{T_{N+1}(x)}{(N + 1)(x - x_{N,k})}\right)^2, \tag{12.104}$$

the fundamental interpolation polynomials.
The Max-product interpolation Hermite-Fejér operators on Chebyshev knots of the first kind (see p. 12 of [5]) are defined by

$$H^{(M)}_{2N+1}(f)(x) = \frac{\bigvee_{k=0}^N h_{N,k}(x) f\left(x_{N,k}\right)}{\bigvee_{k=0}^N h_{N,k}(x)}, \quad \forall N \in \mathbb{N}, \tag{12.105}$$

for $f \in C_+ ([-1, 1]), \forall x \in [-1, 1]$.

By [5], p. 287, we have

$$H_{2N+1}^{(M)} (|\cdot - x|) (x) \le \frac{2\pi}{N+1}, \ \forall x \in [-1, 1], \forall N \in \mathbb{N}. \tag{12.106}$$

And by [3], we get that

$$H_{2N+1}^{(M)} (|\cdot - x|^m) (x) \le \frac{2^m \pi}{N+1}, \ \forall x \in [-1, 1], \forall m, N \in \mathbb{N}. \tag{12.107}$$

Notice $H_{2N+1}^{(M)} (1) = 1$, and $H_{2N+1}^{(M)}$ maps $C_+ ([-1, 1])$ into itself, and it is a positive sublinear operator. Furthermore it holds $\bigvee_{k=0}^{N} h_{N,k} (x) > 0, \ \forall x \in [-1, 1]$. We also have $h_{N,k} (x_{N,k}) = 1$, and $h_{N,k} (x_{N,j}) = 0$, if $k \neq j$, and $H_{2N+1}^{(M)} (f) (x_{N,j}) = f (x_{N,j})$, for all $j \in \{0, 1, ..., N\}$, see [5], p. 282.

We need

Definition 12.38 Let $f \in C_+ ([-1, 1]^k)$, $k \in \mathbb{N} - \{1\}$, and $\overrightarrow{N} = (N_1, ..., N_k) \in \mathbb{N}^k$. We define the multivariate Max-product interpolation Hermite-Fejér operators on Chebyshev knots of the first kind, as follows:

$$H_{2\overrightarrow{N}+1}^{(M)} (f) (x) :=$$

$$\frac{\bigvee_{i_1=0}^{N_1} \bigvee_{i_2=0}^{N_2} \cdots \bigvee_{i_k=0}^{N_k} h_{N_1,i_1} (x_1) h_{N_2,i_2} (x_2) ... h_{N_k,i_k} (x_k) f (x_{N_1,i_1}, x_{N_2,i_2}, ..., x_{N_k,i_k})}{\bigvee_{i_1=0}^{N_1} \bigvee_{i_2=0}^{N_2} \cdots \bigvee_{i_k=0}^{N_k} h_{N_1,i_1} (x_1) h_{N_2,i_2} (x_2) ... h_{N_k,i_k} (x_k)}, \tag{12.108}$$

$\forall x = (x_1, ..., x_k) \in [-1, 1]^k$. Call $N_{\min} := \min\{N_1, ..., N_k\}$.

The operators $H_{2\overrightarrow{N}+1}^{(M)} (f) (x)$ are positive sublinear mapping $C_+ ([-1, 1]^k)$ into itself, and $H_{2\overrightarrow{N}+1}^{(M)} (1) = 1$.

We also have

$$H_{2\overrightarrow{N}+1}^{(M)} (f) (x) :=$$

$$\frac{\bigvee_{i_1=0}^{N_1} \bigvee_{i_2=0}^{N_2} \cdots \bigvee_{i_k=0}^{N_k} h_{N_1,i_1} (x_1) h_{N_2,i_2} (x_2) ... h_{N_k,i_k} (x_k) f (x_{N_1,i_1}, x_{N_2,i_2}, ..., x_{N_k,i_k})}{\prod_{\lambda=1}^{k} \left(\bigvee_{i_\lambda=0}^{N_\lambda} h_{N_\lambda,i_\lambda} (x_\lambda) \right)}, \tag{12.109}$$

$\forall x = (x_1, ..., x_\lambda, ..., x_k) \in [-1, 1]^k$, by the maximum multiplicative principle, see (12.20). Notice that $H_{2\overrightarrow{N}+1}^{(M)} (f) (x_{N_1,i_1}, ..., x_{N_k,i_k}) = f (x_{N_1,i_1}, ..., x_{N_k,i_k})$.

We make

Remark 12.39 The coordinate Max-product interpolation Hermite-Fejér operators on Chebyshev knots of the first kind, are defined as follows ($\lambda = 1, ..., k$):

$$H_{2N_\lambda+1}^{(M)}(g)(x_\lambda) := \frac{\vee_{i_\lambda=0}^{N_\lambda} h_{N_\lambda,i_\lambda}(x_\lambda)\, g\left(x_{N_\lambda,i_\lambda}\right)}{\vee_{i_\lambda=0}^{N_\lambda} h_{N_\lambda,i_\lambda}(x_\lambda)}, \tag{12.110}$$

$\forall\, N_\lambda \in \mathbb{N}$, and $\forall\, x_\lambda \in [-1, 1]$, $\forall\, g \in C_+([-1, 1])$.

Here we have $(\lambda = 1, ..., k;\ x_\lambda \in [-1, 1])$

$$h_{N_\lambda,i_\lambda}(x_\lambda) = \left(1 - x_\lambda \cdot x_{N_\lambda,i_\lambda}\right)\left(\frac{T_{N_\lambda+1}(x_\lambda)}{(N_\lambda+1)\left(x_\lambda - x_{N_\lambda,i_\lambda}\right)}\right)^2, \tag{12.111}$$

where the Chebyshev knots $x_{N_\lambda,i_\lambda} = \cos\left(\frac{(2(N_\lambda-i_\lambda)+1)}{2(N_\lambda+1)}\pi\right) \in (-1, 1)$, $i_\lambda \in \{0, 1, ..., N_\lambda\}, -1 < x_{N_\lambda,0} < x_{N_\lambda,1} < ... < x_{N_\lambda,N_\lambda} < 1$ are the roots of the first kind Chebyshev polynomial $T_{N_\lambda+1}(x_\lambda) = \cos\left((N_\lambda+1)\arccos x_\lambda\right)$, $x_\lambda \in [-1, 1]$.

In case of $f \in C_+\left([-1, 1]^k\right)$ such that $f(x) := g(x_\lambda)$, $\forall\, x \in [-1, 1]^k$ and $g \in C_+([-1, 1])$, we get that

$$H_{2\vec{N}+1}^{(M)}(f)(x) = H_{2N_\lambda+1}^{(M)}(g)(x_\lambda), \tag{12.112}$$

by the maximum multiplicative principle (12.20) and simplification of (12.109).

We present

Theorem 12.40 *Let* $x \in (-1, 1)^k$, $k \in \mathbb{N} - \{1\}$, *be fixed, and let* $f \in C^n\left([-1, 1]^k, \mathbb{R}_+\right)$, $n \in \mathbb{N}$. *We assume that* $f_\alpha(x) = 0$, *for all* $\alpha : |\alpha| = 1, ..., n$. *Here* $|f_\alpha|$ *is assumed to be convex over* $[-1, 1]^k$, *all* $\alpha : |\alpha| = n$. *The set* $[-1, 1]^k$, $k \in \mathbb{N} - \{1\}$ *is endowed with the* l_1-*norm and* $\omega_1(f_\alpha, \cdot)$, *all* $\alpha : |\alpha| = n$ *is with respect to* l_1-*norm. For a sufficiently large* $N_{\min}^* := \min\{N_1^*, ..., N_k^*\}$, *where* $\vec{N}^* := \left(N_1^*, ..., N_k^*\right) \in \mathbb{N}^k$, *the ball in* $\mathbb{R}^k : B\left(x, \frac{1}{(N_{\min}^*+1)}\right) \subset [-1, 1]^k$. *Then*

$$\left|H_{2\vec{N}+1}^{(M)}(f)(x) - f(x)\right| \leq \left(\frac{(2k)^{n+1}\pi}{(n+1)!}\right)\left(\max_{\alpha:|\alpha|=n} \omega_1\left(f_\alpha, \frac{1}{(N_{\min}+1)}\right)\right), \tag{12.113}$$

$\forall\, \vec{N} \in \mathbb{N}^k$, *where* $\vec{N} := (N_1, ..., N_k) \in \mathbb{N}^k$ *and* $N_{\min} := \min\{N_1, ..., N_k\}$, *with* $N_{\min} \geq N_{\min}^*$.

It holds $\lim\limits_{\vec{N}\to(\infty,...,\infty)} H_{2\vec{N}+1}^{(M)}(f)(x) = f(x)$.

Proof By (12.19) we get:

$$\left|H_{2\vec{N}+1}^{(M)}(f)(x) - f(x)\right| \overset{(12.112)}{\leq} \frac{\left(\max_{\alpha:|\alpha|=n} \omega_1(f_\alpha, h)\right)k^n}{h(n+1)!}.$$

$$\left(\sum_{i=1}^{k} H_{2N_i+1}^{(M)} \left(|t_i - x_i|^{n+1} \right) (x_i) \right) \overset{(12.107)}{\leq}$$

$$\frac{\left(\max_{\alpha:|\alpha|=n} \omega_1 \left(f_\alpha, h \right) \right) k^n}{h \, (n+1)!} \left(\sum_{i=1}^{k} \frac{2^{n+1} \pi}{(N_i + 1)} \right) \leq \qquad (12.114)$$

$$\frac{\left(\max_{\alpha:|\alpha|=n} \omega_1 \left(f_\alpha, h \right) \right) k^{n+1}}{h \, (n+1)!} \left(2^{n+1} \pi \right) \frac{1}{(N_{\min} + 1)} =$$

(setting $h := \frac{1}{(N_{\min}+1)}$)

$$\left(\max_{\alpha:|\alpha|=n} \omega_1 \left(f_\alpha, \frac{1}{(N_{\min} + 1)} \right) \right) \left(\frac{2^{n+1} k^{n+1} \pi}{(n+1)!} \right),$$

proving the claim. ∎

We continue with

Theorem 12.41 *Let* $[-1, 1]^k$, $k \in \mathbb{N} - \{1\}$, *with the* l_1-*norm* $\|\cdot\|$ *and* $f \in C_+$ $\left([-1, 1]^k \right) : |f(t) - f(x)|$ *is a convex function in* $t \in [-1, 1]^k$ *for a fixed* $x \in$ $(-1, 1)^k$. *For a sufficiently large* $N_{\min}^* := \min\{N_1^*, ..., N_k^*\}$, *where* $\overrightarrow{N}^* :=$ $\left(N_1^*, ..., N_k^* \right) \in \mathbb{N}^k$, *the ball in* $\mathbb{R}^k : B\left(x, \frac{1}{(N_{\min}^*+1)} \right) \subset [-1, 1]^k$. *Then*

$$\left| H_{2\overrightarrow{N}+1}^{(M)} (f)(x) - f(x) \right| \leq 2k\pi\omega_1 \left(f, \frac{1}{(N_{\min} + 1)} \right), \qquad (12.115)$$

$\forall \, \overrightarrow{N} \in \mathbb{N}^k$, *where* $\overrightarrow{N} := (N_1, ..., N_k) \in \mathbb{N}^k$ *and* $N_{\min} := \min\{N_1, ..., N_k\}$, *with* $N_{\min}$ $\geq N_{\min}^*$.

It holds $\lim_{\overrightarrow{N} \to (\infty,...,\infty)} H_{2\overrightarrow{N}+1}^{(M)} (f)(x) = f(x)$.

Proof By (12.33) we have that

$$\left| H_{2\overrightarrow{N}+1}^{(M)} (f)(x) - f(x) \right| \overset{(12.112)}{\leq} \frac{\omega_1 (f, h)}{h} \left(\sum_{i=1}^{k} H_{2N_i+1}^{(M)} \left(|t_i - x_i| \right) (x_i) \right)$$

$$\overset{(12.106)}{\leq} \frac{\omega_1 (f, h)}{h} \left(\sum_{i=1}^{k} \frac{2\pi}{(N_i + 1)} \right) \leq \frac{\omega_1 (f, h)}{h} (2\pi k) \frac{1}{(N_{\min} + 1)} = \qquad (12.116)$$

(setting $h := \frac{1}{(N_{\min}+1)}$)

$$= \omega_1 \left(f, \frac{1}{(N_{\min} + 1)} \right) (2k\pi) \, ,$$

proving the claim. ∎

We give

Theorem 12.42 *Let Q with the l_1-norm $\|\cdot\|$, be a compact and convex subset of $\mathbb{R}^k$, $k \in \mathbb{N} - \{1\}$, and $f \in C_+ (Q) : |f(t) - f(x)|$ is a convex function in $t \in Q$ for a fixed $x \in Q^o$. The first modulus of continuity $\omega_1 (f, \cdot)$ is with respect to l_1-norm. Let $\{L_N\}_{N \in \mathbb{N}}$ be positive sublinear operators from $C_+ (Q)$ into itself : $L_N (1) = 1$, $\forall N \in \mathbb{N}$. We assume that $L_N (\|t - x\|) (x) > 0$ and the ball in $\mathbb{R}^k : B(x, L_N (\|t - x\|) (x)) \subset Q$, $\forall N \in \mathbb{N}$. Then*

$$|L_N (f) (x) - f(x)| \le \omega_1 (f, L_N (\|t - x\|) (x)) \, , \; \forall N \in \mathbb{N}. \tag{12.117}$$

If $L_N (\|t - x\|) (x) \to 0$, then $L_N (f) (x) \to f(x)$, as $N \to +\infty$.

Proof By (12.33). ∎

We need

Theorem 12.43 *All as in Theorem 12.2, $f \in C^n (Q, \mathbb{R}_+)$. Let $\{L_N\}_{N \in \mathbb{N}}$ be positive sublinear operators mapping $C_+ (Q)$ into itself, such that $L_N (1) = 1$, $\forall N \in \mathbb{N}$. Then*

$$|L_N (f) (x) - f(x)| \le \frac{\left(\max\limits_{\alpha : |\alpha| = n} \omega_1 (f_\alpha, h) \right)}{h (n + 1)!} \left(L_N \left(\|t - x\|^{n+1} \right) \right) (x) \, , \tag{12.118}$$

$\forall N \in \mathbb{N}.$

Proof By (12.12), see Definition 12.3, and by (12.18). ∎

We finally present

Theorem 12.44 *Let Q with the l_1-norm $\|\cdot\|$, be a compact and convex subset of $\mathbb{R}^k$, $k \ge 1$, let $x \in Q^o$ be fixed. Let $f \in C_+^n (Q, \mathbb{R}_+)$, $n \in \mathbb{N}$, and f_α with $\alpha : |\alpha| = n$, has a first modulus of continuity $\omega_1 (f_\alpha, \cdot)$ relative to Q with respect to l_1-norm. Each $|f_\alpha|$ is a convex function over Q, all $\alpha : |\alpha| = n$. Assume further that $f_\alpha (x) = 0$, all $\alpha : |\alpha| = 1, ..., n$. Let $\{L_N\}_{N \in \mathbb{N}}$ be positive sublinear operators from $C_+ (Q)$ into $C_+ (Q) : L_N (1) = 1$, $\forall N \in \mathbb{N}$. We further assume that $L_N \left(\|t - x\|^{n+1} \right) (x) > 0$ and the ball in $\mathbb{R}^k : B \left(x, L_N \left(\|t - x\|^{n+1} \right) (x) \right) \subset Q$, $\forall N \in \mathbb{N}$. Then*

$$|L_N (f) (x) - f(x)| \le \frac{\left(\max\limits_{\alpha : |\alpha| = n} \omega_1 \left(f_\alpha, \left(L_N \left(\|t - x\|^{n+1} \right) \right) (x) \right) \right)}{(n + 1)!} \, , \; \forall N \in \mathbb{N}. \tag{12.119}$$

It holds, as $L_N \left(\|t - x\|^{n+1} \right) (x) \to 0$, then $L_N (f) (x) \to f(x)$, when $N \to +\infty$.

Proof By (12.118). ∎

References

1. G. Anastassiou, *Moments in Probability and Approximation Theory*, Pitman Research Notes in Mathematics Series (Longman Group, New York, 1993)
2. G. Anastassiou, *Approximation by Sublinear Operators* (2017, submitted)
3. G. Anastassiou, *Approximation by Max-Product Operators* (2017, submitted)
4. G. Anastassiou, *Approximations by Multivariate Sublinear and Max-product Operators under Convexity* (2017, submitted)
5. B. Bede, L. Coroianu, S. Gal, *Approximation by Max-Product type Operators* (Springer, Heidelberg, 2016)

Printed in the United States
By Bookmasters